Test Bank

for
Gustafson and Frisk's

College Algebra

Eighth Edition

Australia • Canada • Mexico • Singapore • Spain • United Kingdom • United States

COPYRIGHT © 2004 Brooks/Cole, a division of Thomson Learning, Inc. Thomson Learning™ is a trademark used herein under license.

ALL RIGHTS RESERVED. Instructors of classes adopting *College Algebra*, Eigth Edition by R. David Gustafson and Peter Frisk as an assigned textbook may reproduce material from this publication for classroom use or in a secure electronic network environment that prevents downloading or reproducing the copyrighted material. Otherwise, no part of this work covered by the copyright hereon may be reproduced or used in any form or by any means—graphic, electronic, or mechanical, including but not limited to photocopying, recording, taping, Web distribution, information networks, or information storage and retrieval systems—without the written permission of the publisher.

Printed in the United States of America
1 2 3 4 5 6 7 08 07 06 05 04

Printer: West Group

ISBN: 0-534-40072-8

For more information about our products, contact us at:
Thomson Learning Academic Resource Center
1-800-423-0563

For permission to use material from this text or product, submit a request online at
http://www.thomsonrights.com.
Any additional questions about permissions can be submitted by email to **thomsonrights@thomson.com**.

Thomson Brooks/Cole
10 Davis Drive
Belmont, CA 94002-3098
USA

Asia
Thomson Learning
5 Shenton Way #01-01
UIC Building
Singapore 068808

Australia/New Zealand
Thomson Learning
102 Dodds Street
Southbank, Victoria 3006
Australia

Canada
Nelson
1120 Birchmount Road
Toronto, Ontario M1K 5G4
Canada

Europe/Middle East/South Africa
Thomson Learning
High Holborn House
50/51 Bedford Row
London WC1R 4LR
United Kingdom

Latin America
Thomson Learning
Seneca, 53
Colonia Polanco
11560 Mexico D.F.
Mexico

Spain/Portugal
Paraninfo
Calle/Magallanes, 25
28015 Madrid, Spain

Table of Contents

Chapter 0
Test Form A- Free Response
Test Form B- Free Response
Test Form C- Multiple Choice
Test Form D- Multiple Choice
Test Form E- Mixed (Free Response/Multiple Choice)
Test Form F- Mixed (Free Response/Multiple Choice)
Test Form G- Mixed (Free Response/Multiple Choice)
Test Form H- Mixed (Free Response/Multiple Choice)

Chapter 1
Test Form A- Free Response
Test Form B- Free Response
Test Form C- Multiple Choice
Test Form D- Multiple Choice
Test Form E- Mixed (Free Response/Multiple Choice)
Test Form F- Mixed (Free Response/Multiple Choice)
Test Form G- Mixed (Free Response/Multiple Choice)
Test Form H- Mixed (Free Response/Multiple Choice)

Chapter 2
Test Form A- Free Response
Test Form B- Free Response
Test Form C- Multiple Choice
Test Form D- Multiple Choice
Test Form E- Mixed (Free Response/Multiple Choice)
Test Form F- Mixed (Free Response/Multiple Choice)
Test Form G- Mixed (Free Response/Multiple Choice)
Test Form H- Mixed (Free Response/Multiple Choice)

Chapter 3
Test Form A- Free Response
Test Form B- Free Response
Test Form C- Multiple Choice
Test Form D- Multiple Choice
Test Form E- Mixed (Free Response/Multiple Choice)
Test Form F- Mixed (Free Response/Multiple Choice)
Test Form G- Mixed (Free Response/Multiple Choice)
Test Form H- Mixed (Free Response/Multiple Choice)

Chapter 4
Test Form A- Free Response
Test Form B- Free Response
Test Form C- Multiple Choice
Test Form D- Multiple Choice
Test Form E- Mixed (Free Response/Multiple Choice)
Test Form F- Mixed (Free Response/Multiple Choice)
Test Form G- Mixed (Free Response/Multiple Choice)
Test Form H- Mixed (Free Response/Multiple Choice)

Chapter 5
Test Form A- Free Response
Test Form B- Free Response
Test Form C- Multiple Choice
Test Form D- Multiple Choice
Test Form E- Mixed (Free Response/Multiple Choice)
Test Form F- Mixed (Free Response/Multiple Choice)
Test Form G- Mixed (Free Response/Multiple Choice)
Test Form H- Mixed (Free Response/Multiple Choice)

Chapter 6
Test Form A- Free Response
Test Form B- Free Response
Test Form C- Multiple Choice
Test Form D- Multiple Choice
Test Form E- Mixed (Free Response/Multiple Choice)
Test Form F- Mixed (Free Response/Multiple Choice)
Test Form G- Mixed (Free Response/Multiple Choice)
Test Form H- Mixed (Free Response/Multiple Choice)

Chapter 7
Test Form A- Free Response
Test Form B- Free Response
Test Form C- Multiple Choice
Test Form D- Multiple Choice
Test Form E- Mixed (Free Response/Multiple Choice)
Test Form F- Mixed (Free Response/Multiple Choice)
Test Form G- Mixed (Free Response/Multiple Choice)
Test Form H- Mixed (Free Response/Multiple Choice)

Chapter 8
Test Form A- Free Response
Test Form B- Free Response
Test Form C- Multiple Choice
Test Form D- Multiple Choice
Test Form E- Mixed (Free Response/Multiple Choice)
Test Form F- Mixed (Free Response/Multiple Choice)
Test Form G- Mixed (Free Response/Multiple Choice)
Test Form H- Mixed (Free Response/Multiple Choice)

Chapter 9
Test Form A- Free Response
Test Form B- Free Response
Test Form C- Multiple Choice
Test Form D- Multiple Choice
Test Form E- Mixed (Free Response/Multiple Choice)
Test Form F- Mixed (Free Response/Multiple Choice)
Test Form G- Mixed (Free Response/Multiple Choice)
Test Form H- Mixed (Free Response/Multiple Choice)

Final Exams
Final Exam Form A- Free Response
Final Exam Form B- Multiple Choice
Final Exam Form C- Mixed (Free Response/Multiple Choice)

Note: Each test form includes an answer key and list of problem codes for BCA Testing

Gustafson/ Frisk - College Algebra 8E Chapter 0 Form A

1. How many prime numbers are between -6 and 18 on the number line?

2. Write the inequality as the union of two intervals.

   $x \leq -16$ or $x > 3$

3. Write the expression without using absolute value symbols.

   $|2|$

4. Find the distance between the following two points on the number line.

   -7 and 3

5. Simplify the expression.

   $(x^2)^2 (x^3)^3$

6. Simplify the expression.

   $$\dfrac{x^3 x^5}{x^5 x}$$

7. Simplify the expression.

   $$\dfrac{(8^{-1} z^{-1} y)^{-1}}{(5y^2 z^{-2})^4 (5yz^{-2})^{-1}}$$

8. Let $x = -2$, $y = 0$, $z = 2$ and evaluate the expression.

   $$\dfrac{-(x^2 z^3)}{z^2 - y^2}$$

9. Simplify the radical expression.

$\frac{12}{\sqrt{8}}$

10. Simplify the expression.

$\left(-\frac{125}{27}\right)^{-1/3}$

11. Simplify the expression.

$4y^3\sqrt[3]{64y^3} - 2\sqrt{36y^9}$

12. Rationalize the denominator and simplify.

$\dfrac{4}{\sqrt[3]{4}}$

13. Perform the division and write the answer without using negative exponents.

$\dfrac{-36x^6 y^4 z^9}{9x^9 y^6 z^0}$

14. Give the degree of the polynomial.

$915x^{10} + 3x^{658} + 6$

15. Perform the operations and simplify.

$(2r - 10s)$

$(2r^2 - 8rs - 4s^2)$

16. Rationalize the denominator.

$$\frac{7}{\sqrt{7} - 2}$$

17. Factor the expression completely.

    $4x^2 + 2x^3$

18. Factor the expression completely.

    $2x^3 y^3 z^3 + 4x^2 y^2 z^2 - 8xyz$

19. Factor the expression completely.

    $4ax + 12ay - 5bx - 15by$

20. Factor the expression completely.

    $z^4 - 16$

21. Factor the expression completely.

    $x^2 + 12x + 27$

22. Perform the operations and simplify. Assume that no denominators are 0.

$$\frac{x^2 + 3x}{x - 3} \cdot \frac{x^2 - 9}{x + 4}$$

23. Perform the operations and simplify. Assume that no denominators are 0.

$$\frac{4x}{x^2 - 4} - \frac{4}{x + 2}$$

24. Perform the operations and simplify. Assume that no denominators are 0.

$$\frac{-4}{2x - 9y} + \frac{4}{2x - 3z} - \frac{12z - 36y}{(2x - 9y)(2x - 3z)}$$

25. Write the expression without using negative exponents, and simplify the resulting complex fraction. Assume that no denominators are 0.

$$\frac{8y^{-1}}{2x^{-1} + 13y^{-1}}$$

ANSWER KEY

Gustafson/ Frisk - College Algebra 8E Chapter 0 Form A

1. 7

2. $(-\infty, -16] \cup (3, \infty)$

3. 2

4. 10

5. x^{13}

6. x^2

7. $\dfrac{8^1 \cdot z^7}{\left(5^3 \cdot y^0\right)}$

8. (-8)

9. $\sqrt[4]{2}$

10. $-\dfrac{3}{5}$

11. $20y^4 \cdot \sqrt{y}$

12. $\sqrt[3]{16}$

13. $\dfrac{-4z^9}{\left(x^3 \cdot y^2\right)}$

14. 658

ANSWER KEY

Gustafson/ Frisk - College Algebra 8E Chapter 0 Form A

15. $4r^3 - 36r^2 \cdot s + 72r \cdot s^2 + 40s^3$

16. $\dfrac{(7\sqrt{7} + 14)}{3}$

17. $2x^2 \cdot (2 + x)$

18. $2x \cdot y \cdot z \cdot (x^2 \cdot y^2 \cdot z^2 + 2x \cdot y \cdot z - 4)$

19. $(x + 3y) \cdot (4a - 5b)$

20. $(z^2 + 4) \cdot (z - 2) \cdot (z + 2)$

21. $(x + 9) \cdot (x + 3)$

22. $\dfrac{x \cdot (x + 3)^2}{(x + 4)}$

23. $\dfrac{8}{(x^2 - 4)}$

24. 0

25. $\dfrac{8x}{(2y + 13x)}$

List of Problem Codes for BCA Testing

Gustafson/ Frisk - College Algebra 8E Chapter 0 Form A

1. gfca.00.01.4.39_NoAlgs
2. gfca.00.01.4.65_NoAlgs
3. gfca.00.01.4.69_NoAlgs
4. gfca.00.01.4.85_NoAlgs
5. gfca.00.02.4.41_NoAlgs
6. gfca.00.02.4.63_NoAlgs
7. gfca.00.02.4.79_NoAlgs
8. gfca.00.02.4.89_NoAlgs
9. gfca.00.03.4.113_NoAlgs
10. gfca.00.03.4.49_NoAlgs
11. gfca.00.03.4.85_NoAlgs
12. gfca.00.03.4.97_NoAlgs
13. gfca.00.04.4.101_NoAlgs
14. gfca.00.04.4.11_NoAlgs
15. gfca.00.04.4.68_NoAlgs
16. gfca.00.04.4.81_NoAlgs
17. gfca.00.05.4.11_NoAlgs
18. gfca.00.05.4.16_NoAlgs
19. gfca.00.05.4.24_NoAlgs
20. gfca.00.05.4.40_NoAlgs
21. gfca.00.05.4.51_NoAlgs
22. gfca.00.06.4.35_NoAlgs
23. gfca.00.06.4.57_NoAlgs
24. gfca.00.06.4.70_NoAlgs
25. gfca.00.06.4.94_NoAlgs

Gustafson/ Frisk - College Algebra 8E Chapter 0 Form B

1. Perform the operations and simplify. Assume that no denominators are 0.

$$\frac{x^2 + 3x}{x - 3} \cdot \frac{x^2 - 9}{x + 4}$$

2. Simplify the expression.

$(x^2)^2 (x^3)^3$

3. Rationalize the denominator.

$$\frac{7}{\sqrt{7} - 2}$$

4. Simplify the expression.

$4y^3 \sqrt{64y^3} - 2\sqrt{36y^9}$

5. Perform the operations and simplify. Assume that no denominators are 0.

$$\frac{-4}{2x - 9y} + \frac{4}{2x - 3z} - \frac{12z - 36y}{(2x - 9y)(2x - 3z)}$$

6. Write the inequality as the union of two intervals.

$x \leq -16$ or $x > 3$

7. Simplify the expression. Write the answer without using negative exponents. Assume that the variable is restricted to those numbers for which the expression is defined.

$$\frac{x^3 x^5}{x^5 x}$$

8. How many prime numbers are between -6 and 18 on the number line?

9. Factor the expression completely.

    $4x^2 + 2x^3$

10. Simplify the expression. Write the answer without using negative exponents. Assume that all variables are restricted to those numbers for which the expression is defined.

    $$\frac{(8^{-1}z^{-1}y)^{-1}}{(5y^2z^{-2})^4(5yz^{-2})^{-1}}$$

11. Rationalize the denominator and simplify.

    $$\frac{4}{\sqrt[3]{4}}$$

12. Let $x = -2$, $y = 0$, $z = 2$ and evaluate the expression.

    $$\frac{-(x^2z^3)}{z^2-y^2}$$

13. Simplify the radical expression.

    $\sqrt[12]{8}$

14. Factor the expression completely.

    $x^2 + 12x + 27$

15. Give the degree of polynomial.

    $915x^{10} + 3x^{658} + 6$

16. Perform the operations and simplify.

    $(2r - 10s)$

    $(2r^2 - 8rs - 4s^2)$

17. Write the expression without using absolute value symbols.

    $|2|$

18. Find the distance between the following two points on the number line.

    -7 and 3

19. Perform the operations and simplify. Assume that no denominators are 0.

    $$\frac{4x}{x^2 - 4} - \frac{4}{x + 2}$$

20. Simplify the expression.

    $$\left(-\frac{125}{27}\right)^{-1/3}$$

21. Write the expression without using negative exponents, and simplify the resulting complex fraction. Assume that no denominators are 0.

    $$\frac{8y^{-1}}{2x^{-1} + 13y^{-1}}$$

22. Factor the expression completely.

    $z^4 - 16$

23. Factor the expression completely.

    $4ax + 12ay - 5bx - 15by$

24. Perform the division and write the answer without using negative exponents.

$$\frac{-36x^6 y^4 z^9}{9x^9 y^6 z^0}$$

25. Factor the expression completely.

$2x^3 y^3 z^3 + 4x^2 y^2 z^2 - 8xyz$

ANSWER KEY

Gustafson/ Frisk - College Algebra 8E Chapter 0 Form B

1. $\dfrac{x \cdot (x+3)^2}{(x+4)}$

2. x^{13}

3. $\dfrac{(7\sqrt{7}+14)}{3}$

4. $20y^4 \cdot \sqrt{y}$

5. 0

6. $(-\infty, -16] \cup (3, \infty)$

7. x^2

8. 7

9. $2x^2 \cdot (2+x)$

10. $\dfrac{8^1 \cdot z^7}{\left(5^3 \cdot y^8\right)}$

11. $\sqrt[3]{16}$

12. (-8)

13. $\sqrt[4]{2}$

14. $(x+9) \cdot (x+3)$

15. 658

16. $4r^3 - 36r^2 \cdot s + 72r \cdot s^2 + 40s^3$

17. 2

18. 10

ANSWER KEY

Gustafson/ Frisk - College Algebra 8E Chapter 0 Form B

19. $\dfrac{8}{(x^2-4)}$

20. $-\dfrac{3}{5}$

21. $\dfrac{8x}{(2y+13x)}$

22. $(z^2+4)\cdot(z-2)\cdot(z+2)$

23. $(x+3y)\cdot(4a-5b)$

24. $\dfrac{-4z^9}{(x^3\cdot y^2)}$

25. $2x\cdot y\cdot z\cdot(x^2\cdot y^2\cdot z^2+2x\cdot y\cdot z-4)$

List of Problem Codes for BCA Testing

Gustafson/ Frisk - College Algebra 8E Chapter 0 Form B

1. gfca.00.06.4.35_NoAlgs
2. gfca.00.02.4.41_NoAlgs
3. gfca.00.04.4.81_NoAlgs
4. gfca.00.03.4.85_NoAlgs
5. gfca.00.06.4.70_NoAlgs
6. gfca.00.01.4.65_NoAlgs
7. gfca.00.02.4.63_NoAlgs
8. gfca.00.01.4.39_NoAlgs
9. gfca.00.05.4.11_NoAlgs
10. gfca.00.02.4.79_NoAlgs
11. gfca.00.03.4.97_NoAlgs
12. gfca.00.02.4.89_NoAlgs
13. gfca.00.03.4.113_NoAlgs
14. gfca.00.05.4.51_NoAlgs
15. gfca.00.04.4.11_NoAlgs
16. gfca.00.04.4.68_NoAlgs
17. gfca.00.01.4.69_NoAlgs
18. gfca.00.01.4.85_NoAlgs
19. gfca.00.06.4.57_NoAlgs
20. gfca.00.03.4.49_NoAlgs
21. gfca.00.06.4.94_NoAlgs
22. gfca.00.05.4.40_NoAlgs
23. gfca.00.05.4.24_NoAlgs
24. gfca.00.04.4.101_NoAlgs
25. gfca.00.05.4.16_NoAlgs

1. How many prime numbers are between -9 and 18 on the number line?

   Select the correct answer.

   a. 0
   b. 8
   c. 18
   d. 26
   e. 7
   f. 17

2. Select the correct representation in interval notation of the following inequality.

   $-15 \leq x < 7$

   a. $[-15, 7)$
   b. $(-\infty, 7]$
   c. $[-15, \infty)$
   d. $(-15, 7)$
   e. $[-15, 7]$

3. Write the expression without using absolute value symbols.

   $-|-3|$

   Select the correct answer.

   a. 1
   b. 0
   c. 3
   d. -3

4. Write the expression without using absolute value symbols.

   $|x - 10|$ and $x < -13$

   Select the correct answer.

   a. $-(x - 10)$
   b. $x + 10$
   c. $-(x + 10)$
   d. $x - 10$

5. Calculate the volume of a box that has dimensions of 5,000, 9,300, 5,500 millimeters.

   Select the correct answer.

   a. $2.5575 \times 10^{11} \; mm^3$

   b. $2.5575 \times 10^{10} \; mm^3$

   c. $1.9872 \times 10^{11} \; mm^3$

6. Rationalize the denominator and simplify.

   $$\frac{3}{\sqrt[5]{3}}$$

   Select the correct answer.

   a. $\sqrt[10]{81}$

   b. $\sqrt[5]{84}$

   c. $\sqrt[5]{181}$

   d. $\sqrt[6]{82}$

   e. $\sqrt[5]{81}$

7. Simplify the expression. Write the answer without using negative exponents. Assume that all variables are restricted to those numbers for which the expression is defined.

$$\frac{(8^{-2} z^{-4} y)^{-2}}{(5 y^5 z^{-1})^3 (5 y z^{-1})^{-1}}$$

Select the correct answer.

a. $\dfrac{4.096 z^{10}}{25 y^{16}}$

b. $\dfrac{512 z^9}{25 y^{17}}$

c. $\dfrac{25 y^{16}}{4.096 z^{10}}$

8. Let $x = -3$, $y = 0$, $z = 3$ and evaluate the expression.

$$\frac{-(x^2 z^3)}{z^2 - y^2}$$

Select the correct answer.

a. 0
b. -27
c. -81

9. Simplify the radical expression.

$\sqrt[9]{27}$

Select the correct answer.

a. $\sqrt[3]{27}$

b. $\sqrt[3]{3}$

c. $\sqrt[3]{300}$

d. $\sqrt[9]{3}$

e. $\dfrac{27}{\sqrt{3}}$

10. Simplify the expression.

$$\left(-\frac{343}{8}\right)^{-2/3}$$

Select the correct answer.

a. $\dfrac{98}{49}$

b. $\dfrac{4}{8}$

c. $\dfrac{4}{49}$

d. $\dfrac{49}{4}$

11. Perform division and write the answer without using negative exponents.

    $$\frac{-64x^6y^4z^9}{16x^9y^6z^0}$$

    Select the correct answer.

    a. $\dfrac{4z^9}{x^3y^6}$

    b. $\dfrac{4z^9}{x^3y^2}$

    c. $\dfrac{-4z^4}{x^3y^2}$

    d. $\dfrac{-4z^9}{x^3y^2}$

12. Perform the operations and simplify.

    $(3r - 3s)$

    $(10r^2 - 4rs - 4s^2)$

    Select the correct answer.

    a. $30r^3 + 42r^2s^2 - 6s^4 + 30s^3$
    b. $30r^3 - 42r^2s + 6rs^2 + 12s^3$
    c. $30r^3 + 42r^2s^2 - 6r^2s^2 + 12s^3$
    d. $30r^3 + 42r^2s^2 - 6s^4 - 30s^3$

13. Rationalize the denominator.

$$\frac{7}{\sqrt{6}-2}$$

Select the correct answer.

a. $\dfrac{7\sqrt{6}+\sqrt{14}}{2}$

b. $\dfrac{7\sqrt{6}+14}{2}$

c. $\dfrac{7\sqrt{6}}{2}$

d. $\dfrac{7\sqrt{6}-14}{2}$

14. Perform the division and write the answer without using negative exponents.

$$\frac{50a^{-2}b^3}{10ab^6}$$

Select the correct answer.

a. $5ab^3$

b. $\dfrac{5a}{b^3}$

c. $\dfrac{5}{ab^3}$

d. $\dfrac{5a^2}{b^3}$

15. Factor the expression completely.

    $12x^2 + 6x^3$

    Select the correct answer.

    a. $6x^2(3 + x)$
    b. $6x^2(2 - x)$
    c. $6x^2(2 + x^2)$
    d. $6x^2(2 + x)$

16. Factor the expression completely.

    $5x^3y^3z^3 + 25x^2y^2z^2 - 125xyz$

    Select the correct answer.

    a. $25xyz(x^2y^2z^2 + 5xyz - 25)$
    b. $5xyz(x^2y^2z^2 + 5xyz - 25)$
    c. $5xyz(x^2y^2z^2 - 5xyz + 25)$
    d. $25xyz(x^2y^2z^2 + 5xyz - 5)$

17. Factor the expression completely.

    $4ax + 12ay - 5bx - 15by$

    Select the correct answer.

    a. $(x + 3y)(4a - 5b)$
    b. $(x + 3y)(5b - 4a)$
    c. $(x - 3y)(4a + 5b)$

18. Factor the expression completely.

    $z^4 - 16$

    Select the correct answer.

    a. $(z + 2)(z - 2)$
    b. $(z^4 + 4)(z - 2)(z + 2)$
    c. $(z^4 - 4)(z - 2)(z + 2)$
    d. $(z^2 + 4)(z^2 - 4)$

19. Perform the operations and simplify. Assume that no denominators are 0.

$$\frac{x^2 + 8x}{x - 8} \cdot \frac{x^2 - 64}{x + 9}$$

Select the correct answer.

a. $\dfrac{x^3 + 16x^2 - 16x}{x + 9}$

b. $\dfrac{x^2 + 16x + 64}{x + 9}$

c. $\dfrac{x^3 - 16x^2 + 64x}{x - 9}$

d. $\dfrac{x^3 + 16x^2 + 64x}{x + 9}$

20. Perform the operations and simplify. Assume that no denominators are 0.

$$\frac{25x}{x^2 - 25} - \frac{25}{x + 5}$$

Select the correct answer.

a. $\dfrac{125}{x^2 + 25}$

b. $\dfrac{125}{x^2 - 5}$

c. $\dfrac{25}{x^2 - 25}$

d. $\dfrac{125}{x^2 - 25}$

21. Perform the operations and simplify. Assume that no denominators are 0.

$$\frac{-8}{2x-5y} + \frac{8}{2x-9z} - \frac{72z-40y}{(2x-5y)(2x-9z)}$$

Select the correct answer.

a. $\dfrac{8}{2x-9z}$

b. $\dfrac{72z-40y}{(2x-5y)(9z-2x)}$

c. 0

d. $\dfrac{8}{2x-5y}$

22. Simplify the expression. Assume that all variables represent positive numbers, so that no absolute value symbols are needed.

$$6y^2\sqrt{64y^5} - 2\sqrt{100y^9}$$

Select the correct answer.

a. $28y^4\sqrt{y}$

b. $9y^5\sqrt{y}$

c. $28y^5\sqrt{y}$

d. $28y^4\sqrt[5]{y}$

23. Write the expression without using negative exponents, and simplify the resulting complex fraction. Assume that denominator is not 0.

$$\frac{6y^{-1}}{8x^{-1} + 5y^{-1}}$$

Select the correct answer.

a. $\dfrac{6x}{5y + 8x}$

b. $\dfrac{6x}{8y - 5x}$

c. $\dfrac{5x}{8y + 6x}$

d. $\dfrac{6x}{8y + 5x}$

24. Simplify the expression.

$(x^3)^4 (x^2)^3$

Select the correct answer.

a. x^{12}
b. x^{10}
c. x^{18}

25. Factor the expression completely.

$x^2 + 10x + 24$

Select the correct answer.

a. $(x + 4)(x - 6)$
b. $(x + 4)(x + 4)$
c. $(x - 4)(x - 6)$
d. $(x + 4)(x + 6)$

ANSWER KEY

Gustafson/ Frisk - College Algebra 8E Chapter 0 Form C

1. e
2. a
3. d
4. a
5. a
6. e
7. a
8. b
9. b
10. c
11. d
12. b
13. b
14. b
15. d
16. b
17. a
18. b
19. d
20. d
21. c
22. a
23. d
24. c
25. d

List of Problem Codes for BCA Testing

Gustafson/ Frisk - College Algebra 8E Chapter 0 Form C

1. gfca.00.01.4.39m_NoAlgs
2. gfca.00.01.4.57m_NoAlgs
3. gfca.00.01.4.73m_NoAlgs
4. gfca.00.01.4.83m_NoAlgs
5. gfca.00.02.4.120m_NoAlgs
6. gfca.00.03.4.97m_NoAlgs
7. gfca.00.02.4.79m_NoAlgs
8. gfca.00.02.4.89m_NoAlgs
9. gfca.00.03.4.113m_NoAlgs
10. gfca.00.03.4.49m_NoAlgs
11. gfca.00.04.4.101m_NoAlgs
12. gfca.00.04.4.68m_NoAlgs
13. gfca.00.04.4.81m_NoAlgs
14. gfca.00.04.4.99m_NoAlgs
15. gfca.00.05.4.11m_NoAlgs
16. gfca.00.05.4.16m_NoAlgs
17. gfca.00.05.4.24m_NoAlgs
18. gfca.00.05.4.40m_NoAlgs
19. gfca.00.06.4.35m_NoAlgs
20. gfca.00.06.4.57m_NoAlgs
21. gfca.00.06.4.70m_NoAlgs
22. gfca.00.03.4.85m_NoAlgs
23. gfca.00.06.4.94m_NoAlgs
24. gfca.00.02.4.41m_NoAlgs
25. gfca.00.05.4.51m_NoAlgs

Gustafson/ Frisk - College Algebra 8E Chapter 0 Form D

1. Write the expression without using negative exponents, and simplify the resulting complex fraction. Assume that denominator is not 0.

$$\frac{6y^{-1}}{8x^{-1} + 5y^{-1}}$$

Select the correct answer.

a. $\dfrac{6x}{5y + 8x}$

b. $\dfrac{6x}{8y - 5x}$

c. $\dfrac{5x}{8y + 6x}$

d. $\dfrac{6x}{8y + 5x}$

2. Write the expression without using absolute value symbols.

$-|-3|$

Select the correct answer.

a. 1
b. 0
c. 3
d. -3

3. Perform the division and write the answer without using negative exponents.

$$\frac{50a^2b^3}{10ab^6}$$

Select the correct answer.

a. $5ab^3$

b. $\dfrac{5a}{b^3}$

c. $\dfrac{5}{ab^3}$

d. $\dfrac{5a^2}{b^3}$

4. Perform the operations and simplify. Assume that no denominators are 0.

$$\frac{x^2 + 8x}{x - 8} \cdot \frac{x^2 - 64}{x + 9}$$

Select the correct answer.

a. $\dfrac{x^2 + 16x + 64}{x + 9}$

b. $\dfrac{x^3 + 16x^2 - 16x}{x + 9}$

c. $\dfrac{x^3 - 16x^2 + 64x}{x - 9}$

d. $\dfrac{x^3 + 16x^2 + 64x}{x + 9}$

5. How many prime numbers are between -9 and 18 on the number line?

   Select the correct answer.

   a. 0
   b. 8
   c. 18
   d. 26
   e. 7
   f. 17

6. Perform the operations and simplify.

   $(3r - 3s)$

   $(10r^2 - 4rs - 4s^2)$

   Select the correct answer.

   a. $30r^3 + 42r^2s^2 - 6r^2s^2 + 12s^3$
   b. $30r^3 + 42r^2s^2 - 6s^4 + 30s^3$
   c. $30r^3 + 42r^2s^2 - 6s^4 - 30s^3$
   d. $30r^3 - 42r^2s + 6rs^2 + 12s^3$

7. Select the correct representation in interval notation of the following inequality.

   $-15 \leq x < 7$

   a. $[-15, 7)$
   b. $(-\infty, 7]$
   c. $[-15, \infty)$
   d. $(-15, 7)$
   e. $[-15, 7]$

8. Simplify the expression.

$$\left(-\frac{343}{8}\right)^{-2/3}$$

Select the correct answer.

a. $\dfrac{49}{4}$

b. $\dfrac{4}{8}$

c. $\dfrac{98}{49}$

d. $\dfrac{4}{49}$

9. Let $x = -3$, $y = 0$, $z = 3$ and evaluate the expression.

$$\frac{-(x^2 z^3)}{z^2 - y^2}$$

Select the correct answer.

a. 0
b. -27
c. -81

10. Perform the operations and simplify. Assume that no denominators are 0.

$$\frac{25x}{x^2 - 25} - \frac{25}{x + 5}$$

Select the correct answer.

a. $\dfrac{125}{x^2 + 25}$

b. $\dfrac{125}{x^2 - 5}$

c. $\dfrac{25}{x^2 - 25}$

d. $\dfrac{125}{x^2 - 25}$

11. Factor the expression completely.

    $4ax + 12ay - 5bx - 15by$

    Select the correct answer.

    a. $(x + 3y)(4a - 5b)$
    b. $(x + 3y)(5b - 4a)$
    c. $(x - 3y)(4a + 5b)$

12. Factor the expression completely.

    $5x^3 y^3 z^3 + 25x^2 y^2 z^2 - 125xyz$

    Select the correct answer.

    a. $25xyz(x^2 y^2 z^2 + 5xyz - 25)$
    b. $5xyz(x^2 y^2 z^2 + 5xyz - 25)$
    c. $5xyz(x^2 y^2 z^2 - 5xyz + 25)$
    d. $25xyz(x^2 y^2 z^2 + 5xyz - 5)$

13. Factor the expression completely.

    $12x^2 + 6x^3$

    Select the correct answer.

    a. $6x^2(3 + x)$
    b. $6x^2(2 - x)$
    c. $6x^2(2 + x^2)$
    d. $6x^2(2 + x)$

14. Rationalize the denominator and simplify.

    $$\frac{3}{\sqrt[5]{3}}$$

    Select the correct answer.
    a. $\sqrt[6]{82}$

    b. $\sqrt[5]{84}$

    c. $\sqrt[5]{181}$

    d. $\sqrt[10]{81}$

    e. $\sqrt[5]{81}$

15. Rationalize the denominator.

$$\frac{7}{\sqrt{6} - 2}$$

Select the correct answer.

a. $\dfrac{7\sqrt{6} + \sqrt{14}}{2}$

b. $\dfrac{7\sqrt{6} + 14}{2}$

c. $\dfrac{7\sqrt{6}}{2}$

d. $\dfrac{7\sqrt{6} - 14}{2}$

16. Write the expression without using absolute value symbols.

　|　$x - 10$　| and $x < -13$

Select the correct answer.

a. 　$-(x - 10)$
b. 　$x + 10$
c. 　$-(x + 10)$
d. 　$x - 10$

17. Simplify the expression. Write the answer without using negative exponents. Assume that all variables are restricted to those numbers for which the expression is defined.

$$\frac{(8^{-2}z^{-4}y)^{-2}}{(5y^5z^{-1})^3(5yz^{-1})^{-1}}$$

Select the correct answer.

a. $\dfrac{4.096z^{10}}{25y^{16}}$

b. $\dfrac{512z^9}{25y^{17}}$

c. $\dfrac{25y^{16}}{4.096z^{10}}$

18. Factor the expression completely.

$z^4 - 16$

Select the correct answer.

a. $(z + 2)(z - 2)$
b. $(z^4 + 4)(z - 2)(z + 2)$
c. $(z^4 - 4)(z - 2)(z + 2)$
d. $(z^2 + 4)(z^2 - 4)$

19. Calculate the volume of a box that has dimensions of 5,000, 9,300, 5,500 millimeters.

Select the correct answer.

a. 2.5575×10^{11} mm^3

b. 2.5575×10^{10} mm^3

c. 1.9872×10^{11} mm^3

20. Simplify the expression.

    $(x^3)^4 \cdot (x^2)^3$

    Select the correct answer.

    a. x^{12}
    b. x^{10}
    c. x^{18}

21. Factor the expression completely.

    $x^2 + 10x + 24$

    Select the correct answer.
    a. $(x + 4)(x - 6)$
    b. $(x + 4)(x + 4)$
    c. $(x - 4)(x - 6)$
    d. $(x + 4)(x + 6)$

22. Simplify the radical expression.

    $\sqrt[9]{27}$

    a. $\sqrt[9]{3}$

    b. $\sqrt[3]{300}$

    c. $\sqrt[3]{3}$

    d. $\sqrt[3]{27}$

    e. $\sqrt[27]{3}$

23. Perform division and write the answer without using negative exponents.

$$\frac{-64x^6y^4z^9}{16x^9y^6z^0}$$

Select the correct answer.

a. $\dfrac{4z^9}{x^3y^6}$

b. $\dfrac{4z^9}{x^3y^2}$

c. $\dfrac{-4z^4}{x^3y^2}$

d. $\dfrac{-4z^9}{x^3y^2}$

24. Perform the operations and simplify. Assume that no denominators are 0.

$$\frac{-8}{2x-5y} + \frac{8}{2x-9z} - \frac{72z-40y}{(2x-5y)(2x-9z)}$$

Select the correct answer.

a. $\dfrac{8}{2x-9z}$

b. $\dfrac{72z-40y}{(2x-5y)(9z-2x)}$

c. 0

d. $\dfrac{8}{2x-5y}$

25. Simplify the expression. Assume that all variables represent positive numbers, so that no absolute value symbols are needed.

$$6y^2\sqrt{64y^5} - 2\sqrt{100y^9}$$

Select the correct answer.

a. $28y^4\sqrt{y}$

b. $9y^5\sqrt{y}$

c. $28y^5\sqrt{y}$

d. $28y^4\sqrt[5]{y}$

ANSWER KEY

Gustafson/ Frisk - College Algebra 8E Chapter 0 Form D

1. d
2. d
3. b
4. d
5. e
6. d
7. a
8. d
9. b
10. d
11. a
12. b
13. d
14. e
15. b
16. a
17. a
18. b
19. a
20. c
21. d
22. c
23. d
24. c
25. a

List of Problem Codes for BCA Testing

Gustafson/ Frisk - College Algebra 8E Chapter 0 Form D

1. gfca.00.06.4.94m_NoAlgs
2. gfca.00.01.4.73m_NoAlgs
3. gfca.00.04.4.99m_NoAlgs
4. gfca.00.06.4.35m_NoAlgs
5. gfca.00.01.4.39m_NoAlgs
6. gfca.00.04.4.68m_NoAlgs
7. gfca.00.01.4.57m_NoAlgs
8. gfca.00.03.4.49m_NoAlgs
9. gfca.00.02.4.89m_NoAlgs
10. gfca.00.06.4.57m_NoAlgs
11. gfca.00.05.4.24m_NoAlgs
12. gfca.00.05.4.16m_NoAlgs
13. gfca.00.05.4.11m_NoAlgs
14. gfca.00.03.4.97m_NoAlgs
15. gfca.00.04.4.81m_NoAlgs
16. gfca.00.01.4.83m_NoAlgs
17. gfca.00.02.4.79m_NoAlgs
18. gfca.00.05.4.40m_NoAlgs
19. gfca.00.02.4.120m_NoAlgs
20. gfca.00.02.4.41m_NoAlgs
21. gfca.00.05.4.51m_NoAlgs
22. gfca.00.03.4.113m_NoAlgs
23. gfca.00.04.4.101m_NoAlgs
24. gfca.00.06.4.70m_NoAlgs
25. gfca.00.03.4.85m_NoAlgs

Gustafson/ Frisk - College Algebra 8E Chapter 0 Form E

1. How many prime numbers are between -9 and 18 on the number line?

   Select the correct answer.

   a. 18
   b. 17
   c. 0
   d. 7
   e. 8
   f. 26

2. Write the inequality $-6 \leq x < 11$ using interval notation.

3. Calculate the volume of a box that has dimensions of 5,000 by 8,100 by 4,500 millimeters. Write the answer in scientific notation.

4. Simplify the expression.

   $x^6 x^3$

   Select the correct answer.

   a. x^{18}
   b. x^9
   c. x^8

5. Select the correct distance between the following two points on the number line.

   - 20 and 16

   a. 20
   b. 37
   c. 36
   d. 35
   e. 16

6. Simplify the expression. Write the answer without using negative exponents. Assume that all variables are restricted to those numbers for which the expression is defined.

$$\frac{(8^{-2} z^{-4} y)^{-2}}{(5 y^5 z^{-1})^3 (5 y z^{-1})^{-1}}$$

Select the correct answer.

a. $\dfrac{512 z^9}{25 y^{17}}$

b. $\dfrac{25 y^{16}}{4{,}096 z^{10}}$

c. $\dfrac{4{,}096 z^{10}}{25 y^{16}}$

7. Let $x = -2$, $y = 0$, $z = 2$ and evaluate the expression.

$$\frac{-(x^2 z^3)}{z^2 - y^2}$$

8. Simplify the radical expression.

$\sqrt[9]{27}$

a. $\sqrt[3]{27}$

b. $\sqrt[3]{3}$

c. $\sqrt[3]{300}$

d. $\sqrt[27]{3}$

e. $\sqrt[9]{3}$

9. Simplify the expression.

$$\left(-\frac{27x^9}{64y^3}\right)^{1/3}$$

10. Simplify the expression.

$$-8^{4/3}$$

Select the correct answer.

a. -32
b. -16
c. -48
d. 19
e. -18

11. Rationalize the denominator and simplify.

$$\frac{4}{\sqrt[3]{4}}$$

12. Perform the division.

$$x^2 + x - 1 \overline{)3x^3 - 2x^2 - 8x + 5}$$

13. Give the degree of the polynomial.

$$\sqrt{4 \cdot 83}$$

Select the correct answer.

a. This is not a polynomial
b. No defined degree
c. 0

14. Perform the operations and simplify. Assume that no denominators are 0.

$$\frac{-4}{2x-9y} + \frac{4}{2x-3z} - \frac{12z-36y}{(2x-9y)(2x-3z)}$$

15. Write the expression without using absolute value symbols.

    $|2|$

16. Rationalize the denominator.

    $$\frac{7}{\sqrt{6} - 2}$$

    Select the correct answer.

    a. $\dfrac{7\sqrt{6} - 14}{2}$

    b. $\dfrac{7\sqrt{6} + \sqrt{14}}{2}$

    c. $\dfrac{7\sqrt{6} + 14}{2}$

    d. $\dfrac{7\sqrt{6}}{2}$

17. Perform the division and write the answer without using negative exponents.

    $$\frac{77a^2 b^3}{11ab^6}$$

18. Factor the expression completely.

    $12x^2 + 6x^3$

    Select the correct answer.

    a. $6x^2(2 + x^2)$
    b. $6x^2(2 - x)$
    c. $6x^2(2 + x)$
    d. $6x^2(3 + x)$

19. Factor the expression completely.

    $4x^3 + 4x^2 - 11x - 11$

    Select the correct answer.

    a. $(x + 1)(11 - 4x^2)$
    b. $(1 - x)(4x^2 - 11)$
    c. $(x + 1)(4x^2 - 11)$
    d. $(x - 1)(4x^2 + 11)$

20. Factor the expression completely.

    $4ax + 12ay - 5bx - 15by$

21. Factor the expression completely.

    $(x + y)^3 - 64$

22. Factor the expression completely.

    $(a + b)^2 - 3(a + b) - 4$

    Select the correct answer.

    a. $(a + b - 4)(a + b - 1)$
    b. $(a - b + 4)(a + b + 1)$
    c. $(a - b - 4)(a - b + 1)$
    d. $(a + b - 4)(a + b + 1)$

23. Perform the operations and simplify. Assume that no denominators are 0.

    $$\frac{x^3 + 512}{x^2 - 36} \div \left(\frac{x^2 + 17x + 72}{x^2 + 6x} \div \frac{x^2 + 2x - 48}{x^2 - 8x + 64} \right)$$

24. Perform the operations and simplify. Assume that no denominators are 0.

$$\frac{25x}{x^2 - 25} - \frac{25}{x + 5}$$

Select the correct answer.

a. $\dfrac{25}{x^2 - 25}$

b. $\dfrac{125}{x^2 + 25}$

c. $\dfrac{125}{x^2 - 5}$

d. $\dfrac{125}{x^2 - 25}$

25. Write the expression without using negative exponents, and simplify the resulting complex fraction. Assume that denominator is not 0.

$$\frac{6y^{-1}}{8x^{-1} + 5y^{-1}}$$

Select the correct answer.

a. $\dfrac{5x}{8y + 6x}$

b. $\dfrac{6x}{8y + 5x}$

c. $\dfrac{6x}{8y - 5x}$

d. $\dfrac{6x}{5y + 8x}$

ANSWER KEY

Gustafson/ Frisk - College Algebra 8E Chapter 0 Form E

1. d

2. $[-6, 11)$

3. $1.8225 \cdot 10^{11}$

4. b

5. c

6. c

7. (-8)

8. b

9. $\dfrac{-3x^3}{(4y)}$

10. b

11. $\sqrt[3]{16}$

12. $3x - 5$

13. c

14. 0

15. 2

16. c

ANSWER KEY

Gustafson/ Frisk - College Algebra 8E Chapter 0 Form E

17. $\dfrac{7a}{b^3}$

18. c

19. c

20. $(x + 3y) \cdot (4a - 5b)$

21. $(x + y - 4) \cdot (x^2 + 2x \cdot y + 16 + y^2 + 4x + 4y)$

22. d

23. $\dfrac{(x^2 + 8x)}{(x + 9)}$

24. d

25. b

List of Problem Codes for BCA Testing

Gustafson/ Frisk - College Algebra 8E Chapter 0 Form E

1. gfca.00.01.4.39m_NoAlgs
2. gfca.00.01.4.57_NoAlgs
3. gfca.00.02.4.120_NoAlgs
4. gfca.00.02.4.33m_NoAlgs
5. gfca.00.01.4.85m_NoAlgs
6. gfca.00.02.4.79m_NoAlgs
7. gfca.00.02.4.89_NoAlgs
8. gfca.00.03.4.113m_NoAlgs
9. gfca.00.03.4.33_NoAlgs
10. gfca.00.03.4.37m_NoAlgs
11. gfca.00.03.4.97_NoAlgs
12. gfca.00.04.4.111_NoAlgs
13. gfca.00.04.4.15m_NoAlgs
14. gfca.00.06.4.70_NoAlgs
15. gfca.00.01.4.69_NoAlgs
16. gfca.00.04.4.81m_NoAlgs
17. gfca.00.04.4.99_NoAlgs
18. gfca.00.05.4.11m_NoAlgs
19. gfca.00.05.4.21m_NoAlgs
20. gfca.00.05.4.24_NoAlgs
21. gfca.00.05.4.85_NoAlgs
22. gfca.00.05.4.95m_NoAlgs
23. gfca.00.06.4.45_NoAlgs
24. gfca.00.06.4.57m_NoAlgs
25. gfca.00.06.4.94m_NoAlgs

1. Factor the expression completely.

   $(a + b)^2 - 3(a + b) - 4$

   Select the correct answer.
   a. $(a + b - 4)(a + b - 1)$
   b. $(a - b + 4)(a + b + 1)$
   c. $(a - b - 4)(a - b + 1)$
   d. $(a + b - 4)(a + b + 1)$

2. Select the correct distance between the following two points on the number line.

   -20 and 16

   a. 36
   b. 37
   c. 35
   d. 16
   e. 20

3. Give the degree of the polynomial.

   $\sqrt{4 \cdot 83}$

   Select the correct answer.
   a. This is not a polynomial
   b. No defined degree
   c. 0

4. Simplify the expression.

   $$\left(-\frac{27x^9}{64y^3} \right)^{1/3}$$

5. Perform the division.

   $$x^2 + x - 1 \overline{\smash{\big)}\, 3x^3 - 2x^2 - 8x + 5}$$

6. Write the expression without using absolute value symbols.

   $|2|$

7. Simplify the expression.

   $x^6 x^3$

   Select the correct answer.

   a. x^{18}
   b. x^9
   c. x^8

8. Perform the operations and simplify. Assume that no denominators are 0.

   $$\frac{-4}{2x-9y} + \frac{4}{2x-3z} - \frac{12z-36y}{(2x-9y)(2x-3z)}$$

9. Perform the division and write the answer without using negative exponents.

   $$\frac{77a^2b^3}{11ab^6}$$

10. Perform the operations and simplify. Assume that no denominators are 0.

    $$\frac{x^3+512}{x^2-36} \div \left(\frac{x^2+17x+72}{x^2+6x} \div \frac{x^2+2x-48}{x^2-8x+64} \right)$$

11. Write the inequality $-6 \leq x < 11$ using interval notation.

12. Let $x = -2$, $y = 0$, $z = 2$ and evaluate the expression.

    $$\frac{-(x^2 z^3)}{z^2 - y^2}$$

13. Perform the operations and simplify. Assume that no denominators are 0.

$$\frac{25x}{x^2 - 25} - \frac{25}{x+5}$$

Select the correct answer.

a. $\dfrac{25}{x^2 - 25}$

b. $\dfrac{125}{x^2 + 25}$

c. $\dfrac{125}{x^2 - 5}$

d. $\dfrac{125}{x^2 - 25}$

14. Rationalize the denominator.

$$\frac{7}{\sqrt{6} - 2}$$

Select the correct answer.

a. $\dfrac{7\sqrt{6} - 14}{2}$

b. $\dfrac{7\sqrt{6} + \sqrt{14}}{2}$

c. $\dfrac{7\sqrt{6} + 14}{2}$

d. $\dfrac{7\sqrt{6}}{2}$

15. Factor the expression completely.

$(x + y)^3 - 64$

16. Calculate the volume of a box that has dimensions of 5,000 by 8,100 by 4,500 millimeters. Write the answer in scientific notation.

17. Factor the expression completely.

    $4ax + 12ay - 5bx - 15by$

18. How many prime numbers are between -9 and 18 on the number line?

    Select the correct answer.

    a. 18
    b. 17
    c. 0
    d. 7
    e. 8
    f. 26

19. Factor the expression completely.

    $12x^2 + 6x^3$

    Select the correct answer.

    a. $6x^2(2 + x^2)$
    b. $6x^2(2 - x)$
    c. $6x^2(2 + x)$
    d. $6x^2(3 + x)$

20. Simplify the radical expression.

    $\sqrt[9]{27}$

    a. $\sqrt[9]{3}$

    b. $\sqrt[3]{27}$

    c. $\sqrt[27]{3}$

    d. $\sqrt[3]{3}$

    e. $\sqrt[3]{300}$

21. Rationalize the denominator and simplify.

    $\dfrac{4}{\sqrt[3]{4}}$

22. Write the expression without using negative exponents, and simplify the resulting complex fraction. Assume that denominator is not 0.

$$\frac{6y^{-1}}{8x^{-1} + 5y^{-1}}$$

Select the correct answer.

a. $\dfrac{5x}{8y + 6x}$

b. $\dfrac{6x}{8y + 5x}$

c. $\dfrac{6x}{8y - 5x}$

d. $\dfrac{6x}{5y + 8x}$

23. Simplify the expression.

$$-8^{4/3}$$

Select the correct answer.

a. -32
b. -16
c. -48
d. 19
e. -18

24. Factor the expression completely.

$4x^3 + 4x^2 - 11x - 11$

Select the correct answer.

a. $(x + 1)(11 - 4x^2)$
b. $(1 - x)(4x^2 - 11)$
c. $(x + 1)(4x^2 - 11)$
d. $(x - 1)(4x^2 + 11)$

25. Simplify the expression. Write the answer without using negative exponents. Assume that all variables are restricted to those numbers for which the expression is defined.

$$\frac{(8^{-2}z^{-4}y)^{-2}}{(5y^5z^{-1})^3(5yz^{-1})^{-1}}$$

Select the correct answer.

a. $\dfrac{512z^9}{25y^{17}}$

b. $\dfrac{25y^{16}}{4,096z^{10}}$

c. $\dfrac{4,096z^{10}}{25y^{16}}$

ANSWER KEY

Gustafson/ Frisk - College Algebra 8E Chapter 0 Form F

1. d

2. a

3. c

4. $\dfrac{-3x^3}{(4y)}$

5. $3x - 5$

6. 2

7. b

8. 0

9. $\dfrac{7a}{b^3}$

10. $\dfrac{(x^2+8x)}{(x+9)}$

11. $[-6, 11)$

12. (-8)

13. d

ANSWER KEY

Gustafson/ Frisk - College Algebra 8E Chapter 0 Form F

14. c

15. $(x + y - 4) \cdot (x^2 + 2x \cdot y + 16 + y^2 + 4x + 4y)$

16. $1.8225 \cdot 10^{11}$

17. $(x + 3y) \cdot (4a - 5b)$

18. d

19. c

20. d

21. $\sqrt[3]{16}$

22. b

23. b

24. c

25. c

List of Problem Codes for BCA Testing

Gustafson/ Frisk - College Algebra 8E Chapter 0 Form F

1. gfca.00.05.4.95m_NoAlgs
2. gfca.00.01.4.85m_NoAlgs
3. gfca.00.04.4.15m_NoAlgs
4. gfca.00.03.4.33_NoAlgs
5. gfca.00.04.4.111_NoAlgs
6. gfca.00.01.4.69_NoAlgs
7. gfca.00.02.4.33m_NoAlgs
8. gfca.00.06.4.70_NoAlgs
9. gfca.00.04.4.99_NoAlgs
10. gfca.00.06.4.45_NoAlgs
11. gfca.00.01.4.57_NoAlgs
12. gfca.00.02.4.89_NoAlgs
13. gfca.00.06.4.57m_NoAlgs
14. gfca.00.04.4.81m_NoAlgs
15. gfca.00.05.4.85_NoAlgs
16. gfca.00.02.4.120_NoAlgs
17. gfca.00.05.4.24_NoAlgs
18. gfca.00.01.4.39m_NoAlgs
19. gfca.00.05.4.11m_NoAlgs
20. gfca.00.03.4.113m_NoAlgs
21. gfca.00.03.4.97_NoAlgs
22. gfca.00.06.4.94m_NoAlgs
23. gfca.00.03.4.37m_NoAlgs
24. gfca.00.05.4.21m_NoAlgs
25. gfca.00.02.4.79m_NoAlgs

Gustafson/ Frisk - College Algebra 8E Chapter 0 Form G

1. Perform the operations and simplify. Assume that no denominators are 0

$$\frac{x^3 + 512}{x^2 - 36} \div \left(\frac{x^2 + 17x + 72}{x^2 + 6x} \div \frac{x^2 + 2x - 48}{x^2 - 8x + 64} \right)$$

2. Express the number -180,000,000 in scientific notation.

   Select the correct answer.

   a. -1.8×10^8

   b. -1.8×10^9

   c. -1.8×10^7

   d. -18×10^9

3. Let $x = -3$, $y = 0$, $z = 3$ and evaluate the expression.

   $$\frac{-(x^2 z^3)}{z^2 - y^2}$$

   Select the correct answer.

   a. - 27
   b. 0
   c. - 81

4. Write the expression without using absolute value symbols.

   $-|-3|$

   Select the correct answer.

   a. 0
   b. - 3
   c. 3
   d. 1

5. Perform the operations and simplify. Assume that no denominators are 0.

$$\frac{-8}{2x-5y} + \frac{8}{2x-9z} - \frac{72z-40y}{(2x-5y)(2x-9z)}$$

Select the correct answer.

a. $\dfrac{8}{2x-5y}$

b. 0

c. $\dfrac{72z-40y}{(2x-5y)(9z-2x)}$

d. $\dfrac{8}{2x-9z}$

6. Write the expression without using absolute value symbols.

|2|

7. Factor the expression completely.

$(a+b)^2 - 3(a+b) - 4$

Select the correct answer.

a. $(a+b-4)(a+b+1)$
b. $(a-b-4)(a-b+1)$
c. $(a-b+4)(a+b+1)$
d. $(a+b-4)(a+b-1)$

8. Simplify the expression. Write your answer without using negative exponents.

$\left(-\dfrac{125}{27}\right)^{-1/3}$

9. Factor the expression completely.

$4ax + 12ay - 5bx - 15by$

10. Factor the expression completely.

    $(x + y)^3 - 64$

11. Perform the operations and simplify. Assume that no denominators are 0.

    $$\frac{x^2 + 8x}{x - 8} \cdot \frac{x^2 - 64}{x + 9}$$

    Select the correct answer.

    a. $\dfrac{x^3 + 16x^2 + 64x}{x + 9}$

    b. $\dfrac{x^2 + 16x + 64}{x + 9}$

    c. $\dfrac{x^3 + 16x^2 - 16x}{x + 9}$

    d. $\dfrac{x^3 - 16x^2 + 64x}{x - 9}$

12. Rationalize the denominator.

    $$\frac{7}{\sqrt{6} - 2}$$

    Select the correct answer.

    a. $\dfrac{7\sqrt{6} - 14}{2}$

    b. $\dfrac{7\sqrt{6} + \sqrt{14}}{2}$

    c. $\dfrac{7\sqrt{6} + 14}{2}$

    d. $\dfrac{7\sqrt{6}}{2}$

13. Simplify the expression.

    $(x^3)^4 (x^2)^3$

    Select the correct answer.

    a. x^{18}
    b. x^{10}
    c. x^{12}

14. Factor the expression completely.

    $12x^2 + 6x^3$

    Select the correct answer.

    a. $6x^2(2+x)$
    b. $6x^2(2-x)$
    c. $6x^2(2+x^2)$
    d. $6x^2(3+x)$

15. Perform the division and write the answer without using negative exponents.

    $$\frac{50a^2b^3}{10ab^6}$$

    Select the correct answer.

    a. $5ab^3$

    b. $\dfrac{5a}{b^3}$

    c. $\dfrac{5}{ab^3}$

    d. $\dfrac{5a^2}{b^3}$

16. Simplify the expression.

    $-8^{4/3}$

    Select the correct answer.

    a. -16
    b. -48
    c. -18
    d. -32
    e. 19

17. Simplify the expression. Write the answer without using negative exponents. Assume that the variable is restricted to those numbers for which the expression is defined.

    $$\frac{x^3 x^5}{x^5 x}$$

18. Perform the operations and simplify.

    $(2r - 10s)$

    $(2r^2 - 8rs - 4s^2)$

19. Perform division and write the answer without using negative exponents.

$$\frac{-64x^6y^4z^9}{16x^9y^6z^0}$$

Select the correct answer.

a. $\dfrac{4z^9}{x^3y^2}$

b. $\dfrac{-4z^4}{x^3y^2}$

c. $\dfrac{-4z^9}{x^3y^2}$

d. $\dfrac{4z^9}{x^3y^6}$

20. Factor the expression completely.

$x^2 + 10x + 24$

Select the correct answer.

a. $(x + 4)(x - 6)$
b. $(x + 4)(x + 6)$
c. $(x - 4)(x - 6)$
d. $(x + 4)(x + 4)$

21. How many prime numbers are between -9 and 18 on the number line?

Select the correct answer.

a. 7
b. 18
c. 8
d. 26
e. 17
f. 0

22. Find the distance between the following two points on the number line.

-7 and 3

23. Rationalize the denominator and simplify.

$$\frac{3}{\sqrt[5]{3}}$$

Select the correct answer.

a. $\sqrt[5]{181}$

b. $\sqrt[6]{82}$

c. $\sqrt[5]{84}$

d. $\sqrt[10]{81}$

e. $\sqrt[5]{81}$

24. Simplify the expression. Assume that all variables represent positive numbers, so that no absolute value symbols are needed.

$$6y^2\sqrt{64y^5} - 2\sqrt{100y^9}$$

Select the correct answer.

a. $9y^5\sqrt{y}$

b. $28y^4\sqrt[5]{y}$

c. $28y^4\sqrt{y}$

d. $28y^5\sqrt{y}$

25. Write the expression without using negative exponents, and simplify the resulting complex fraction. Assume that no denominators are 0.

$$\frac{8y^{-1}}{2x^{-1} + 13y^{-1}}$$

ANSWER KEY

Gustafson/ Frisk - College Algebra 8E Chapter 0 Form G

1. $\dfrac{(x^2+8x)}{(x+9)}$

2. a

3. a

4. b

5. b

6. 2

7. a

8. $-\dfrac{3}{5}$

9. $(x+3y)\cdot(4a-5b)$

10. $(x+y-4)\cdot(x^2+2x\cdot y+16+y^2+4x+4y)$

11. a

12. c

13. a

14. a

15. b

16. a

17. x^2

18. $4r^3-36r^2\cdot s+72r\cdot s^2+40s^3$

19. c

20. b

21. a

22. 10

ANSWER KEY

Gustafson/ Frisk - College Algebra 8E Chapter 0 Form G

23. e

24. c

25. $\dfrac{8x}{(2y+13x)}$

List of Problem Codes for BCA Testing

Gustafson/ Frisk - College Algebra 8E Chapter 0 Form G

1. gfca.00.06.4.45_NoAlgs
2. gfca.00.02.4.97m_NoAlgs
3. gfca.00.02.4.89m_NoAlgs
4. gfca.00.01.4.73m_NoAlgs
5. gfca.00.06.4.70m_NoAlgs
6. gfca.00.01.4.69_NoAlgs
7. gfca.00.05.4.95m_NoAlgs
8. gfca.00.03.4.49_NoAlgs
9. gfca.00.05.4.24_NoAlgs
10. gfca.00.05.4.85_NoAlgs
11. gfca.00.06.4.35m_NoAlgs
12. gfca.00.04.4.81m_NoAlgs
13. gfca.00.02.4.41m_NoAlgs
14. gfca.00.05.4.11m_NoAlgs
15. gfca.00.04.4.99m_NoAlgs
16. gfca.00.03.4.37m_NoAlgs
17. gfca.00.02.4.63_NoAlgs
18. gfca.00.04.4.68_NoAlgs
19. gfca.00.04.4.101m_NoAlgs
20. gfca.00.05.4.51m_NoAlgs
21. gfca.00.01.4.39m_NoAlgs
22. gfca.00.01.4.85_NoAlgs
23. gfca.00.03.4.97m_NoAlgs
24. gfca.00.03.4.85m_NoAlgs
25. gfca.00.06.4.94_NoAlgs

1. Perform the operations and simplify. Assume that no denominators are 0.

   $$\frac{-8}{2x-5y} + \frac{8}{2x-9z} - \frac{72z-40y}{(2x-5y)(2x-9z)}$$

   Select the correct answer.

   a. $\dfrac{8}{2x-5y}$

   b. 0

   c. $\dfrac{72z-40y}{(2x-5y)(9z-2x)}$

   d. $\dfrac{8}{2x-9z}$

2. Write the expression without using absolute value symbols.

   $-|-3|$

   Select the correct answer.

   a. 0
   b. -3
   c. 3
   d. 1

3. Perform the division and write the answer without using negative exponents.

$$\frac{50a^2b^3}{10ab^6}$$

Select the correct answer.

a. $5ab^3$

b. $\dfrac{5a}{b^3}$

c. $\dfrac{5}{ab^3}$

d. $\dfrac{5a^2}{b^3}$

4. Write the expression without using absolute value symbols.

$|2|$

5. Factor the expression completely.

$x^2 + 10x + 24$

Select the correct answer.
a. $(x + 4)(x - 6)$
b. $(x - 4)(x - 6)$
c. $(x + 4)(x + 6)$
d. $(x + 4)(x + 4)$

6. Simplify the expression. Assume that all variables represent positive numbers, so that no absolute value symbols are needed.

$$6y^2\sqrt{64y^5} - 2\sqrt{100y^9}$$

Select the correct answer.

a. $9y^5\sqrt{y}$

b. $28y^4\sqrt[5]{y}$

c. $28y^4\sqrt{y}$

d. $28y^5\sqrt{y}$

7. Find the distance between the following two points on the number line.

   -7 and 3

8. Simplify the expression. Write your answer without using negative exponents.

$$\left(-\frac{125}{27}\right)^{-1/3}$$

9. Factor the expression completely.

   $4ax + 12ay - 5bx - 15by$

10. Write the expression without using negative exponents, and simplify the resulting complex fraction. Assume that no denominators are 0.

$$\frac{8y^{-1}}{2x^{-1} + 13y^{-1}}$$

11. Express the number -180,000,000 in scientific notation.

    Select the correct answer.

    a. -1.8×10^8

    b. -1.8×10^9

    c. -1.8×10^7

    d. -18×10^9

12. Rationalize the denominator and simplify.

    $$\frac{3}{\sqrt[5]{3}}$$

    Select the correct answer.

    a. $\sqrt[5]{181}$

    b. $\sqrt[6]{82}$

    c. $\sqrt[5]{84}$

    d. $\sqrt[10]{81}$

    e. $\sqrt[5]{81}$

13. Factor the expression completely.

    $(a + b)^2 - 3(a + b) - 4$

    Select the correct answer.

    a. $(a + b - 4)(a + b + 1)$
    b. $(a - b - 4)(a - b + 1)$
    c. $(a - b + 4)(a + b + 1)$
    d. $(a + b - 4)(a + b - 1)$

14. How many prime numbers are between -9 and 18 on the number line?

    Select the correct answer.

    a. 7
    b. 18
    c. 8
    d. 26
    e. 17
    f. 0

15. Factor the expression completely.

    $(x + y)^3 - 64$

16. Rationalize the denominator.

    $$\frac{7}{\sqrt{6} - 2}$$

    Select the correct answer.

    a. $\dfrac{7\sqrt{6}}{2}$

    b. $\dfrac{7\sqrt{6} + 14}{2}$

    c. $\dfrac{7\sqrt{6} + \sqrt{14}}{2}$

    d. $\dfrac{7\sqrt{6} - 14}{2}$

17. Let $x = -3$, $y = 0$, $z = 3$ and evaluate the expression.

    $$\frac{-(x^2 z^3)}{z^2 - y^2}$$

    Select the correct answer.

    a. -27
    b. 0
    c. -81

18. Perform the operations and simplify. Assume that no denominators are 0.

$$\frac{x^2 + 8x}{x - 8} \cdot \frac{x^2 - 64}{x + 9}$$

Select the correct answer.

a. $\dfrac{x^3 + 16x^2 + 64x}{x + 9}$

b. $\dfrac{x^2 + 16x + 64}{x + 9}$

c. $\dfrac{x^3 + 16x^2 - 16x}{x + 9}$

d. $\dfrac{x^3 - 16x^2 + 64x}{x - 9}$

19. Simplify the expression.

$$-8^{4/3}$$

Select the correct answer.

a. -16
b. -48
c. -18
d. -32
e. 19

20. Simplify the expression. Write the answer without using negative exponents. Assume that the variable is restricted to those numbers for which the expression is defined.

$$\frac{x^3 x^5}{x^5 x}$$

21. Perform the operations and simplify.

    $(2r - 10s)$

    $(2r^2 - 8rs - 4s^2)$

22. Perform the operations and simplify. Assume that no denominators are 0.

    $$\frac{x^3 + 512}{x^2 - 36} \div \left(\frac{x^2 + 17x + 72}{x^2 + 6x} \div \frac{x^2 + 2x - 48}{x^2 - 8x + 64} \right)$$

23. Simplify the expression.

    $(x^3)^4 (x^2)^3$

    Select the correct answer.

    a. x^{18}
    b. x^{10}
    c. x^{12}

24. Perform division and write the answer without using negative exponents.

    $$\frac{-64x^6 y^4 z^9}{16x^9 y^6 z^0}$$

    Select the correct answer.

    a. $\dfrac{4z^9}{x^3 y^2}$

    b. $\dfrac{-4z^4}{x^3 y^2}$

    c. $\dfrac{-4z^9}{x^3 y^2}$

    d. $\dfrac{4z^9}{x^3 y^6}$

25. Factor the expression completely.

$12x^2 + 6x^3$

Select the correct answer.

a. $6x^2(2 + x)$
b. $6x^2(2 - x)$
c. $6x^2(2 + x^2)$
d. $6x^2(3 + x)$

ANSWER KEY

Gustafson/ Frisk - College Algebra 8E Chapter 0 Form H

1. b

2. b

3. b

4. 2

5. c

6. c

7. 10

8. $-\dfrac{3}{5}$

9. $(x + 3y) \cdot (4a - 5b)$

10. $\dfrac{8x}{(2y + 13x)}$

11. a

12. e

13. a

14. a

15. $(x + y - 4) \cdot (x^2 + 2x \cdot y + 16 + y^2 + 4x + 4y)$

16. b

17. a

18. a

19. a

20. x^2

ANSWER KEY

Gustafson/ Frisk - College Algebra 8E Chapter 0 Form H

21. $4r^3 - 36r^2 \cdot s + 72r \cdot s^2 + 40s^3$

22. $\dfrac{\left(x^2 + 8x\right)}{(x+9)}$

23. a

24. c

25. a

List of Problem Codes for BCA Testing

Gustafson/ Frisk - College Algebra 8E Chapter 0 Form H

1. gfca.00.06.4.70m_NoAlgs
2. gfca.00.01.4.73m_NoAlgs
3. gfca.00.04.4.99m_NoAlgs
4. gfca.00.01.4.69_NoAlgs
5. gfca.00.05.4.51m_NoAlgs
6. gfca.00.03.4.85m_NoAlgs
7. gfca.00.01.4.85_NoAlgs
8. gfca.00.03.4.49_NoAlgs
9. gfca.00.05.4.24_NoAlgs
10. gfca.00.06.4.94_NoAlgs
11. gfca.00.02.4.97m_NoAlgs
12. gfca.00.03.4.97m_NoAlgs
13. gfca.00.05.4.95m_NoAlgs
14. gfca.00.01.4.39m_NoAlgs
15. gfca.00.05.4.85_NoAlgs
16. gfca.00.04.4.81m_NoAlgs
17. gfca.00.02.4.89m_NoAlgs
18. gfca.00.06.4.35m_NoAlgs
19. gfca.00.03.4.37m_NoAlgs
20. gfca.00.02.4.63_NoAlgs
21. gfca.00.04.4.68_NoAlgs
22. gfca.00.06.4.45_NoAlgs
23. gfca.00.02.4.41m_NoAlgs
24. gfca.00.04.4.101m_NoAlgs
25. gfca.00.05.4.11m_NoAlgs

Gustafson/ Frisk - College Algebra 8E Chapter 1 Form A

1. Solve the equation by factoring.

   $$4x^2 - 28x = -24$$

2. Solve the formula for the indicated variable.

   $$\frac{x^2}{m^2} + \frac{y^2}{f^2} = 1 \; ; \; y$$

3. Find the absolute value.

   $$\left| \frac{1}{1+i} \right|$$

4. Solve the inequality and write the answer in interval notation.

   $$\frac{4}{3}x - \frac{3}{2}x \leq \frac{3}{8}\left(x + \frac{4}{3}\right) + \frac{1}{3}$$

5. John drove to a distant city in 4 hours. When he returned, there was less traffic, and the trip took only 2 hours. If John averaged 22 mph faster on the return trip, how fast did he drive each way?

6. Solve the inequality. Write the answer in interval notation.

   $$\frac{4}{x-7} \leq 2$$

7. A rectangle is 2 times as long as it is wide. If the area is 18 square feet, find its perimeter.

8. A full-price ticket for a college basketball game costs $5, and a student ticket costs $4.50. If 700 tickets were sold, and the total receipts were $3,370.00, how many tickets were student tickets?

9. Solve the equation for real values of the variable by factoring.

   $6m^{\frac{2}{3}} - 36m^{\frac{1}{3}} + 48 = 0$

10. Use the square root property to solve the equation.

    $x^2 - 14x + 49 = 64$

11. Find the values of *x* and *y*.

    $x + 97i = y - yi$

12. Solve the inequality. Express the solution set in interval notation.

    $0 < |4x + 1| < 5$

13. Solve the formula for the *x*.

    $r = \dfrac{k - x}{w}$

14. Solve the equation.

    $x(x - 7) - 14 = (x - 1)^2$

15. A college student earns $50 per day delivering advertising brochures door-to-door, plus 50 cents for each person he interviews. How many people did he interview on a day when he earned $78?

16. Solve the inequality and write the answer in interval notation.

    $-4x - 13 > -9$

17. Solve the equation for x.

    $\left| \dfrac{3x - 20}{2} \right| = 13$

18. Solve the equation.

    $3x + 9 = x + 13$

19. A cyclist rides from DeKalb to Rockford, a distance of 120 miles. His return trip takes 1 hours longer, because his speed decreases by 10 miles per hour. How fast does he ride each way?

20. Find all real solutions of the equation.

    $\sqrt{y + 1} = 5 - y$

21. Do the operation and express the answer in a + bi form.

    $(3 + \sqrt{-16})(5 - \sqrt{-25})$

22. Solve the equation for x.

    $|x + 9| = |9 - x|$

23. Solve the equation for real values of the variable by factoring.

    $8p - 80p^{\frac{1}{2}} + 168 = 0$

24. Solve the equation by completing the square.

    $x^2 - 6x - 55 = 0$

25. A piece of tin, $y = 15$ inches on a side, is to have four equal squares cut from its corners, as in illustration. If the edges are then to be folded up to make a box with a floor area of 25 square inches, find the depth of the box.

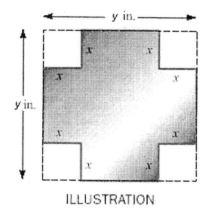

ILLUSTRATION

ANSWER KEY

Gustafson/ Frisk - College Algebra 8E Chapter 1 Form A

1. 1, 6
2. $f \cdot \sqrt{\left(1 - \frac{x}{m}\right) \cdot \left(1 + \frac{x}{m}\right)}, \ -f \cdot \sqrt{\left(1 - \frac{x}{m}\right) \cdot \left(1 + \frac{x}{m}\right)}$
3. $\dfrac{1}{\sqrt{2}}$
4. $[-1.538462, \infty)$
5. 22, 44
6. $(-\infty, 7) \cup [9, \infty)$
7. 18
8. 260
9. 64, 8
10. 15, −1
11. $(-97, -97)$
12. $\left(\dfrac{-6}{4}, \dfrac{-1}{4}\right) \cup \left(\dfrac{-1}{4}, 1\right)$
13. $x = k - r \cdot w$
14. −3
15. 56
16. $(-\infty, -1)$
17. $\dfrac{46}{3}, \ -2$
18. 2
19. 30, 40
20. 3
21. $35 + 5i$
22. 0
23. 9, 49
24. −5, 11
25. 5

List of Problem Codes for BCA Testing

Gustafson/ Frisk - College Algebra 8E Chapter 1 Form A

1. gfca.01.03.4.16_NoAlgs
2. gfca.01.03.4.67_NoAlgs
3. gfca.01.05.4.63_NoAlgs
4. gfca.01.07.4.31_NoAlgs
5. gfca.01.02.4.43_NoAlgs
6. gfca.01.07.4.78_NoAlgs
7. gfca.01.04.4.04_NoAlgs
8. gfca.01.02.4.21_NoAlgs
9. gfca.01.06.4.17_NoAlgs
10. gfca.01.03.4.26_NoAlgs
11. gfca.01.05.4.10_NoAlgs
12. gfca.01.08.4.61_NoAlgs
13. gfca.01.01.4.78_NoAlgs
14. gfca.01.01.4.48_NoAlgs
15. gfca.01.02.4.12_NoAlgs
16. gfca.01.07.4.17_NoAlgs
17. gfca.01.08.4.25_NoAlgs
18. gfca.01.01.4.18_NoAlgs
19. gfca.01.04.4.11_NoAlgs
20. gfca.01.06.4.39_NoAlgs
21. gfca.01.05.4.25_NoAlgs
22. gfca.01.08.4.36_NoAlgs
23. gfca.01.06.4.23_NoAlgs
24. gfca.01.03.4.39_NoAlgs
25. gfca.01.04.4.07_NoAlgs

Gustafson/ Frisk - College Algebra 8E Chapter 1 Form B

1. Solve the equation by factoring.

    $4x^2 - 28x = -24$

2. A college student earns $50 per day delivering advertising brochures door-to-door, plus 50 cents for each person he interviews. How many people did he interview on a day when he earned $78?

3. Solve the equation.

    $3x + 9 = x + 13$

4. Find the absolute value.

    $\left| \dfrac{1}{1+i} \right|$

5. Solve the equation by completing the square.

    $x^2 - 6x - 55 = 0$

6. Solve the equation for real values of the variable by factoring.

    $6m^{\frac{2}{3}} - 36m^{\frac{1}{3}} + 48 = 0$

7. Solve the inequality and write the answer in interval notation.

    $-4x - 13 > -9$

8. Do the operation and express the answer in *a + bi* form.

    $(3 + \sqrt{-16})(5 - \sqrt{-25})$

9. Solve the equation.

$$x(x-7) - 14 = (x-1)^2$$

10. John drove to a distant city in 4 hours. When he returned, there was less traffic, and the trip took only 2 hours. If John averaged 22 mph faster on the return trip, how fast did he drive each way?

11. Solve the equation for x.

$$|x + 9| = |9 - x|$$

12. Use the square root property to solve the equation.

$$x^2 - 14x + 49 = 64$$

13. A rectangle is 2 times as long as it is wide. If the area is 18 square feet, find its perimeter.

14. Solve the equation for real values of the variable by factoring.

$$8p - 80p^{\frac{1}{2}} + 168 = 0$$

15. Solve the equation for x.

$$\left| \frac{3x - 20}{2} \right| = 13$$

16. A cyclist rides from DeKalb to Rockford, a distance of 120 miles. His return trip takes 1 hours longer, because his speed decreases by 10 miles per hour. How fast does he ride each way?

17. Solve the formula for the indicated variable.

$$\frac{x^2}{m^2} + \frac{y^2}{f^2} = 1 \; ; \; y$$

18. Solve the inequality and write the answer in interval notation.

$$\frac{4}{3}x - \frac{3}{2}x \leq \frac{3}{8}\left(x + \frac{4}{3}\right) + \frac{1}{3}$$

19. Find all real solutions of the equation.

$$\sqrt{y+1} = 5 - y$$

20. Solve the formula for the x.

$$r = \frac{k-x}{w}$$

21. Solve the inequality. Write the answer in interval notation.

$$\frac{4}{x-7} \leq 2$$

22. A piece of tin, $y = 15$ inches on a side, is to have four equal squares cut from its corners, as in illustration. If the edges are then to be-folded up to make a box with a floor area of 25 square inches, find the depth of the box.

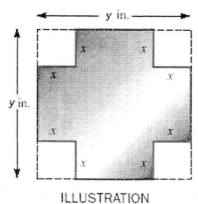

ILLUSTRATION

23. Find the values of x and y.

$x + 97i = y - yi$

24. Solve the inequality. Express the solution set in interval notation.

$0 < |\,4x + 1\,| < 5$

25. A full-price ticket for a college basketball game costs $5, and a student ticket costs $4.50. If 700 tickets were sold, and the total receipts were $3,370.00, how many tickets were student tickets?

ANSWR KEY

Gustafson/ Frisk - College Algebra 8E Chapter 1 Form B

1. 1, 6
2. 56
3. 2
4. $\dfrac{1}{\sqrt{2}}$
5. −5, 11
6. 64, 8
7. $(-\infty, -1)$
8. $35 + 5i$
9. −3
10. 22, 44
11. 0
12. 15, −1
13. 18
14. 9, 49
15. $\dfrac{46}{3}, -2$
16. 30; 40
17. $f \cdot \sqrt{\left(1 - \dfrac{x}{m}\right) \cdot \left(1 + \dfrac{x}{m}\right)}, -f \cdot \sqrt{\left(1 - \dfrac{x}{m}\right) \cdot \left(1 + \dfrac{x}{m}\right)}$
18. $[-1.538462, \infty)$
19. 3
20. $x = k - r \cdot w$
21. $(-\infty, 7) \cup [9, \infty)$
22. 5
23. $(-97, -97)$
24. $\left(\dfrac{-6}{4}, \dfrac{-1}{4}\right) \cup \left(\dfrac{-1}{4}, 1\right)$
25. 260

List of Problem Codes for BCA Testing

Gustafson/ Frisk - College Algebra 8E Chapter 1 Form B

1. gfca.00.06.4.35_NoAlgs
2. gfca.00.02.4.41_NoAlgs
3. gfca.00.04.4.81_NoAlgs
4. gfca.00.03.4.85_NoAlgs
5. gfca.00.06.4.70_NoAlgs
6. gfca.00.01.4.65_NoAlgs
7. gfca.00.02.4.63_NoAlgs
8. gfca.00.01.4.39_NoAlgs
9. gfca.00.05.4.11_NoAlgs
10. gfca.00.02.4.79_NoAlgs
11. gfca.00.03.4.97_NoAlgs
12. gfca.00.02.4.89_NoAlgs
13. gfca.00.03.4.113_NoAlgs
14. gfca.00.05.4.51_NoAlgs
15. gfca.00.04.4.11_NoAlgs
16. gfca.00.04.4.68_NoAlgs
17. gfca.00.01.4.69_NoAlgs
18. gfca.00.01.4.85_NoAlgs
19. gfca.00.06.4.57_NoAlgs
20. gfca.00.03.4.49_NoAlgs
21. gfca.00.06.4.94_NoAlgs
22. gfca.00.05.4.40_NoAlgs
23. gfca.00.05.4.24_NoAlgs
24. gfca.00.04.4.101_NoAlgs
25. gfca.00.05.4.16_NoAlgs

1. A full-price ticket for a college basketball game costs $5.00, and a student ticket costs $4.00. If 458 tickets were sold, and the total receipts were $2,063.00, how many tickets were student tickets?

   Select the correct answer.

   a. 227 tickets
   b. 234 tickets
   c. 224 tickets
   d. 230 tickets
   e. 219 tickets

2. John drove to a distant city in 4 hours. When he returned, there was less traffic, and the trip took only 2 hours. If John averaged 40 mph faster on the return trip, how fast did he drive each way?

   Select the correct answer.

   a. 48 mph, 88 mph
   b. 40 mph, 80 mph
   c. 27 mph, 67 mph
   d. 44 mph, 84 mph
   e. 51 mph, 91 mph

3. Solve the equation by factoring.

   $$9x^2 - 54x = -72$$

   Select the correct answer(s).

   a. $x = 2$
   b. $x = 7$
   c. $x = 4$
   d. $x = -1$

4. A rectangle is 2 times as long as it is wide. If the area is 8 square feet, find its perimeter.

   Select the correct answer.

   a. 24 ft
   b. 12 ft
   c. 22 ft
   d. 17 ft

5. Use the square root property to solve the equation.

    $x^2 - 16x + 64 = 49$

    Select the correct answer(s).

    a. $x = 7$
    b. $x = 8$
    c. $x = 16$
    d. $x = 1$
    e. $x = 15$

6. Solve the equation by completing the square.

    $x^2 - 6x - 40 = 0$

    Select the correct answers.

    a. $x = -4, x = 10$
    b. $x = 7, x = 10$
    c. $x = 3, x = 8$
    d. $x = -4, x = 3$

7. Solve the equation.

    $3x + 13 = x + 11$

    Select the correct answer.

    a. $x = -1$
    b. $x = 48$
    c. $x = -6$
    d. $x = 51$

8. Solve the equation.

    $x(x - 6) - 15 = (x - 1)^2$

    Select the correct answer.

    a. $x = 3$
    b. $x = -13$
    c. $x = -4$
    d. $x = 1$

9. Solve the formula for the x.

$$p = \frac{n - x}{q}$$

Select the correct answer.

a. $x = pq - n$
b. $x = n - pq$
c. $x = nq - p$
d. $x = p - nq$

10. A college student earns $45 per day delivering advertising brochures door-to-door, plus 50 cents for each person he interviews. How many people did he interview on a day when he earned $113?

Select the correct answer.

a. 141
b. 138
c. 135
d. 137
e. 136

11. Solve the formula for the indicated variable.

$$\frac{x^2}{d^2} + \frac{y^2}{f^2} = 1 \; ; \; y$$

Select the correct answer.

a. $y = \sqrt{f\left(1 - \frac{x}{d}\right)^2}$, $y = -\sqrt{f\left(1 - \frac{x}{d}\right)^2}$

b. $y = f\sqrt{\left(1 - \frac{x}{d}\right)\left(1 + \frac{x}{d}\right)}$, $y = -f\sqrt{\left(1 - \frac{x}{d}\right)\left(1 + \frac{x}{d}\right)}$

c. $y = \sqrt{f(1 - xd)(1 + xd)}$, $y = -\sqrt{f(1 - xd)(1 + xd)}$

12. A piece of tin, y = 12 inches on a side, is to have four equal squares cut from its corners, as in the illustration. If the edges are then to be-folded up to make a box with a floor area of 16 square inches, find the depth of the box.

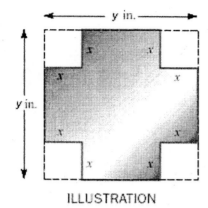

ILLUSTRATION

Select the correct answer.

a.   4 in
b.   8 in
c.   7 in
d.   9 in

13. A cyclist rides from DeKalb to Rockford, a distance of 120 miles. His return trip takes 1 hour longer, because his speed decreases by 10 miles per hour. How fast does he ride each way?

    Select the correct answer.

    a. 40 mph going and 40 mph returning
    b. 30 mph going and 120 mph returning
    c. 60 mph going and 30 mph returning
    d. 40 mph going and 30 mph returning

14. Do the operation and express the answer in a + bi form.

    $(2 - \sqrt{-9}) - (9i - 6)$

    Select the correct answer.

    a. $8 + 12i$
    b. $8 - 12i$
    c. $-8 + 12i$
    d. $-8 - 12i$

15. Do the operation and express the answer in a + bi form.

    $\dfrac{1}{6 + i}$

    Select the correct answer.

    a. $\dfrac{6}{7} + \dfrac{i}{7}$

    b. $\dfrac{6}{37} - \dfrac{i}{37}$

    c. $-\dfrac{6}{37} + \dfrac{i}{37}$

    d. $-\dfrac{6}{7} - \dfrac{i}{7}$

16. Find the absolute value.

$$\left| \frac{1}{4+i} \right|$$

Select the correct answer.

a. $\dfrac{1}{\sqrt{17}}$

b. 17

c. $\sqrt{17}$

d. $\dfrac{1}{17}$

17. Solve the equation for real values of the variable by factoring.

$$2m^{\frac{2}{3}} - 18m^{\frac{1}{3}} + 28 = 0$$

Select the correct answer(s).

a. $x = 7$
b. $x = 2$
c. $x = 8$
d. $x = 343$

18. Solve the equation for real values of the variable by factoring.

$$9p - 63p^{\frac{1}{2}} + 90 = 0$$

Select the correct answer(s).

a. $x = 25$
b. $x = 5$
c. $x = 4$
d. $x = 2$

19. Find all real solutions of the equation.

    $\sqrt{y + 7} = 5 - y$

    Select the correct answer(s).

    a. $x = 2$
    b. $x = 5$
    c. $x = 0$
    d. $x = 7$

20. Solve the inequality.

    $-2x - 11 > -1$

    Select the correct answer.

    a. $[5, \infty)$
    b. $(-\infty, 5)$
    c. $(-\infty, 5]$
    d. $(-\infty, -5)$

21. Solve the inequality.

    $\frac{4}{3}x - \frac{3}{2}x \leq \frac{3}{8}(x + \frac{4}{3}) + \frac{1}{3}$

    Select the correct answer.

    a. $(-\infty, -\frac{20}{13}]$
    b. $[-\frac{20}{13}, \infty)$
    c. $(-\frac{20}{13}, \infty)$
    d. $(-\infty, -\frac{20}{13})$

22. Solve the inequality.

$$\frac{8}{x-1} \leq 2$$

Select the correct answer.

a. $(-\infty, 1) \cup [5, \infty)$
b. $(-\infty, 1] \cup [5, \infty)$
c. $(-\infty, 1) \cup (5, \infty)$
d. $(-\infty, 1] \cup (5, \infty)$
e. $(1, 5]$

23. Solve the equation for x.

$$\left| \frac{3x-16}{2} \right| = 11$$

Select the correct answer.

a. $x_1 = -\frac{38}{3}, x_2 = -2$

b. $x_1 = \frac{38}{3}, x_2 = -2$

c. $x_1 = -\frac{38}{3}, x_2 = \frac{2}{3}$

24. Solve the equation for x.

$|x + 9| = |9 - x|$

Select the correct answer.

a. $x_1 = 9, x_2 = 0$
b. $x_1 = 0, x_2 = -1$
c. $x = -9$
d. $x = 0$

25. Solve the inequality. Express the solution set in interval notation.

$0 < |4x + 3| < 7$

Select the correct answer.

a. $\left(-\dfrac{5}{2}, -\dfrac{3}{4}\right) \cup \left(-\dfrac{3}{4}, 1\right)$

b. $\left(-\infty, -\dfrac{3}{4}\right) \cup (1, \infty)$

c. $\left(-\dfrac{3}{4}, 1\right)$

d. $(-\infty, -1) \cup \left(\dfrac{3}{4}, \infty\right)$

ANSWER KEY

Gustafson/ Frisk - College Algebra 8E Chapter 1 Form C

1. a
2. b
3. a,c
4. b
5. d,e
6. a
7. a
8. c
9. b
10. e
11. b
12. a
13. d
14. b
15. b
16. a
17. c,d
18. a,c
19. a
20. d
21. b
22. a
23. b
24. d
25. a

List of Problem Codes for BCA Testing

Gustafson Frisk - College Algebra 8E Chapter 1 Form C

1. gfca.00.01.4.39m_NoAlgs
2. gfca.00.01.4.57m_NoAlgs
3. gfca.00.01.4.73m_NoAlgs
4. gfca.00.01.4.83m_NoAlgs
5. gfca.00.02.4.120m_NoAlgs
6. gfca.00.03.4.97m_NoAlgs
7. gfca.00.02.4.79m_NoAlgs
8. gfca.00.02.4.89m_NoAlgs
9. gfca.00.03.4.113m_NoAlgs
10. gfca.00.03.4.49m_NoAlgs
11. gfca.00.04.4.101m_NoAlgs
12. gfca.00.04.4.68m_NoAlgs
13. gfca.00.04.4.81m_NoAlgs
14. gfca.00.04.4.99m_NoAlgs
15. gfca.00.05.4.11m_NoAlgs
16. gfca.00.05.4.16m_NoAlgs
17. gfca.00.05.4.24m_NoAlgs
18. gfca.00.05.4.40m_NoAlgs
19. gfca.00.06.4.35m_NoAlgs
20. gfca.00.06.4.57m_NoAlgs
21. gfca.00.06.4.70m_NoAlgs
22. gfca.00.03.4.85m_NoAlgs
23. gfca.00.06.4.94m_NoAlgs
24. gfca.00.02.4.41m_NoAlgs
25. gfca.00.05.4.51m_NoAlgs

Gustafson/ Frisk - College Algebra 8E Chapter 1 Form D

1. Solve the formula for the indicated variable.

   $$\frac{x^2}{d^2} + \frac{y^2}{f^2} = 1 \; ; \; y$$

   Select the correct answer.

   a. $y = \sqrt{f\left(1-\frac{x}{d}\right)^2}$, $y = -\sqrt{f\left(1-\frac{x}{d}\right)^2}$

   b. $y = f\sqrt{\left(1-\frac{x}{d}\right)\left(1+\frac{x}{d}\right)}$, $y = -f\sqrt{\left(1-\frac{x}{d}\right)\left(1+\frac{x}{d}\right)}$

   c. $y = \sqrt{f(1-xd)(1+xd)}$, $y = -\sqrt{f(1-xd)(1+xd)}$

2. A piece of tin, $y = 12$ inches on a side, is to have four equal squares cut from its corners, as in the illustration. If the edges are then to be folded up to make a box with a floor area of 16 square inches, find the depth of the box.

   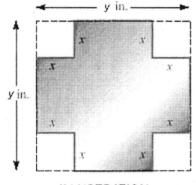

   ILLUSTRATION

   Select the correct answer.

   a. 4 in
   b. 8 in
   c. 7 in
   d. 9 in

3. Solve the equation.

$$x(x-6) - 15 = (x-1)^2$$

Select the correct answer.

a. $x = 3$
b. $x = -13$
c. $x = -4$
d. $x = 1$

4. Solve the inequality.

$$\frac{8}{x-1} \leq 2$$

Select the correct answer.

a. $(-\infty, 1) \cup [5, \infty)$
b. $(-\infty, 1] \cup [5, \infty)$
c. $(-\infty, 1) \cup (5, \infty)$
d. $(-\infty, 1] \cup (5, \infty)$
e. $(1, 5]$

5. Do the operation and express the answer in $a + bi$ form.

$$(2 - \sqrt{-9}) - (9i - 6)$$

Select the correct answer.

a. $8 - 12i$
b. $8 + 12i$
c. $-8 + 12i$
d. $-8 - 12i$

6. A college student earns $45 per day delivering advertising brochures door-to-door, plus 50 cents for each person he interviews. How many people did he interview on a day when he earned $113?

   Select the correct answer.

   a. 136
   b. 135
   c. 138
   d. 141
   e. 137

7. Find the absolute value.

   $$\left| \frac{1}{4+i} \right|$$

   Select the correct answer.

   a. $\dfrac{1}{\sqrt{17}}$

   b. 17

   c. $\sqrt{17}$

   d. $\dfrac{1}{17}$

8. Solve the inequality. Express the solution set in interval notation.

   $0 < |4x + 3| < 7$

   Select the correct answer.

   a. $\left(-\dfrac{3}{4}, 1\right)$

   b. $(-\infty, -1) \cup \left(\dfrac{3}{4}, \infty\right)$

   c. $\left(-\dfrac{5}{2}, -\dfrac{3}{4}\right) \cup \left(-\dfrac{3}{4}, 1\right)$

   d. $\left(-\infty, -\dfrac{3}{4}\right) \cup (1, \infty)$

9. Solve the equation by factoring.

$$9x^2 - 54x = -72$$

Select the correct answer(s).

a. $x = -1$
b. $x = 2$
c. $x = 7$
d. $x = 4$

10. Solve the equation for real values of the variable by factoring.

$$2m^{\frac{2}{3}} - 18m^{\frac{1}{3}} + 28 = 0$$

Select the correct answer(s).

a. $x = 7$
b. $x = 2$
c. $x = 8$
d. $x = 343$

11. Do the operation and express the answer in $a + bi$ form.

$$\frac{1}{6 + i}$$

Select the correct answer.

a. $\frac{6}{7} + \frac{i}{7}$

b. $\frac{6}{37} - \frac{i}{37}$

c. $-\frac{6}{37} + \frac{i}{37}$

d. $-\frac{6}{7} - \frac{i}{7}$

12. Solve the equation for real values of the variable by factoring.

    $$9p - 63p^{\frac{1}{2}} + 90 = 0$$

    Select the correct answer(s).

    a. $x = 25$
    b. $x = 5$
    c. $x = 4$
    d. $x = 2$

13. A full-price ticket for a college basketball game costs $5.00, and a student ticket costs $4.00. If 458 tickets were sold, and the total receipts were $2,063.00, how many tickets were student tickets?

    Select the correct answer.

    a. 227 tickets
    b. 234 tickets
    c. 224 tickets
    d. 230 tickets
    e. 219 tickets

14. Solve the formula for the x.

    $$p = \frac{n - x}{q}$$

    Select the correct answer.

    a. $x = pq - n$
    b. $x = n - pq$
    c. $x = nq - p$
    d. $x = p - nq$

15. Find all real solutions of the equation.

    $$\sqrt{y + 7} = 5 - y$$

    Select the correct answer(s).

    a. $x = 2$
    b. $x = 5$
    c. $x = 0$
    d. $x = 7$

16. Solve the equation for x.

    Select the correct answer.

    $|x + 9| = |9 - x|$

    a. $x_1 = 9, x_2 = 0$
    b. $x_1 = 0, x_2 = -1$
    c. $x = -9$
    d. $x = 0$

17. Solve the inequality.

    $$\frac{4}{3}x - \frac{3}{2}x \leq \frac{3}{8}\left(x + \frac{4}{3}\right) + \frac{1}{3}$$

    Select the correct answer.

    a. $\left(-\infty, -\frac{20}{13}\right]$

    b. $\left[-\frac{20}{13}, \infty\right)$

    c. $\left(-\frac{20}{13}, \infty\right)$

    d. $\left(-\infty, -\frac{20}{13}\right)$

18. Solve the equation for x.

    $$\left|\frac{3x - 16}{2}\right| = 11$$

    Select the correct answer.

    a. $x_1 = -\frac{38}{3}, x_2 = -2$

    b. $x_1 = \frac{38}{3}, x_2 = -2$

    c. $x_1 = -\frac{38}{3}, x_2 = \frac{2}{3}$

19. Solve the equation.

$3x + 13 = x + 11$

Select the correct answer.

a. $x = 51$
b. $x = -6$
c. $x = 48$
d. $x = -1$

20. A rectangle is 2 times as long as it is wide. If the area is 8 square feet, find its perimeter.

Select the correct answer.

a. 24 ft

b. 12 ft

c. 22 ft

d. 17 ft

21. John drove to a distant city in 4 hours. When he returned, there was less traffic, and the trip took only 2 hours. If John averaged 40 mph faster on the return trip, how fast did he drive each way?

Select the correct answer.

a. 48 mph, 88 mph
b. 40 mph, 80 mph
c. 27 mph, 67 mph
d. 44 mph, 84 mph
e. 51 mph, 91 mph

22. Solve the inequality.

$-2x - 11 > -1$

Select the correct answer.

a. $[5, \infty)$

b. $(-\infty, 5)$

c. $(-\infty, 5]$

d. $(-\infty, -5)$

23. Use the square root property to solve the equation.

    $x^2 - 16x + 64 = 49$

    Select the correct answer(s).

    a. $x = 7$
    b. $x = 8$
    c. $x = 16$
    d. $x = 1$
    e. $x = 15$

24. A cyclist rides from DeKalb to Rockford, a distance of 120 miles. His return trip takes 1 hour longer, because his speed decreases by 10 miles per hour. How fast does he ride each way?

    Select the correct answer.

    a.   40 mph going and 40 mph returning
    b.   30 mph going and 120 mph returning
    c.   60 mph going and 30 mph returning
    d.   40 mph going and 30 mph returning

25. Solve the equation by completing the square.

    $x^2 - 6x - 40 = 0$

    Select the correct answers.

    a. $x = 7, x = 10$
    b. $x = -4, x = 10$
    c. $x = -4, x = 3$
    d. $x = 3, x = 8$

ANSWER KEY

Gustafson/ Frisk - College Algebra 8E Chapter 1 Form D

1. b
2. a
3. c
4. a
5. a
6. a
7. a
8. c
9. b,d
10. c,d
11. b
12. a,c
13. a
14. b
15. a
16. d
17. b
18. b
19. d
20. b
21. b
22. d
23. d,e
24. d
25. b

List of Problem Codes for BCA Testing

Gustafson/ Frisk - College Algebra 8E Chapter 1 Form D

1. gfca.01.03.4.67m_NoAlgs
2. gfca.01.04.4.07m_NoAlgs
3. gfca.01.01.4.48m_NoAlgs
4. gfca.01.07.4.78m_NoAlgs
5. gfca.01.05.4.18m_NoAlgs
6. gfca.01.02.4.12m_NoAlgs
7. gfca.01.05.4.63m_NoAlgs
8. gfca.01.08.4.61m_NoAlgs
9. gfca.01.03.4.16m_NoAlgs
10. gfca.01.06.4.17m_NoAlgs
11. gfca.01.05.4.33m_NoAlgs
12. gfca.01.06.4.23m_NoAlgs
13. gfca.01.02.4.21m_NoAlgs
14. gfca.01.01.4.78m_NoAlgs
15. gfca.01.06.4.39m_NoAlgs
16. gfca.01.08.4.36m_NoAlgs
17. gfca.01.07.4.31m_NoAlgs
18. gfca.01.08.4.25m_NoAlgs
19. gfca.01.01.4.18m_NoAlgs
20. gfca.01.04.4.04m_NoAlgs
21. gfca.01.02.4.43m_NoAlgs
22. gfca.01.07.4.17m_NoAlgs
23. gfca.01.03.4.26m_NoAlgs
24. gfca.01.04.4.11m_NoAlgs
25. gfca.01.03.4.39m_NoAlgs

1. Solve the equation.

   $3x + 13 = x + 11$

   Select the correct answer.

   a. $x = 51$
   b. $x = -1$
   c. $x = 48$
   d. $x = -6$

2. Solve the equation.

   $x(x - 6) - 15 = (x - 1)^2$

   Select the correct answer.

   a. $x = 3$
   b. $x = -4$
   c. $x = -13$
   d. $x = 1$

3. Solve the formula for the r.

   $$\frac{1}{r} = \frac{1}{m} + \frac{1}{j}$$

4. A full-price ticket for a college basketball game costs $5, and a student ticket costs $4.50. If 700 tickets were sold, and the total receipts were $3,370.00, how many tickets were student tickets?

5. Of the 401 tickets sold to a movie, 34 were full-price tickets costing $6.00 each. If the gate receipts were $1,855.50, what did a student ticket cost?

   Select the correct answer.

   a. $4.50
   b. $4.00
   c. $6.50
   d. $5.00
   e. $5.50

6. Solve the inequality. Express the solution set in interval notation.

   $0 < |4x + 1| < 5$

7. The Norman window with dimensions as shown in illustration is a rectangle topped by a semicircle. If the area of the window is 7.57 square feet, find its height *h*.

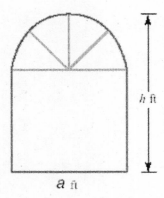

a = 2

8. Solve the equation by factoring.

$$4x^2 - 28x = -24$$

9. Use the square root property to solve the equation

$$x^2 - 16x + 64 = 49$$

Select the correct answer(s).

a. $x = 15$
b. $x = 1$
c. $x = 8$
d. $x = 7$
e. $x = 16$

10. Solve the inequality. Write the answer in interval notation.

$$\frac{4}{x-7} \leq 2$$

11. Solve the formula for the indicated variable.

$$\frac{x^2}{d^2} + \frac{y^2}{f^2} = 1 \; ; \; y$$

Select the correct answer.

a. $y = \sqrt{f\left(1-\frac{x}{d}\right)^2}$, $y = -\sqrt{f\left(1-\frac{x}{d}\right)^2}$

b. $y = f\sqrt{\left(1-\frac{x}{d}\right)\left(1+\frac{x}{d}\right)}$, $y = -f\sqrt{\left(1-\frac{x}{d}\right)\left(1+\frac{x}{d}\right)}$

c. $y = \sqrt{f(1-xd)(1+xd)}$, $y = -\sqrt{f(1-xd)(1+xd)}$

12. Use the discriminant to determine the nature of the roots of the equation.

$$3x^2 + 3x + 6 = 0$$

Select the correct answer.

a. real
b. not real

13. A rectangle is 2 times as long as it is wide. If the area is 8 square feet, find its perimeter.

Select the correct answer.

a. 17 ft
b. 22 ft
c. 24 ft
d. 12 ft

14. A cyclist rides from DeKalb to Rockford, a distance of 120 miles. His return trip takes 1 hour longer, because his speed decreases by 10 miles per hour. How fast does he ride each way?

15. If two opposite sides of a square are increased by 8 meters and the other sides are decreased by 3 meters, the area of the rectangle that is formed is 102 square meters. Find the area of the original square.

    Select the correct answer.

    a. 121 square meters
    b. 84 square meters
    c. 100 square meters
    d. 81 square meters

16. Do the operation and express the answer in *a + bi* form.

    $(8 - \sqrt{-25}) - (2i - 4)$

17. Do the operation and express the answer in *a + bi* form.

    $$\frac{1}{6 + i}$$

    Select the correct answer.

    a. $\frac{6}{37} - \frac{i}{37}$
    b. $\frac{6}{7} + \frac{i}{7}$
    c. $-\frac{6}{37} + \frac{i}{37}$
    d. $-\frac{6}{7} - \frac{i}{7}$

18. Find the absolute value.

    $$\left| \frac{1}{1 + i} \right|$$

19. Find all real solutions of the equation.

    $\sqrt{x+240} = 9\sqrt{x}$

20. Solve the equation for x.

    $\left|\dfrac{3x-20}{2}\right| = 13$

21. Find all real solutions of the equation.

    $\sqrt{y+7} = 5 - y$

    Select the correct answer(s).

    a. $x = 2$
    b. $x = 5$
    c. $x = 7$
    d. $x = 0$

22. Find all real solutions of the equation.

    $\sqrt{\sqrt{x+22} - \sqrt{x-29}} = \sqrt{3}$

23. Solve the inequality.

    $-2x - 11 > -1$

    Select the correct answer.

    a. $(-\infty, -5)$
    b. $(-\infty, 5)$
    c. $[5, \infty)$
    d. $(-\infty, 5]$

24. Solve the inequality.

    $9x^2 + 24x > -16$

    Select the correct answer.

    a. $(-\infty, \frac{4}{3}) \cup (\frac{4}{3}, \infty)$

    b. $(-\infty, -\frac{4}{3}) \cup [\frac{4}{3}, \infty)$

    c. $(-\infty, -\frac{4}{3}) \cup (-\frac{4}{3}, \infty)$

    d. $(-\infty, -\frac{4}{3}] \cup [\frac{4}{3}, \infty)$

    e. $(-\infty, -\frac{4}{3}) \cup (\frac{4}{3}, \infty)$

25. Solve the equation for x.

    $|x + 9| = |9 - x|$

ANSWER KEY

Gustafson/ Frisk - College Algebra 8E Chapter 1 Form E

1. b
2. b
3. $r = \dfrac{m \cdot j}{(m + j)}$
4. 260
5. a
6. $\left(\dfrac{-6}{4}, \dfrac{-1}{4}\right) \cup \left(\dfrac{-1}{4}, 1\right)$
7. 4
8. 1, 6
9. a, b
10. $(-\infty, 7) \cup [9, \infty)$
11. b
12. b
13. d
14. 30, 40
15. d
16. 12 − 7i
17. a
18. $\dfrac{1}{\sqrt{2}}$
19. 3
20. $\dfrac{46}{3}, -2$
21. a

ANSWER KEY

Gustafson/ Frisk - College Algebra 8E Chapter 1 Form E

22. 78

23. a

24. c

25. 0

List of Problem Codes for BCA Testing

Gustafson/ Frisk - College Algebra 8E Chapter 1 Form E

1. gfca.00.01.4.39m_NoAlgs
2. gfca.00.01.4.57_NoAlgs
3. gfca.00.02.4.120_NoAlgs
4. gfca.00.02.4.33m_NoAlgs
5. gfca.00.01.4.85m_NoAlgs
6. gfca.00.02.4.79m_NoAlgs
7. gfca.00.02.4.89_NoAlgs
8. gfca.00.03.4.113m_NoAlgs
9. gfca.00.03.4.33_NoAlgs
10. gfca.00.03.4.37m_NoAlgs
11. gfca.00.03.4.97_NoAlgs
12. gfca.00.04.4.111_NoAlgs
13. gfca.00.04.4.15m_NoAlgs
14. gfca.00.06.4.70_NoAlgs
15. gfca.00.01.4.69_NoAlgs
16. gfca.00.04.4.81m_NoAlgs
17. gfca.00.04.4.99_NoAlgs
18. gfca.00.05.4.11m_NoAlgs
19. gfca.00.05.4.21m_NoAlgs
20. gfca.00.05.4.24_NoAlgs
21. gfca.00.05.4.85_NoAlgs
22. gfca.00.05.4.95m_NoAlgs
23. gfca.00.06.4.45_NoAlgs
24. gfca.00.06.4.57m_NoAlgs
25. gfca.00.06.4.94m_NoAlgs

1. Use the square root property to solve the equation.

   $$x^2 - 16x + 64 = 49$$

   Select the correct answer(s).

   a. $x = 8$
   b. $x = 15$
   c. $x = 1$
   d. $x = 7$
   e. $x = 16$

2. Solve the equation for x.

   $$\left| \frac{3x - 20}{2} \right| = 13$$

3. A rectangle is 2 times as long as it is wide. If the area is 8 square feet, find its perimeter.

   Select the correct answer.

   a. 24 ft
   b. 12 ft
   c. 17 ft
   d. 22 ft

4. Find all real solutions of the equation.

   $$\sqrt{x + 240} = 9\sqrt{x}$$

5. Solve the formula for the r.

   $$\frac{1}{r} = \frac{1}{m} + \frac{1}{j}$$

6. Solve the equation for x.

   $|x + 9| = |9 - x|$

7. Find the absolute value.

$$\left| \frac{1}{1+i} \right|$$

8. Solve the equation by factoring.

$$4x^2 - 28x = -24$$

9. Solve the formula for the indicated variable.

$$\frac{x^2}{d^2} + \frac{y^2}{f^2} = 1 \; ; \; y$$

   Select the correct answer.

   a. $y = f\sqrt{\left(1-\frac{x}{d}\right)\left(1+\frac{x}{d}\right)}$, $y = -f\sqrt{\left(1-\frac{x}{d}\right)\left(1+\frac{x}{d}\right)}$

   b. $y = \sqrt{f(1-xd)(1+xd)}$, $y = -\sqrt{f(1-xd)(1+xd)}$

   c. $y = \sqrt{f\left(1-\frac{x}{d}\right)^2}$, $y = -\sqrt{f\left(1-\frac{x}{d}\right)^2}$

10. Do the operation and express the answer in $a + bi$ form.

$$(8 - \sqrt{-25}) - (2i - 4)$$

11. Solve the equation.

    $3x + 13 = x + 11$

    Select the correct answer.

    a. $x = 51$
    b. $x = -1$
    c. $x = 48$
    d. $x = -6$

12. Solve the inequality. Write the answer in interval notation.

$$\frac{4}{x-7} \leq 2$$

13. Do the operation and express the answer in *a + bi* form.

$$\frac{1}{6+i}$$

Select the correct answer.

a. $\dfrac{6}{7} + \dfrac{i}{7}$

b. $-\dfrac{6}{7} - \dfrac{i}{7}$

c. $-\dfrac{6}{37} + \dfrac{i}{37}$

d. $\dfrac{6}{37} - \dfrac{i}{37}$

14. The Norman window with dimensions as shown in illustration is a rectangle topped by a semicircle. If the area of the window is 7.57 square feet, find its height *h*.

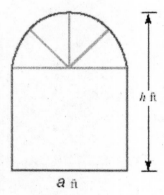

$a = 2$

15. If two opposite sides of a square are increased by 8 meters and the other sides are decreased by 3 meters, the area of the rectangle that is formed is 102 square meters. Find the area of the original square.

    Select the correct answer.

    a. 121 square meters
    b. 84 square meters
    c. 100 square meters
    d. 81 square meters

16. Use the discriminant to determine the nature of the roots of the equation.

    $3x^2 + 3x + 6 = 0$

    Select the correct answer.
    a. real
    b. not real

17. A full-price ticket for a college basketball game costs $5, and a student ticket costs $4.50. If 700 tickets were sold, and the total receipts were $3,370.00, how many tickets were student tickets?

18. Find all real solutions of the equation.

    $\sqrt{\sqrt{x+22} - \sqrt{x-29}} = \sqrt{3}$

19. Solve the equation

    $x(x-6) - 15 = (x-1)^2$

    Select the correct answer.

    a. $x = 3$
    b. $x = -4$
    c. $x = -13$
    d. $x = 1$

20. Solve the inequality. Express the solution set in interval notation.

    $0 < |4x + 1| < 5$

21. Solve the inequality.

    $-2x - 11 > -1$

    Select the correct answer.

    a. $(-\infty, -5)$
    b. $(-\infty, 5)$
    c. $[5, \infty)$
    d. $(-\infty, 5]$

22. Solve the inequality.

    $9x^2 + 24x > -16$

    Select the correct answer.

    a. $(-\infty, \frac{4}{3}) \cup (\frac{4}{3}, \infty)$
    b. $(-\infty, -\frac{4}{3}) \cup [\frac{4}{3}, \infty)$
    c. $(-\infty, -\frac{4}{3}) \cup (-\frac{4}{3}, \infty)$
    d. $(-\infty, -\frac{4}{3}] \cup [\frac{4}{3}, \infty)$
    e. $(-\infty, -\frac{4}{3}) \cup (\frac{4}{3}, \infty)$

23. A cyclist rides from DeKalb to Rockford, a distance of 120 miles. His return trip takes 1 hours longer, because his speed decreases by 10 miles per hour. How fast does he ride each way?

24. Of the 401 tickets sold to a movie, 34 were full-price tickets costing $6.00 each. If the gate receipts were $1,855.50, what did a student ticket cost?

    Select the correct answer.

    a. $4.50
    b. $4.00
    c. $6.50
    d. $5.00
    e. $5.50

25. Find all real solutions of the equation.

$$\sqrt{y + 7} = 5 - y$$

Select the correct answer(s).

a. $x = 7$
b. $x = 2$
c. $x = 0$
d. $x = 5$

ANSWER KEY

Gustafson/ Frisk - College Algebra 8E Chapter 1 Form F

1. b,c

2. $\dfrac{46}{3}, -2$

3. b

4. 3

5. $r = \dfrac{m \cdot j}{(m + j)}$

6. 0

7. $\dfrac{1}{\sqrt{2}}$

8. 1, 6

9. a

10. $12 - 7i$

11. b

12. $(-\infty, 7) \cup [9, \infty)$

13. d

14. 4

15. d

16. b

17. 260

18. 78

19. b

20. $\left(\dfrac{-6}{4}, \dfrac{-1}{4}\right) \cup \left(\dfrac{-1}{4}, 1\right)$

ANSWER KEY

Gustafson/ Frisk - College Algebra 8E Chapter 1 Form F

21. a

22. c

23. 30.40

24. a

25. b

List of Problem Codes for BCA Testing

Gustafson/ Frisk - College Algebra 8E Chapter 1 Form F

1. gfca.00.05.4.95m_NoAlgs
2. gfca.00.01.4.85m_NoAlgs
3. gfca.00.04.4.15m_NoAlgs
4. gfca.00.03.4.33_NoAlgs
5. gfca.00.04.4.111_NoAlgs
6. gfca.00.01.4.69_NoAlgs
7. gfca.00.02.4.33m_NoAlgs
8. gfca.00.06.4.70_NoAlgs
9. gfca.00.04.4.99_NoAlgs
10. gfca.00.06.4.45_NoAlgs
11. gfca.00.01.4.57_NoAlgs
12. gfca.00.02.4.89_NoAlgs
13. gfca.00.06.4.57m_NoAlgs
14. gfca.00.04.4.81m_NoAlgs
15. gfca.00.05.4.85_NoAlgs
16. gfca.00.02.4.120_NoAlgs
17. gfca.00.05.4.24_NoAlgs
18. gfca.00.01.4.39m_NoAlgs
19. gfca.00.05.4.11m_NoAlgs
20. gfca.00.03.4.113m_NoAlgs
21. gfca.00.03.4.97_NoAlgs
22. gfca.00.06.4.94m_NoAlgs
23. gfca.00.03.4.37m_NoAlgs
24. gfca.00.05.4.21m_NoAlgs
25. gfca.00.02.4.79m_NoAlgs

Gustafson/ Frisk - College Algebra 8E Chapter 1 Form G

1. Solve the equation.

   $3x + 13 = x + 11$

   Select the correct answer.

   a. $x = 48$
   b. $x = -6$
   c. $x = 51$
   d. $x = -1$

2. Solve the formula for the x.

   $$p = \frac{n - x}{q}$$

   Select the correct answer.

   a. $x = pq - n$
   b. $x = n - pq$
   c. $x = p - nq$
   d. $x = nq - p$

3. Solve the formula for the r.

   $$\frac{1}{r} = \frac{1}{m} + \frac{1}{j}$$

4. A college student earns $45 per day delivering advertising brochures door-to-door, plus 50 cents for each person he interviews. How many people did he interview on a day when he earned $113

   Select the correct answer.

   a. 136
   b. 135
   c. 138
   d. 137
   e. 141

5. Solve the inequality. Express the solution set in interval notation.

   $0 < |\, 4x + 1 \,| < 5$

Gustafson/ Frisk - College Algebra 8E Chapter 1 Form G

6. John drove to a distant city in 4 hours. When he returned, there was less traffic, and the trip took only 2 hours. If John averaged 40 mph faster on the return trip, how fast did he drive each way?

    Select the correct answer.

    a. 44 mph, 84 mph
    b. 40 mph, 80 mph
    c. 48 mph, 88 mph
    d. 27 mph, 67 mph
    e. 51 mph, 91 mph

7. The Norman window with dimensions as shown in illustration is a rectangle topped by a semicircle. If the area of the window is 7.57 square feet, find its height h.

    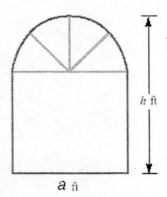

    $a = 2$

8. Use the square root property to solve the equation.

    $$x^2 - 16x + 64 = 49$$

    Select the correct answer(s).

    a. $x = 8$
    b. $x = 7$
    c. $x = 15$
    d. $x = 16$
    e. $x = 1$

Gustafson/ Frisk - College Algebra 8E Chapter 1 Form G

9. Solve the equation by completing the square.

    $x^2 - 6x - 40 = 0$

    Select the correct answers.

    a. $x = -4, x = 3$
    b. $x = -4, x = 10$
    c. $x = 3, x = 8$
    d. $x = 7, x = 10$

10. Solve the formula for the indicated variable.

    $\dfrac{x^2}{m^2} + \dfrac{y^2}{r^2} = 1 \; ; \; y$

11. Use the discriminant to determine the nature of the roots of the equation.

    $3x^2 + 3x + 6 = 0$

    Select the correct answer.

    a. not real
    b. real

12. A rectangle is 2 times as long as it is wide. If the area is 18 square feet, find its perimeter.

13. A cyclist rides from DeKalb to Rockford, a distance of 120 miles. His return trip takes 1 hours longer, because his speed decreases by 10 miles per hour. How fast does he ride each way?

    Select the correct answer.

    a. 60 mph going and 30 mph returning
    b. 40 mph going and 30 mph returning
    c. 40 mph going and 40 mph returning
    d. 30 mph going and 120 mph returning

14. If two opposite sides of a square are increased by 8 meters and the other sides are decreased by 3 meters, the area of the rectangle that is formed is 102 square meters. Find the area of the original square.

    Select the correct answer.

    a. 84 square meters
    b. 100 square meters
    c. 81 square meters
    d. 121 square meters

15. Do the operation and express the answer in $a + bi$ form.

    $(2 - \sqrt{-9}) - (9i - 6)$

    Select the correct answer.

    a. $8 + 12i$
    b. $-8 - 12i$
    c. $-8 + 12i$
    d. $8 - 12i$

16. Do the operation and express the answer in $a + bi$ form.

    $$\frac{1}{9 + i}$$

17. Find the absolute value.

    $$\left| \frac{1}{4 + i} \right|$$

    Select the correct answer.

    a. $\dfrac{1}{17}$
    b. $\dfrac{1}{\sqrt{17}}$
    c. 17
    d. $\sqrt{17}$

18. Solve the equation for real values of the variable by factoring.

$$2m^{\frac{2}{3}} - 18m^{\frac{1}{3}} + 28 = 0$$

Select the correct answer(s).

a. $x = 343$
b. $x = 2$
c. $x = 7$
d. $x = 8$

19. Solve the equation for real values of the variable by factoring.

$$9p - 63p^{\frac{1}{2}} + 90 = 0$$

Select the correct answer(s).

a. $x = 25$
b. $x = 4$
c. $x = 5$
d. $x = 2$

20. Find all real solutions of the equation.

$$\sqrt{y + 1} = 5 - y$$

21. Solve the inequality.

$-2x - 11 > -1$

Select the correct answer.

a. $(-\infty, -5)$
b. $(-\infty, 5)$
c. $[5, \infty)$
d. $(-\infty, 5]$

22. Solve the inequality and write the answer in interval notation.

$$\frac{4}{3}x - \frac{3}{2}x \leq \frac{3}{8}\left(x + \frac{4}{3}\right) + \frac{1}{3}$$

23. Solve the inequality.

$$\frac{8}{x-1} \leq 2$$

Select the correct answer.

a. $(-\infty, 1) \cup [5, \infty)$
b. $(1, 5]$
c. $(-\infty, 1] \cup (5, \infty)$
d. $(-\infty, 1] \cup [5, \infty)$
e. $(-\infty, 1) \cup (5, \infty)$

24. Solve the equation for x.

$$\left|\frac{3x-16}{2}\right| = 11$$

Select the correct answer.

a. $x_1 = -\frac{38}{3}, \; x_2 = -2$

b. $x_1 = -\frac{38}{3}, \; x_2 = \frac{2}{3}$

c. $x_1 = \frac{38}{3}, \; x_2 = -2$

25. Solve the equation for x.

$|x + 9| = |9 - x|$

Select the correct answer.

a. $x_1 = 0, x_2 = -1$
b. $x_1 = 9, x_2 = 0$
c. $x = 0$
d. $x = -9$

ANSWER KEY

Gustafson/ Frisk - College Algebra 8E Chapter 1 Form G

1. D
2. B
3. $r = \dfrac{m \cdot j}{(m+j)}$
4. a
5. $\left(\dfrac{-6}{4}, \dfrac{-1}{4}\right) \cup \left(\dfrac{-1}{4}, 1\right)$
6. b
7. 4
8. c,e
9. b
10. $f \cdot \sqrt{\left(1 - \dfrac{x}{m}\right) \cdot \left(1 + \dfrac{x}{m}\right)}, \ -f \cdot \sqrt{\left(1 - \dfrac{x}{m}\right) \cdot \left(1 + \dfrac{x}{m}\right)}$
11. a
12. 18
13. b
14. c
15. d
16. $\dfrac{9}{82} - \dfrac{i}{82}$
17. b
18. a,d
19. a,b
20. 3
21. a

ANSWER KEY

Gustafson/ Frisk - College Algebra 8E Chapter 1 Form G

22. $[-1.538462, \infty)$

23. a

24. c

25. c

List of Problem Codes for BCA Testing

Gustafson/ Frisk - College Algebra 8E Chapter 1 Form G

1. gfca.00.06.4.45_NoAlgs
2. gfca.00.02.4.97m_NoAlgs
3. gfca.00.02.4.89m_NoAlgs
4. gfca.00.01.4.73m_NoAlgs
5. gfca.00.06.4.70m_NoAlgs
6. gfca.00.01.4.69_NoAlgs
7. gfca.00.05.4.95m_NoAlgs
8. gfca.00.03.4.49_NoAlgs
9. gfca.00.05.4.24_NoAlgs
10. gfca.00.05.4.85_NoAlgs
11. gfca.00.06.4.35m_NoAlgs
12. gfca.00.04.4.81m_NoAlgs
13. gfca.00.02.4.41m_NoAlgs
14. gfca.00.05.4.11m_NoAlgs
15. gfca.00.04.4.99m_NoAlgs
16. gfca.00.03.4.37m_NoAlgs
17. gfca.00.02.4.63_NoAlgs
18. gfca.00.04.4.68_NoAlgs
19. gfca.00.04.4.101m_NoAlgs
20. gfca.00.05.4.51m_NoAlgs
21. gfca.00.01.4.39m_NoAlgs
22. gfca.00.01.4.85_NoAlgs
23. gfca.00.03.4.97m_NoAlgs
24. gfca.00.03.4.85m_NoAlgs
25. gfca.00.06.4.94_NoAlgs

1. The Norman window with dimensions as shown in illustration is a rectangle topped by a semicircle. If the area of the window is 7.57 square feet, find its height h.

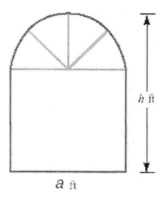

$a = 2$

2. Solve the inequality. Express the solution set in interval notation.

   $0 < |4x + 1| < 5$

3. Solve the equation for x.

   $|x + 9| = |9 - x|$

   Select the correct answer.

   a. $x_1 = 0, x_2 = -1$
   b. $x_1 = 9, x_2 = 0$
   c. $x = 0$
   d. $x = -9$

4. John drove to a distant city in 4 hours. When he returned, there was less traffic, and the trip took only 2 hours. If John averaged 40 mph faster on the return trip, how fast did he drive each way?

   Select the correct answer.

   a. 44 mph, 84 mph
   b. 40 mph, 80 mph
   c. 48 mph, 88 mph
   d. 27 mph, 67 mph
   e. 51 mph, 91 mph

5. Solve the inequality.

$$\frac{8}{x-1} \leq 2$$

Select the correct answer.

a. $(-\infty, 1) \cup [5, \infty)$
b. $(1, 5]$
c. $(-\infty, 1] \cup (5, \infty)$
d. $(-\infty, 1] \cup [5, \infty)$
e. $(-\infty, 1) \cup (5, \infty)$

6. Solve the inequality.

$-2x - 11 > -1$

Select the correct answer.

a. $(-\infty, 5)$
b. $(-\infty, 5]$
c. $[5, \infty)$
d. $(-\infty, -5)$

7. Find all real solutions of the equation.

$$\sqrt{y+1} = 5 - y$$

8. Use the square root property to solve the equation.

$$x^2 - 16x + 64 = 49$$

Select the correct answer(s).

a. $x = 8$
b. $x = 7$
c. $x = 15$
d. $x = 16$
e. $x = 1$

9. Solve the inequality and write the answer in interval notation.

$$\frac{4}{3}x - \frac{3}{2}x \leq \frac{3}{8}\left(x + \frac{4}{3}\right) + \frac{1}{3}$$

10. A college student earns $45 per day delivering advertising brochures door-to-door, plus 50 cents for each person he interviews. How many people did he interview on a day when he earned $113?

Select the correct answer.

a. 138
b. 137
c. 135
d. 136
e. 141

11. A cyclist rides from DeKalb to Rockford, a distance of 120 miles. His return trip takes 1 hours longer, because his speed decreases by 10 miles per hour. How fast does he ride each way?

Select the correct answer.

a. 60 mph going and 30 mph returning
b. 40 mph going and 30 mph returning
c. 40 mph going and 40 mph returning
d. 30 mph going and 120 mph returning

12. Solve the equation for real values of the variable by factoring.

$$9p - 63p^{\frac{1}{2}} + 90 = 0$$

Select the correct answer(s).

a. $x = 25$
b. $x = 2$
c. $x = 5$
d. $x = 4$

13. Solve the formula for the r.

$$\frac{1}{r} = \frac{1}{m} + \frac{1}{j}$$

14. A rectangle is 2 times as long as it is wide. If the area is 18 square feet, find its perimeter.

15. Do the operation and express the answer in $a + bi$ form.

    $$\frac{1}{9 + i}$$

16. Solve the equation for real values of the variable by factoring.

    $$2m^{\frac{2}{3}} - 18m^{\frac{1}{3}} + 28 = 0$$

    Select the correct answer(s).

    a. $x = 8$
    b. $x = 343$
    c. $x = 2$
    d. $x = 7$

17. Solve the formula for the indicated variable.

    $$\frac{x^2}{m^2} + \frac{y^2}{f^2} = 1 \; ; \; y$$

18. Solve the equation.

    $3x + 13 = x + 11$

    Select the correct answer.

    a. $x = 48$
    b. $x = -6$
    c. $x = 51$
    d. $x = -1$

19. Do the operation and express the answer in *a + bi* form.

$$(2 - \sqrt{-9}) - (9i - 6)$$

Select the correct answer.

a. $8 + 12i$
b. $-8 - 12i$
c. $-8 + 12i$
d. $8 - 12i$

20. Find the absolute value.

$$\left| \frac{1}{4 + i} \right|$$

Select the correct answer.

a. $\dfrac{1}{17}$

b. $\dfrac{1}{\sqrt{17}}$

c. 17

d. $\sqrt{17}$

21. Solve the equation for *x*.

$$\left| \frac{3x - 16}{2} \right| = 11$$

Select the correct answer.

a. $x_1 = \dfrac{38}{3}, \; x_2 = -2$

b. $x_1 = -\dfrac{38}{3}, \; x_2 = \dfrac{2}{3}$

c. $x_1 = -\dfrac{38}{3}, \; x_2 = -2$

22. Solve the equation by completing the square.

$$x^2 - 6x - 40 = 0$$

Select the correct answers.

a. $x = 7, x = 10$
b. $x = 3, x = 8$
c. $x = -4, x = 10$
d. $x = -4, x = 3$

23. Solve the formula for the x.

$$p = \frac{n - x}{q}$$

Select the correct answer.

a. $x = pq - n$
b. $x = n - pq$
c. $x = p - nq$
d. $x = nq - p$

24. If two opposite sides of a square are increased by 8 meters and the other sides are decreased by 3 meters, the area of the rectangle that is formed is 102 square meters. Find the area of the original square.

Select the correct answer.

a. 84 square meters
b. 100 square meters
c. 81 square meters
d. 121 square meters

25. Use the discriminant to determine the nature of the roots of the equation.

$$3x^2 + 3x + 6 = 0$$

Select the correct answer.

a. not real
b. real

ANSWER KEY

Gustafson/ Frisk - College Algebra 8E Chapter 1 Form H

1. 4
2. $\left(\dfrac{-6}{4}, \dfrac{-1}{4}\right) \cup \left(\dfrac{-1}{4}, 1\right)$
3. c
4. b
5. a
6. d
7. 3
8. c,e
9. $[-1.538462, \infty)$
10. d
11. b
12. a,d
13. $r = \dfrac{m \cdot j}{(m + j)}$
14. 18
15. $\dfrac{9}{82} - \dfrac{i}{82}$
16. A,b
17. $f \cdot \sqrt{\left(1 - \dfrac{x}{m}\right) \cdot \left(1 + \dfrac{x}{m}\right)}, \ -f \cdot \sqrt{\left(1 - \dfrac{x}{m}\right) \cdot \left(1 + \dfrac{x}{m}\right)}$
18. D
19. D
20. b
21. a
22. c
23. b
24. c
25. a

List of Problem Codes for BCA Testing

Gustafson/ Frisk - College Algebra 8E Chapter 1 Form H

1. gfca.00.06.4.70m_NoAlgs
2. gfca.00.01.4.73m_NoAlgs
3. gfca.00.04.4.99m_NoAlgs
4. gfca.00.01.4.69_NoAlgs
5. gfca.00.05.4.51m_NoAlgs
6. gfca.00.03.4.85m_NoAlgs
7. gfca.00.01.4.85_NoAlgs
8. gfca.00.03.4.49_NoAlgs
9. gfca.00.05.4.24_NoAlgs
10. gfca.00.06.4.94_NoAlgs
11. gfca.00.02.4.97m_NoAlgs
12. gfca.00.03.4.97m_NoAlgs
13. gfca.00.05.4.95m_NoAlgs
14. gfca.00.01.4.39m_NoAlgs
15. gfca.00.05.4.85_NoAlgs
16. gfca.00.04.4.81m_NoAlgs
17. gfca.00.02.4.89m_NoAlgs
18. gfca.00.06.4.35m_NoAlgs
19. gfca.00.03.4.37m_NoAlgs
20. gfca.00.02.4.63_NoAlgs
21. gfca.00.04.4.68_NoAlgs
22. gfca.00.06.4.45_NoAlgs
23. gfca.00.02.4.41m_NoAlgs
24. gfca.00.04.4.101m_NoAlgs
25. gfca.00.05.4.11m_NoAlgs

Gustafson/Frisk - College Algebra 8E Chapter 2 Form A

1. Graph the equation.

   $2(x - y) = 3x + 2$

   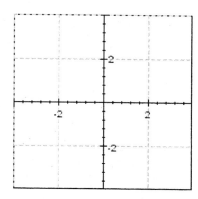

2. Find the distance between the point

   $P(\sqrt{5}, \sqrt{44})$ and $O(0,0)$.

3. One endpoint $P(3, 3)$ and the midpoint $M(6, 4)$ of line segment PQ are given. Find the coordinates of the other endpoint, Q.

4. Find the slope of the line passing through the pair of points.

   $P(-7, 18); Q(11, 20);$

5. Find the slope of the line.

   $10x + 40y = 6$

6. Find the slope of the line.

   $5(y + x) = 5(x - 7)$

7. Find the equation in general form of the circle with center at (1 , 1) and $r = 3$.

8. Find the slope of the line.

    $20x - 18 = 40(y + x)$

9. When a college started an aviation program, the administration agreed to predict enrollments using a straight line method. If the enrollment during the first year was 12, and the enrollment during year 6 was 32, find the rate of growth per year (the slope of the line). See Illustration.

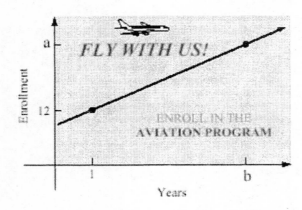

$a = 32$, $b = 6$

10. Use point-slope form to write the equation of the line with the properties:

    $m = 3$, passing through $P(2, 1)$

11. A line passes through the two points $P(6,6)$, and $Q(-1, -1)$. Write the equation in slope-intercept form.

12. Use the slope-intercept form to write the equation of the line passing through the point P(12, 1) and having the slope $m = -\frac{1}{2}$. Express the answer in general form.

13. Write the equation of the line that passes through the point P (0; 0), and is parallel to the line $y = 7x - 7$. Write the answer in slope-intercept form.

14. Find the equation of the line parallel to the line $x = 4$ and passing through the midpoint of the segment joining (1, 2) and (5, -8).

15. Find the y-intercepts of the graph.

    $y = x^4 - 49x^2$

16. Graph the equation. Be sure to find any intercepts and symmetries.

    $y = |x - 2|$

    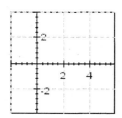

17. Graph the equation.

$y = \sqrt{x} - 1$

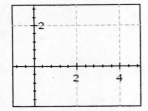

18. Find the *y*-intercept of the function.

$5x + 5y = 4$

19. Graph the equation.

$4x^2 + 4y^2 + 4y = 15$

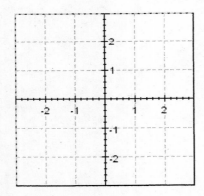

20. Solve the proportion.

$\dfrac{x}{14} = \dfrac{3}{x+1}$

21. Find the constant of proportionality for the stated conditions: *z* is directly proportional to *t*, and *t* = 9 when *z* = 63.

22. Given that y is directly proportional to x and y = 15 when x = 11, find y when x = $\frac{66}{3}$.

23. Given that m varies jointly with the square of n and the square root of q, and m = 26 when n = 9 and q = 7, find m when n = 27 and q = 63.

24. The power, in watts, dissipated as heat in a resistor varies directly with the square of the voltage and inversely with the resistance. If 80 volts are placed across a 80-ohm resistor, it will dissipate 80 watts. What voltage across a 10-ohm resistor will dissipate 10 watts?

25. Graph the equation.

    $2x - y = 4$

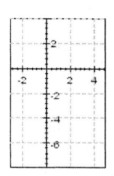

ANSWER KEY

Gustafson/Frisk - College Algebra 8E Chapter 2 Form A

1.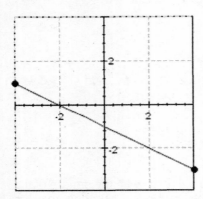

2. 7

3. $(9, 5)$

4. $\dfrac{1}{9}$

5. $-\dfrac{1}{4}$

6. $m = 0$

7. $x^2 + y^2 - 2x - 2y - 7 = 0$

8. $m = -\dfrac{20}{40}$

9. 4

10. $3x - y = 5$

11. $y = x$

12. $x + 2y = 14$

13. $y = 7x$

14. $x = 3$

15. $(0, 0)$

16.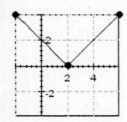

ANSWER KEY

Gustafson/Frisk - College Algebra 8E Chapter 2 Form A

17.
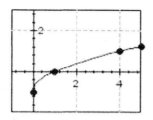

18. $(0, 0, 8)$

19.
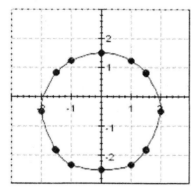

20. $-7, 6$

21. 7

22. 30

23. 702

24. 10

25.
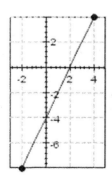

List of Problem Codes for BCA Testing

Gustafson/Frisk - College Algebra 8E Chapter 2 Form A

1. gfca.02.01.4.47_NoAlgs
2. gfca.02.01.4.68_NoAlgs
3. gfca.02.01.4.87_NoAlgs
4. gfca.02.02.4.17_NoAlgs
5. gfca.02.02.4.26_NoAlgs
6. gfca.02.02.4.29_NoAlgs
7. gfca.02.04.4.65_NoAlgs
8. gfca.02.02.4.30_NoAlgs
9. gfca.02.02.4.71_NoAlgs
10. gfca.02.03.4.07_NoAlgs
11. gfca.02.03.4.15_NoAlgs
12. gfca.02.03.4.30_NoAlgs
13. gfca.02.03.4.59_NoAlgs
14. gfca.02.03.4.77_NoAlgs
15. gfca.02.04.4.22_NoAlgs
16. gfca.02.04.4.51_NoAlgs
17. gfca.02.04.4.59_NoAlgs
18. gfca.02.01.4.35_NoAlgs
19. gfca.02.04.4.80_NoAlgs
20. gfca.02.05.4.13_NoAlgs
21. gfca.02.05.4.18_NoAlgs
22. gfca.02.05.4.23_NoAlgs
23. gfca.02.05.4.26_NoAlgs
24. gfca.02.05.4.35_NoAlgs
25. gfca.02.01.4.33_NoAlgs

Gustafson/Frisk - College Algebra 8E Chapter 2 Form B

1. Write the equation of the line that passes through the point $P(0, 0)$, and is parallel to the line $y = 7x - 7$. Write the answer in slope-intercept form.

2. Find the distance between the point

   $P(\sqrt{5}, \sqrt{44})$ and $O(0, 0)$.

3. Graph the equation. Be sure to find any intercepts and symmetries.

   $y = |x - 2|$

   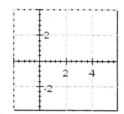

4. When a college started an aviation program, the administration agreed to predict enrollments using a straight line method. If the enrollment during the first year was 12, and the enrollment during year 6 was 32, find the rate of growth per year (the slope of the line). See Illustration.

   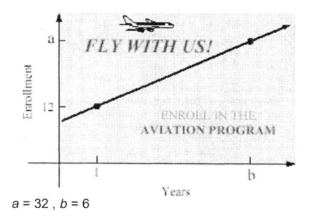

   $a = 32$, $b = 6$

5. Find the constant of proportionality for the stated conditions: z is directly proportional to t, and $t = 9$ when $z = 63$.

6. One endpoint $P(3, 3)$ and the midpoint $M(6, 4)$ of line segment PQ are given. Find the coordinates of the other endpoint, Q.

7. Find the equation in general form of the circle with center at (1, 1) and $r = 3$.

8. Graph the equation.

    $4x^2 + 4y^2 + 4y = 15$

    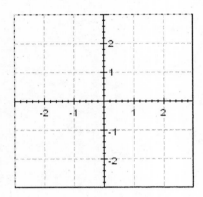

9. Find the equation of the line parallel to the line $x = 4$ and passing through the midpoint of the segment joining (1, 2) and (5, -8).

10. Use point-slope form to write the equation of the line with the properties:

    $m = 3$, passing through $P(2, 1)$

11. Find the y-intercept of the function.

    $5x + 5y = 4$

12. Find the slope of the line passing through the pair of points.

    $P(-7, 18)$; $Q(11, 20)$;

13. Find the slope of the line.

    $20x - 18 = 40(y + x)$

14. Graph the equation.

    $2(x - y) = 3x + 2$

    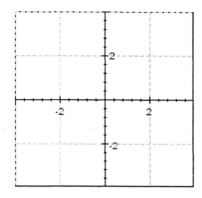

15. Find the slope of the line.

    $10x + 40y = 6$

16. A line passes through the two points $P(6,6)$, and $Q(-1, -1)$. Write the equation in slope-intercept form.

17. Use the slope-intercept form to write the equation of the line passing through the point $P(12, 1)$ and having the slope $m = -\dfrac{1}{2}$. Express the answer in general form.

18. Solve the proportion.

    $\dfrac{x}{14} = \dfrac{3}{x + 1}$

19. Given that m varies jointly with the square of n and the square root of q, and $m = 26$ when $n = 9$ and $q = 7$, find m when $n = 27$ and $q = 63$.

20. The power, in watts, dissipated as heat in a resistor varies directly with the square of the voltage and inversely with the resistance. If 80 volts are placed across a 80-ohm resistor, it will dissipate 80 watts. What voltage across a 10-ohm resistor will dissipate 10 watts?

21. Find the y-intercepts of the graph.

    $y = x^4 - 49x^2$

22. Graph the equation.

    $y = \sqrt{x} - 1$

    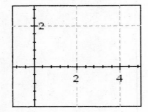

23. Find the slope of the line.

    $5(y + x) = 5(x - 7)$

24. Given that y is directly proportional to x and y = 15 when x = 11, find y when $x = \dfrac{66}{3}$.

25. Graph the equation.

    $2x - y = 4$

    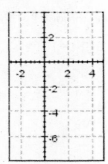

ANSWER KEY

Gustafson/Frisk - College Algebra 8E Chapter 2 Form B

1. $y = 7x$

2. 7

3.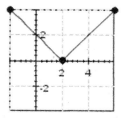

4. 4

5. 7

6. $(9, 5)$

7. $x^2 + y^2 - 2x - 2y - 7 = 0$

8.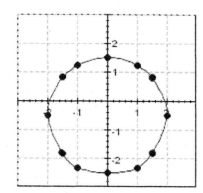

9. $x = 3$

10. $3x - y = 5$

11. $(0, 0, 8)$

12. $\dfrac{1}{9}$

13. $m = -\dfrac{20}{40}$

ANSWER KEY

Gustafson/Frisk - College Algebra 8E Chapter 2 Form B

14.
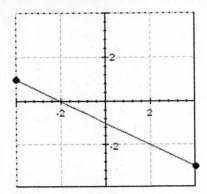

15. $-\dfrac{1}{4}$

16. $y = x$

17. $x + 2y = 14$

18. $-7, 6$

19. 702

20. 10

21. $(0, 0)$

22.
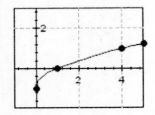

23. $m = 0$

24. 30

25.
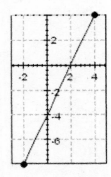

List of Problem Codes for BCA Testing

Gustafson/Frisk - College Algebra 8E Chapter 2 Form B

1. gfca.02.03.4.59_NoAlgs
2. gfca.02.01.4.68_NoAlgs
3. gfca.02.04.4.51_NoAlgs
4. gfca.02.02.4.71_NoAlgs
5. gfca.02.05.4.18_NoAlgs
6. gfca.02.01.4.87_NoAlgs
7. gfca.02.04.4.65_NoAlgs
8. gfca.02.04.4.80_NoAlgs
9. gfca.02.03.4.77_NoAlgs
10. gfca.02.03.4.07_NoAlgs
11. gfca.02.01.4.35_NoAlgs
12. gfca.02.02.4.17_NoAlgs
13. gfca.02.02.4.30_NoAlgs
14. gfca.02.01.4.47_NoAlgs
15. gfca.02.02.4.26_NoAlgs
16. gfca.02.03.4.15_NoAlgs
17. gfca.02.03.4.30_NoAlgs
18. gfca.02.05.4.13_NoAlgs
19. gfca.02.05.4.26_NoAlgs
20. gfca.02.05.4.35_NoAlgs
21. gfca.02.04.4.22_NoAlgs
22. gfca.02.04.4.59_NoAlgs
23. gfca.02.02.4.29_NoAlgs
24. gfca.02.05.4.23_NoAlgs
25. gfca.02.01.4.33_NoAlgs

Gustafson/Frisk - College Algebra 8E Chapter 2 Form C

1. Given that P varies jointly with r and s and P = 6 when r = 9 and s = 2, find P when r = 54 and s = 18.

    Select the correct answer.

    a.   P = 18
    b.   P = 54
    c.   P = 324
    d.   P = 6

2. Use point-slope form to write the equation of the line with the properties.

    m = 5, passing through P(3, 3)

    Select the correct answer.

    a.   5x + y = -12
    b.   y - 5x = 12
    c.   y = 5x - 12
    d.   -y = 5x - 12

3. Determine whether the lines with the given slopes are parallel, perpendicular, or neither.

    $m_1 = \sqrt{200};\ m_2 = 5\sqrt{8}$

    Select the correct answer.

    a.   parallel
    b.   neither
    c.   perpendicular

4. Find the slope of the line.

    8 (y + x) = 8 (x - 17);

    Select the correct answer.

    a.   m = -1
    b.   m = 0
    c.   m = 12
    d.   m = 1

5. Find the graph of the equation.

   $4(x - y) = 3x + 3$

   a.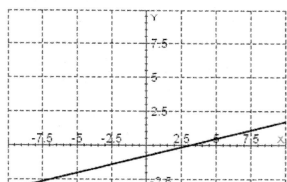

   b.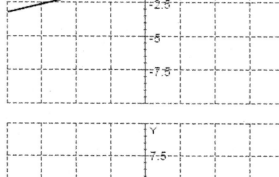

6. Use the slope-intercept form to write the equation of the line passing through the point $P(12, 0)$ and having the slope $m = -\dfrac{1}{4}$. Express the answer in general form.

   Select the correct answer.

   a. $x + 4y = 12$
   b. $x - 4y = -9$
   c. $x - 4y = 12$
   d. $x + 4y = -12$

7. Find the slope of the line.

    $4x + 15y = 10$

    Select the correct answer.

    a. $m = \dfrac{15}{4}$

    b. $m = -\dfrac{4}{15}$

    c. $m = \dfrac{4}{15}$

    d. $m = \dfrac{4}{10}$

8. Find the equation in general form of the circle with center at (3 , 2) and $r = 6$.

    Select the correct answer.

    a. $x^2 + y^2 - 6x - 4y - 49 = 0$
    b. $x^2 + y^2 - 6x - 4y - 23 = 0$
    c. $x^2 + y^2 + 6x + 4y - 23 = 0$
    d. $x^2 + y^2 + 6x + 4y - 49 = 0$

9. Graph the equation.

   $y = |x - 4|$

   Select the correct answer.

   a.
   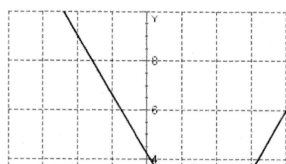

   b.
   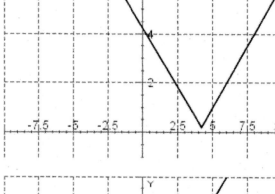

10. Find the distance between the point $P(\sqrt{5}, \sqrt{95})$ and $Q(0,0)$.

    Select the correct answer.

    a. $d(PQ) = 9$
    b. $d(PQ) = 12$
    c. $d(PQ) = 14$
    d. $d(PQ) = 10$

11. Find the *y*-intercept of the function.

$3x + 4y = 4$

Select the correct answer.

a. (0, 1)
b. (0, -1)
c. (0, 4)
d. (0, -4)

12. Solve the proportion.

$$\frac{x}{21} = \frac{2}{x+1}$$

Select the correct answer.

a. $x_1 = 7, x_2 = 6$
b. $x_1 = -7, x_2 = 6$
c. $x_1 = 7, x_2 = -6$
d. no solutions

13. Find the midpoint of the line segment *PQ*.

P(4, 7), *Q*(2, 9)

Select the correct answer.

a. *M* (5, 8)
b. *M* (3, 8)
c. *M* (8, 11)
d. *M* (3, 7)

14. Graph the equation

$y = \sqrt{x} - 2$

Select the correct answer.

a.
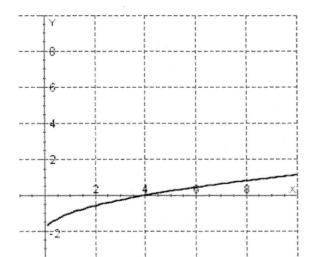

b.
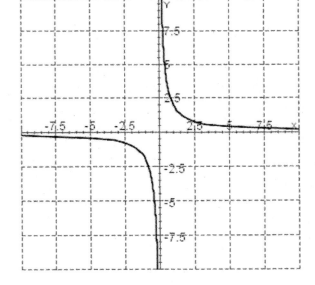

15. When a college started an aviation program, the administration agreed to predict enrollments using a straight line method. If the enrollment during the first year was 12, and the enrollment during year 6 was 32, find the rate of growth per year (the slope of the line). See Illustration.

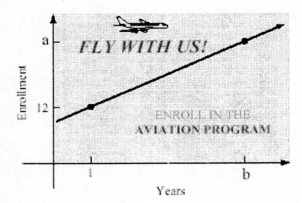

$a = 32$, $b = 6$

Select the correct answer.

a. $m = 3$
b. $m = 4$
c. $m = 8$
d. $m = 5$

16. Find the equation of the line parallel to the line $x = 2$ and passing through the midpoint of the segment joining (1, 9) and (5, -5).

Select the correct answer.

a. $x = 1$
b. $x = -4$
c. $x = 3$
d. $y = 2$

17. Graph the equation.

    $2x^2 + 2y^2 + 2y = 16$

    Select the correct answer.

    a.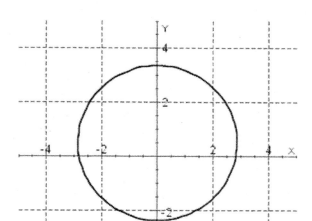

    b.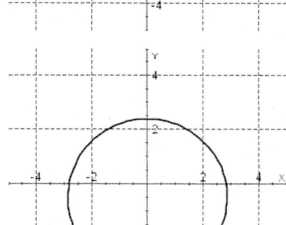

18. Find the graph of the equation.

3x - y = 4

a.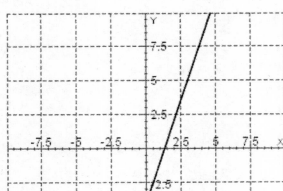

b.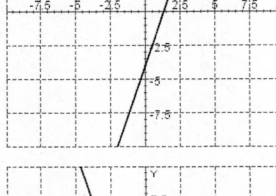

19. A line passes through the two points P(5,5), and Q(-3, -3). Write the equation in slope-intercept form.

Select the correct answer.

a. y = 5x
b. y = 5
c. x = -3
d. y = x

20. Given that w is directly proportional to z and w = 7 when z = 5, find w when z = 40.

    Select the correct answer.

    a.   w = 56
    b.   w = - 56
    c.   w = 8
    d.   w = 40

21. Find the constant of proportionality for the stated conditions. y is directly proportional to x, and x = 7 when y = 35.

    Select the correct answer.

    a.   k = 5
    b.   k = - 5
    c.   k = 7
    d.   k = 35

22. Find the x-intercepts of the graph.

    $y = x^2 - 49$

    Select the correct answer.

    a.   (7, 0), (- 7, 0)
    b.   (7, 0)
    c.   (49, 0)
    d.   (49, 0), (- 49, 0)

23. Find the slope of the line passing through the pair of points.

    P (-19,-5); Q (9,6);

    Select the correct answer.

    a. $m = -\dfrac{11}{28}$

    b. $m = \dfrac{21}{28}$

    c. $m = \dfrac{11}{29}$

    d. $m = \dfrac{11}{28}$

24. Write the equation of the line that passes through the point $P(0, 0)$ and is perpendicular to the line.

    $y = -4x + 5$

    Select the correct answer.

    a. $y = \dfrac{1}{4} x$

    b. $y = -\dfrac{1}{4} x + 5$

    c. $y = 4x$

    d. $y = -\dfrac{1}{5} x$

25. The power, in watts, dissipated as heat in a resistor varies directly with the square of the voltage and inversely with the resistance. If 40 volts are placed across a 40-ohm resistor, it will dissipate 40 watts. What voltage across a 10-ohm resistor will dissipate 640 watts?

    Select the correct answer.

    a. $V = -80$ volts
    b. $V = 6,400$ volts
    c. $V = 80$ volts
    d. $V = 640$ volts

ANSWER KEY

Gustafson/Frisk - College Algebra 8E Chapter 2 Form C

1. c
2. c
3. a
4. b
5. a
6. a
7. b
8. b
9. a
10. d
11. a
12. b
13. b
14. a
15. b
16. c
17. b
18. a
19. d
20. a
21. a
22. a
23. d
24. a
25. c

List of Problem Codes for BCA Testing

Gustafson/Frisk - College Algebra 8E Chapter 2 Form C

1. gfca.02.05.4.25m_NoAlgs
2. gfca.02.03.4.07m_NoAlgs
3. gfca.02.02.4.39m_NoAlgs
4. gfca.02.02.4.29m_NoAlgs
5. gfca.02.01.4.47m_NoAlgs
6. gfca.02.03.4.30m_NoAlgs
7. gfca.02.02.4.26m_NoAlgs
8. gfca.02.04.4.65m_NoAlgs
9. gfca.02.04.4.51m_NoAlgs
10. gfca.02.01.4.68m_NoAlgs
11. gfca.02.01.4.35m_NoAlgs
12. gfca.02.05.4.13m_NoAlgs
13. gfca.02.01.4.79m_NoAlgs
14. gfca.02.04.4.59m_NoAlgs
15. gfca.02.02.4.71m_NoAlgs
16. gfca.02.03.4.77m_NoAlgs
17. gfca.02.04.4.80m_NoAlgs
18. gfca.02.01.4.33m_NoAlgs
19. gfca.02.03.4.15m_NoAlgs
20. gfca.02.05.4.24m_NoAlgs
21. gfca.02.05.4.17m_NoAlgs
22. gfca.02.04.4.11m_NoAlgs
23. gfca.02.02.4.17m_NoAlgs
24. gfca.02.03.4.65m_NoAlgs
25. gfca.02.05.4.35m_NoAlgs

1. The power, in watts, dissipated as heat in a resistor varies directly with the square of the voltage and inversely with the resistance. If 40 volts are placed across a 40-ohm resistor, it will dissipate 40 watts. What voltage across a 10-ohm resistor will dissipate 640 watts?

   Select the correct answer.

   a. $V = 6{,}400$ volts
   b. $V = 80$ volts
   c. $V = 640$ volts
   d. $V = -80$ volts

2. Find the midpoint of the line segment PQ.

   $P(4, 7)$, $Q(2, 9)$

   Select the correct answer.

   a. $M(5, 8)$
   b. $M(3, 8)$
   c. $M(8, 11)$
   d. $M(3, 7)$

3. Find the constant of proportionality for the stated conditions. y is directly proportional to x, and $x = 7$ when $y = 35$.

   Select the correct answer.

   a. $k = 7$
   b. $k = 35$
   c. $k = 5$
   d. $k = -5$

4. Determine whether the lines with the given slopes are parallel, perpendicular, or neither.

   $m_1 = \sqrt{200}$; $m_2 = 5\sqrt{8}$

   Select the correct answer.

   a. parallel
   b. neither
   c. perpendicular

5. When a college started an aviation program, the administration agreed to predict enrollments using a straight line method. If the enrollment during the first year was 12, and the enrollment during year 6 was 32, find the rate of growth per year (the slope of the line). See Illustration.

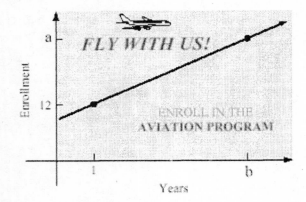

$a = 32$, $b = 6$

Select the correct answer.

a. $m = 3$
b. $m = 4$
c. $m = 8$
d. $m = 5$

6. Use the slope-intercept form to write the equation of the line passing through the point $P(12, 0)$ and having the slope $m = -\dfrac{1}{4}$. Express the answer in general form.

Select the correct answer.

a. $x + 4y = 12$
b. $x + 4y = -12$
c. $x - 4y = 12$
d. $x - 4y = -9$

7. Graph the equation.

   y = | x - 4 |

   Select the correct answer.

   a.
   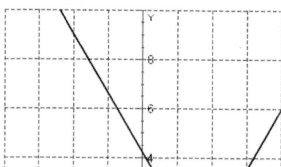

   b.
   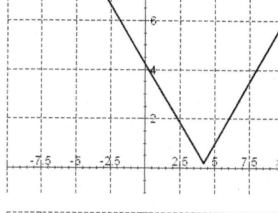

8. Find the slope of the line.

   8 (y + x) = 8 (x - 17);

   Select the correct answer.

   a. m = -1
   b. m = 0
   c. m = 12
   d. m = 1

9. Find the slope of the line.

   $4x + 15y = 10$

   Select the correct answer.

   a. $m = -\dfrac{4}{15}$

   b. $m = \dfrac{4}{10}$

   c. $m = \dfrac{4}{15}$

   d. $m = \dfrac{15}{4}$

10. Given that P varies jointly with r and s and P = 6 when r = 9 and s = 2, find P when r = 54 and s = 18.

    Select the correct answer.

    a.   P = 18
    b.   P = 54
    c.   P = 324
    d.   P = 6

11. Graph the equation.

$$y = \sqrt{x} - 2$$

Select the correct answer.

a.
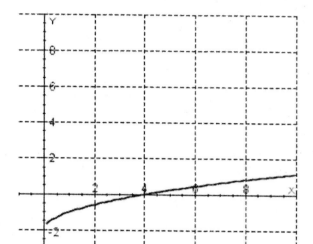

b.
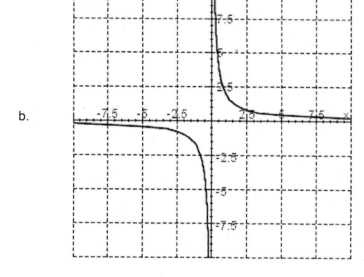

12. Find the graph of the equation.

    $3x - y = 4$

a.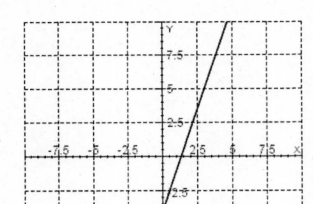

b.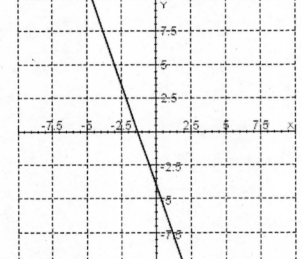

13. Solve the proportion.

$$\frac{x}{21} = \frac{2}{x+1}$$

Select the correct answer.

a. $x_1 = 7, x_2 = 6$
b. $x_1 = -7, x_2 = 6$
c. $x_1 = 7, x_2 = -6$
d. no solutions

14. Find the slope of the line passing through the pair of points.

P (-19,-5); Q (9,6);

Select the correct answer.

a. $m = -\frac{11}{28}$

b. $m = \frac{21}{28}$

c. $m = \frac{11}{29}$

d. $m = \frac{11}{28}$

15. Find the x-intercepts of the graph.

$y = x^2 - 49$

Select the correct answer.

a. (7, 0), (- 7, 0)
b. (49, 0)
c. (49, 0), (- 49, 0)
d. (7, 0)

16. Graph the equation.

    $2x^2 + 2y^2 + 2y = 16$

    Select the correct answer.

    a.

    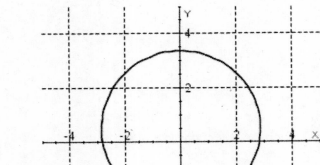

    b.

    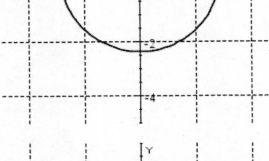

17. Use point-slope form to write the equation of the line with the properties.
    $m = 5$, passing through $P(3, 3)$

    Select the correct answer.

    a. $-y = 5x - 12$
    b. $5x + y = -12$
    c. $y = 5x - 12$
    d. $y - 5x = 12$

18. Write the equation of the line that passes through the point $P(0, 0)$ and is perpendicular to the line.

    $y = -4x + 5$

    Select the correct answer.

    a. $y = \dfrac{1}{4} x$

    b. $y = -\dfrac{1}{4} x + 5$

    c. $y = 4x$

    d. $y = -\dfrac{1}{5} x$

19. Find the distance between the point $P(\sqrt{5}, \sqrt{95})$ and $Q(0,0)$.

    Select the correct answer.

    a. $d(PQ) = 9$
    b. $d(PQ) = 12$
    c. $d(PQ) = 10$
    d. $d(PQ) = 14$

20. Find the equation of the line parallel to the line $x = 2$ and passing through the midpoint of the segment joining $(1, 9)$ and $(5, -5)$.

    Select the correct answer.

    a. $x = 1$
    b. $x = -4$
    c. $x = 3$
    d. $y = 2$

21. Find the graph of the equation.

    $4(x - y) = 3x + 3$

    a.
    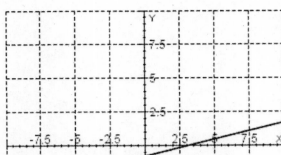

    b.
    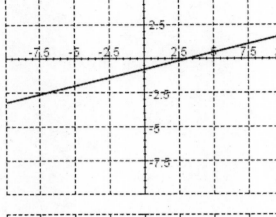

22. Find the equation in general form of the circle with center at (3, 2) and $r = 6$

    Select the correct answer.

    a. $x^2 + y^2 - 6x - 4y - 49 = 0$
    b. $x^2 + y^2 - 6x - 4y - 23 = 0$
    c. $x^2 + y^2 + 6x + 4y - 23 = 0$
    d. $x^2 + y^2 + 6x + 4y - 49 = 0$

23. A line passes through the two points P(5,5), and Q(-3, -3). Write the equation in slope-intercept form.

    Select the correct answer.

    a. $y = 5x$
    b. $y = 5$
    c. $x = -3$
    d. $y = x$

24. Given that w is directly proportional to z and w = 7 when z = 5, find w when z = 40.

    Select the correct answer.

    a. $w = 56$
    b. $w = -56$
    c. $w = 8$
    d. $w = 40$

25. Find the y-intercept of the function.

    $3x + 4y = 4$

    Select the correct answer.

    a. (0, 1)
    b. (0, -1)
    c. (0, 4)
    d. (0, -4)

ANSWER KEY

Gustafson/Frisk - College Algebra 8E Chapter 2 Form D

1. b
2. b
3. c
4. a
5. b
6. a
7. a
8. b
9. a
10. c
11. a
12. a
13. b
14. d
15. a
16. b
17. c
18. a
19. c
20. c
21. a
22. b
23. d
24. a
25. a

List of Problem Codes for BCA Testing

Gustafson/Frisk - College Algebra 8E Chapter 0 Form D

1. gfca.02.05.4.35m_NoAlgs
2. gfca.02.01.4.79m_NoAlgs
3. gfca.02.05.4.17m_NoAlgs
4. gfca.02.02.4.39m_NoAlgs
5. gfca.02.02.4.71m_NoAlgs
6. gfca.02.03.4.30m_NoAlgs
7. gfca.02.04.4.51m_NoAlgs
8. gfca.02.02.4.29m_NoAlgs
9. gfca.02.02.4.26m_NoAlgs
10. gfca.02.05.4.25m_NoAlgs
11. gfca.02.04.4.59m_NoAlgs
12. gfca.02.01.4.33m_NoAlgs
13. gfca.02.05.4.13m_NoAlgs
14. gfca.02.02.4.17m_NoAlgs
15. gfca.02.04.4.11m_NoAlgs
16. gfca.02.04.4.80m_NoAlgs
17. gfca.02.03.4.07m_NoAlgs
18. gfca.02.03.4.65m_NoAlgs
19. gfca.02.01.4.68m_NoAlgs
20. gfca.02.03.4.77m_NoAlgs
21. gfca.02.01.4.47m_NoAlgs
22. gfca.02.04.4.65m_NoAlgs
23. gfca.02.03.4.15m_NoAlgs
24. gfca.02.05.4.24m_NoAlgs
25. gfca.02.01.4.35m_NoAlgs

1. Graph the equation.

    $2x - y = 4$

    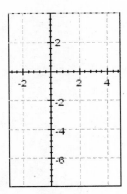

2. Find the y-intercept of the function.

    $3x + 4y = 4$

    Select the correct answer.

    a. $(0, -4)$
    b. $(0, 1)$
    c. $(0, 4)$
    d. $(0, -1)$

3. Find the distance between the point

    $P(\sqrt{5}, \sqrt{44})$ and $O(0,0)$.

4. One endpoint $P(3, 3)$ and the midpoint $M(6, 4)$ of line segment PQ are given. Find the coordinates of the other endpoint, Q.

5. Find the slope of the line passing through the pair of points.

   P (-19,-5); Q (9,6);

   Select the correct answer.

   a. $m = -\dfrac{11}{28}$

   b. $m = \dfrac{11}{29}$

   c. $m = \dfrac{11}{28}$

   d. $m = \dfrac{21}{28}$

6. Find the slope of the line.

   $4x + 15y = 10$

   Select the correct answer.

   a. $m = -\dfrac{4}{15}$

   b. $m = \dfrac{4}{15}$

   c. $m = \dfrac{4}{10}$

   d. $m = \dfrac{15}{4}$

7. Find the slope of the line.

   $20x - 18 = 40\,(y + x)$

8. Determine whether the line through the points P (8,10), Q (10,8) and the line through R (4,5), S(5,4) are parallel, perpendicular, or neither.

   Select the correct answer.

   a. parallel
   b. perpendicular
   c. neither

9. When a college started an aviation program, the administration agreed to predict enrollments using a straight line method. If the enrollment during the first year was 12, and the enrollment during year 6 was 32, find the rate of growth per year (the slope of the line). See Illustration.

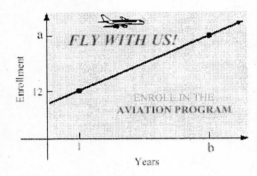

$a = 32$, $b = 6$

Select the correct answer.

a. $m = 3$
b. $m = 4$
c. $m = 8$
d. $m = 5$

10. A line passes through the two points $P(6,6)$, and $Q(-1, -1)$. Write the equation in slope-intercept form.

11. Use the slope-intercept form to write the equation of the line passing through the point $P(12, 0)$ and having the slope $m = -\dfrac{1}{4}$. Express the answer in general form.

Select the correct answer.

a. $x - 4y = 12$
b. $x + 4y = -12$
c. $x - 4y = -9$
d. $x + 4y = 12$

12. Find the y-intercept of the line determined by the equation.

$-3x + 10y = 6$

13. The power, in watts, dissipated as heat in a resistor varies directly with the square of the voltage and inversely with the resistance. If 80 volts are placed across a 80-ohm resistor, it will dissipate 80 watts. What voltage across a 10-ohm resistor will dissipate 10 watts?

14. Find the y-intercepts of the graph.

    $y = x^4 - 49x^2$

15. Graph the equation.

    $y = |x - 4|$

    Select the correct answer.

a.

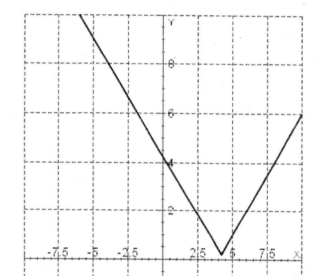

b.

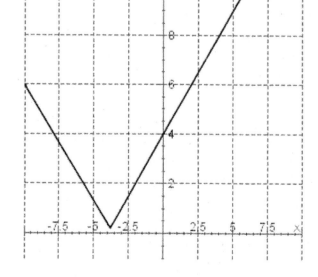

16. Graph the equation.

$$y = \sqrt{x} - 1$$

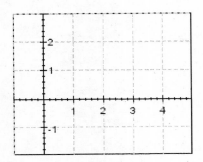

17. Find the equation in general form of the circle with center at (3 , 2) and $r = 6$.

Select the correct answer.

a. $x^2 + y^2 - 6x - 4y - 49 = 0$
b. $x^2 + y^2 + 6x + 4y - 23 = 0$
c. $x^2 + y^2 + 6x + 4y - 49 = 0$
d. $x^2 + y^2 - 6x - 4y - 23 = 0$

18. Solve the proportion.

$$\frac{x}{14} = \frac{3}{x+1}$$

19. Find the constant of proportionality for the stated conditions. z is directly proportional to t, and $t = 9$ when $z = 63$.

20. Graph the equation.

$4x^2 + 4y^2 + 4y = 15$

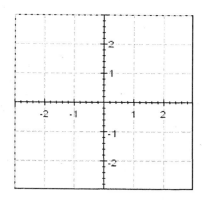

21. Solve the proportion.

$\dfrac{16}{x} = \dfrac{2}{7}$

22. Given that w is directly proportional to z and $w = 7$ when $z = 5$, find w when $z = 40$.

Select the correct answer.

a. $w = 40$
b. $w = -56$
c. $w = 56$
d. $w = 8$

23. Given that P varies jointly with r and s and $P = 6$ when $r = 9$ and $s = 2$, find P when $r = 54$ and $s = 18$.

Select the correct answer.

a. $P = 18$
b. $P = 54$
c. $P = 6$
d. $P = 324$

24. An antique table is expected to appreciate $30 each year. If the table will be worth $280 in 2 years, what will it be worth in 7 years?

    Select the correct answer.

    a. $490
    b. $210
    c. $430
    d. $550

25. Write the equation of the line that passes through the point $P(0, 0)$ and is perpendicular to the line.

    $y = -4x + 5$

    Select the correct answer.

    a. $y = -\frac{1}{5}x$

    b. $y = -\frac{1}{4}x + 5$

    c. $y = \frac{1}{4}x$

    d. $y = 4x$

ANSWER KEY

Gustafson/Frisk - College Algebra 8E Chapter 2 Form E

1.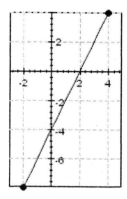

2. b

3. 7

4. $(9, 5)$

5. c

6. a

7. $m = -\dfrac{20}{40}$

8. a

9. b

10. $y = x$

11. d

12. $(0, 0, 6)$

13. 10

14. $(0, 0)$

15. a

ANSWER KEY

Gustafson/Frisk - College Algebra 8E Chapter 2 Form E

16.

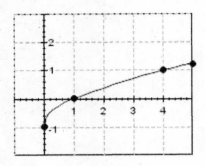

17. d

18. −7, 6

19. 7

20.

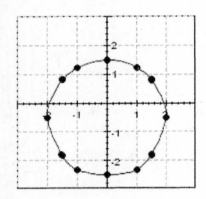

21. 56

22. c

23. d

24. c

25. c

List of Problem Codes for BCA Testing

Gustafson/Frisk - College Algebra 8E Chapter 2 Form E

1. gfca.02.01.4.33_NoAlgs
2. gfca.02.01.4.35m_NoAlgs
3. gfca.02.01.4.68_NoAlgs
4. gfca.02.01.4.87_NoAlgs
5. gfca.02.02.4.17m_NoAlgs
6. gfca.02.02.4.26m_NoAlgs
7. gfca.02.02.4.30_NoAlgs
8. gfca.02.02.4.47m_NoAlgs
9. gfca.02.02.4.71_NoAlgs
10. gfca.02.03.4.15_NoAlgs
11. gfca.02.03.4.30m_NoAlgs
12. gfca.02.03.4.42_NoAlgs
13. gfca.02.05.4.35_NoAlgs
14. gfca.02.04.4.22_NoAlgs
15. gfca.02.04.4.51m_NoAlgs
16. gfca.02.04.4.59_NoAlgs
17. gfca.02.04.4.65m_NoAlgs
18. gfca.02.05.4.13_NoAlgs
19. gfca.02.05.4.18_NoAlgs
20. gfca.02.04.4.80_NoAlgs
21. gfca.02.05.4.11_NoAlgs
22. gfca.02.05.4.24m_NoAlgs
23. gfca.02.05.4.25m_NoAlgs
24. gfca.02.03.4.87m_NoAlgs
25. gfca.02.03.4.65m_NoAlgs

Gustafson/Frisk - College Algebra 8E Chapter 2 Form F

1. Find the slope of the line passing through the pair of points.

   P (-19,-5); Q (9,6);

   Select the correct answer.

   a. $m = -\dfrac{11}{28}$

   b. $m = \dfrac{11}{29}$

   c. $m = \dfrac{11}{28}$

   d. $m = \dfrac{21}{28}$

2. One endpoint P (3, 3) and the midpoint M (6, 4) of line segment PQ are given. Find the coordinates of the other endpoint, Q.

3. The power, in watts, dissipated as heat in a resistor varies directly with the square of the voltage and inversely with the resistance. If 80 volts are placed across a 80-ohm resistor, it will dissipate 80 watts. What voltage across a 10-ohm resistor will dissipate 10 watts?

4. When a college started an aviation program, the administration agreed to predict enrollments using a straightline method. If the enrollment during the first year was 12, and the enrollment during year 6 was 32, find the rate of growth per year (the slope of the line). See Illustration.

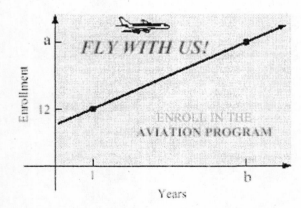

   $a = 32$, $b = 6$

Page 1

5. Solve the proportion.

$$\frac{x}{14} = \frac{3}{x+1}$$

6. Graph the equation.

$2x - y = 4$

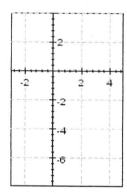

7. Find the distance between the point

$P(\sqrt{5}, \sqrt{44})$ and $O(0,0)$.

8. Given that P varies jointly with r and s and $P = 6$ when $r = 9$ and $s = 2$, find P when $r = 54$ and $s = 18$.

Select the correct answer.

a. $P = 18$
b. $P = 54$
c. $P = 6$
d. $P = 324$

9. Graph the equation.

   $y = |x - 4|$

   Select the correct answer.

   a.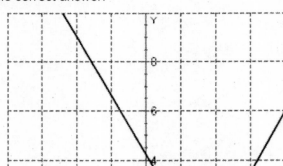

   b.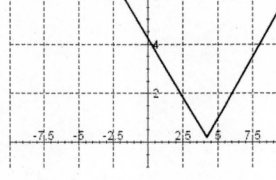

10. Use the slope-intercept form to write the equation of the line passing through the point $P(12, 0)$ and having the slope $m = -\dfrac{1}{4}$. Express the answer in general form.

    Select the correct answer.

    a. $x - 4y = 12$
    b. $x + 4y = -12$
    c. $x - 4y = -9$
    d. $x + 4y = 12$

11. Graph the equation.

    $y = \sqrt{x} - 1$

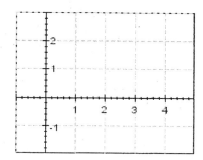

12. Find the constant of proportionality for the stated conditions. z is directly proportional to t, and $t = 9$ when $z = 63$.

13. A line passes through the two points $P(6,6)$, and $Q(-1, -1)$. Write the equation in slope-intercept form.

14. Find the equation in general form of the circle with center at $(3, 2)$ and $r = 6$.

    Select the correct answer.

    a. $x^2 + y^2 - 6x - 4y - 49 = 0$
    b. $x^2 + y^2 + 6x + 4y - 23 = 0$
    c. $x^2 + y^2 + 6x + 4y - 49 = 0$
    d. $x^2 + y^2 - 6x - 4y - 23 = 0$

15. An antique table is expected to appreciate $30 each year. If the table will be worth $280 in 2 years, what will it be worth in 7 years?

    Select the correct answer.

    a. $490
    b. $210
    c. $430
    d. $550

16. Find the slope of the line.

    $20x - 18 = 40(y + x)$

17. Determine whether the line through the points $P(8,10)$, $Q(10,8)$ and the line through $R(4,5)$, $S(5,4)$ are parallel, perpendicular, or neither.

    Select the correct answer.

    a. parallel
    b. perpendicular
    c. neither

18. Find the y-intercept of the function.

    $3x + 4y = 4$

    Select the correct answer.

    a. (0, -4)
    b. (0, 1)
    c. (0, 4)
    d. (0, -1)

19. Find the y-intercept of the line determined by the equation.

    $-3x + 10y = 6$

20. Solve the proportion.

    $\dfrac{16}{x} = \dfrac{2}{7}$

21. Given that w is directly proportional to z and $w = 7$ when $z = 5$, find w when $z = 40$.

    Select the correct answer.

    a. $w = 8$
    b. $w = 40$
    c. $w = -56$
    d. $w = 56$

22. Find the y-intercepts of the graph.

    $y = x^4 - 49x^2$

23. Find the slope of the line.

    $4x + 15y = 10$

    Select the correct answer.

    a. $m = -\dfrac{4}{15}$

    b. $m = \dfrac{4}{15}$

    c. $m = \dfrac{4}{10}$

    d. $m = \dfrac{15}{4}$

24. Graph the equation.

    $4x^2 + 4y^2 + 4y = 15$

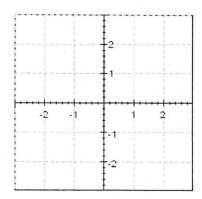

25. Write the equation of the line that passes through the point $P(0, 0)$ and is perpendicular to the line.

$y = -4x + 5$

Select the correct answer.

a. $\quad y = -\dfrac{1}{5} x$

b. $\quad y = -\dfrac{1}{4} x + 5$

c. $\quad y = \dfrac{1}{4} x$

d. $\quad y = 4x$

ANSWER KEY

Gustafson/Frisk - College Algebra 8E Chapter 2 Form F

1. c
2. $(9, 5)$
3. 10
4. 4
5. $-7, 6$
6.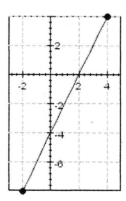
7. 7
8. d
9. a
10. d
11.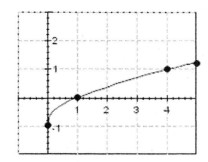
12. 7
13. $y = x$
14. d
15. c
16. $m = -\dfrac{20}{40}$

ANSWER KEY

Gustafson/Frisk - College Algebra 8E Chapter 2 Form F

17. a

18. b

19. $(0, 0.6)$

20. 56

21. d

22. $(0, 0)$

23. a

24.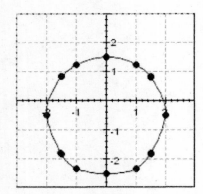

25. c

List of Problem Codes for BCA Testing

Gustafson/Frisk - College Algebra 8E Chapter 2 Form F

1. gfca.02.02.4.17m_NoAlgs
2. gfca.02.01.4.87_NoAlgs
3. gfca.02.05.4.35_NoAlgs
4. gfca.02.02.4.71_NoAlgs
5. gfca.02.05.4.13_NoAlgs
6. gfca.02.01.4.33_NoAlgs
7. gfca.02.01.4.68_NoAlgs
8. gfca.02.05.4.25m_NoAlgs
9. gfca.02.04.4.51m_NoAlgs
10. gfca.02.03.4.30m_NoAlgs
11. gfca.02.04.4.59_NoAlgs
12. gfca.02.05.4.18_NoAlgs
13. gfca.02.03.4.15_NoAlgs
14. gfca.02.04.4.65m_NoAlgs
15. gfca.02.03.4.87m_NoAlgs
16. gfca.02.02.4.30_NoAlgs
17. gfca.02.02.4.47m_NoAlgs
18. gfca.02.01.4.35m_NoAlgs
19. gfca.02.03.4.42_NoAlgs
20. gfca.02.05.4.11_NoAlgs
21. gfca.02.05.4.24m_NoAlgs
22. gfca.02.04.4.22_NoAlgs
23. gfca.02.02.4.26m_NoAlgs
24. gfca.02.04.4.80_NoAlgs
25. gfca.02.03.4.65m_NoAlgs

1. Graph the equation.

   2x - y = 4

   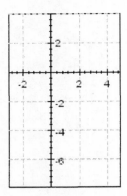

2. Find the equation of the line parallel to the line x = 4 and passing through the midpoint of the segment joining (1, 2) and (5, -8).

3. Find the y-intercept of the function.

   3x + 4y = 4

   Select the correct answer.

   a.   (0, 1)
   b.   (0, 4)
   c.   (0, - 4)
   d.   (0, - 1)

4. Find the graph of the equation

   $4(x - y) = 3x + 3$

   a.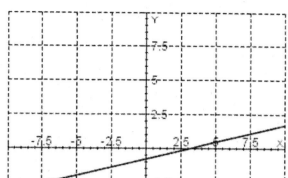

   b.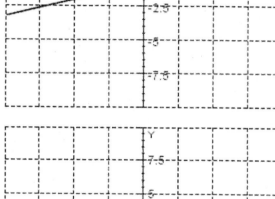

5. Find the distance between the point

   $P(\sqrt{5}, \sqrt{95})$ and $Q(0, 0)$.

   Select the correct answer.

   a. $d(PQ) = 9$
   b. $d(PQ) = 12$
   c. $d(PQ) = 14$
   d. $d(PQ) = 10$

6. Find the slope of the line passing through the pair of points:

$P(-19,-5); Q(9,6);$

Select the correct answer.

a. $m = \dfrac{11}{28}$

b. $m = \dfrac{11}{29}$

c. $m = \dfrac{21}{28}$

d. $m = -\dfrac{11}{28}$

7. Find the slope of the line.

$5(y + x) = 5(x - 7)$

8. Find the slope of the line.

$20x - 18 = 40(y + x)$

9. Determine whether the lines with the given slopes are parallel, perpendicular, or neither.

$m_1 = \sqrt{200} \; ; \; m_2 = 5\sqrt{8}$

Select the correct answer.

a. neither
b. parallel
c. perpendicular

10. Determine whether the line through the points $P(8,10)$, $Q(10,8)$ and the line through $R(4,5)$, $S(5,4)$ are parallel, perpendicular, or neither.

Select the correct answer.

a. perpendicular
b. parallel
c. neither

11. Use point-slope form to write the equation of the line with the properties:

    $m = 5$, passing through $P(3, 3)$

    Select the correct answer.

    a. $y - 5x = 12$
    b. $-y = 5x - 12$
    c. $5x + y = -12$
    d. $y = 5x - 12$

12. A line passes through the two points $P(5,5)$, and $Q(-3, -3)$. Write the equation in slope-intercept form.

    Select the correct answer.

    a. $y = 5$
    b. $x = -3$
    c. $y = 5x$
    d. $y = x$

13. Use the slope-intercept form to write the equation of the line passing through the point $P(12, 0)$ and having the slope $m = -\dfrac{1}{4}$. Express the answer in general form.

    Select the correct answer.

    a. $x + 4y = 12$
    b. $x + 4y = -12$
    c. $x - 4y = 12$
    d. $x - 4y = -9$

14. Write the equation of the line that passes through the point $P(0; 0)$, and is parallel to the line $y = 7x - 7$. Write the answer in slope-intercept form.

15. Graph the equation.

$y = |x - 4|$

Select the correct answer.

a.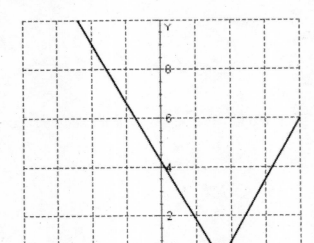

b.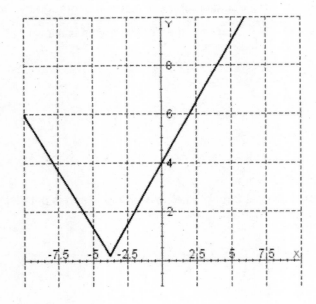

16. Graph the equation.

    $y^2 = 4x$

    Select the correct answer.

    a.
    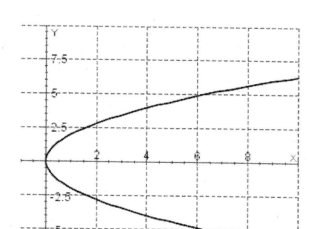

    b.
    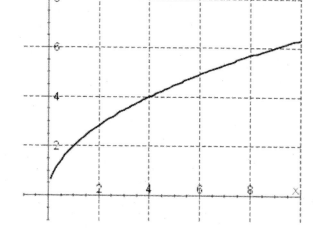

17. Graph the equation

    $y = \sqrt{x} - 1$

    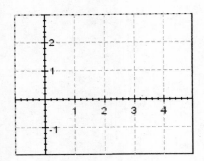

18. Find the equation in general form of the circle with center at (1 , 1) and $r = 3$

19. Find the x-intercepts of the graph.

    $y = x^2 - 49$

    Select the correct answer.

    a. (7, 0), (- 7, 0)
    b. (49, 0), (- 49, 0)
    c. (49, 0)
    d. (7, 0)

20. Graph the equation

    $2x^2 + 2y^2 + 2y = 16$

    Select the correct answer.

a.
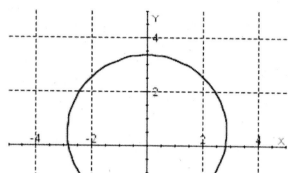

b.
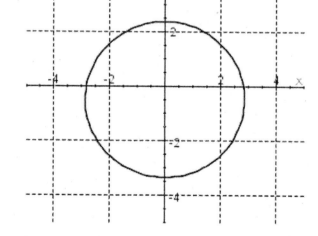

21. Solve the proportion.

$$\frac{x}{21} = \frac{2}{x+1}$$

Select the correct answer.

a. $x_1 = 7, x_2 = -6$
b. $x_1 = -7, x_2 = 6$
c. $x_1 = 7, x_2 = 6$
d. no solutions

22. Find the constant of proportionality for the stated conditions: y is directly proportional to x, and $x = 7$ when $y = 35$.

Select the correct answer.

a. $k = 35$
b. $k = -5$
c. $k = 5$
d. $k = 7$

23. Given that w is directly proportional to z and $w = 7$ when $z = 5$, find w when $z = 40$.

Select the correct answer.

a. $w = -56$
b. $w = 8$
c. $w = 56$
d. $w = 40$

24. Given that P varies jointly with r and s and $P = 6$ when $r = 9$ and $s = 2$, find P when $r = 54$ and $s = 18$.

Select the correct answer.

a. $P = 54$
b. $P = 324$
c. $P = 6$
d. $P = 18$

25. The power, in watts, dissipated as heat in a resistor varies directly with the square of the voltage and inversely with the resistance. If 80 volts are placed across a 80-ohm resistor, it will dissipate 80 watts. What voltage across a 10-ohm resistor will dissipate 10 watts?

ANSWER KEY

Gustafson/Frisk - College Algebra 8E Chapter 2 Form G

1.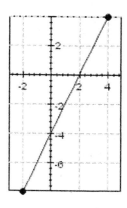

2. $x = 3$

3. a

4. a

5. d

6. a

7. $m = 0$

8. $m = -\dfrac{20}{40}$

9. b

10. b

11. d

12. d

13. a

14. $y = 7x$

15. a

16. a

ANSWER KEY

Gustafson/Frisk - College Algebra 8E Chapter 2 Form G

17.

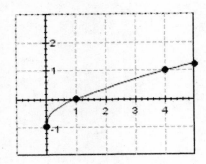

18. $x^2 + y^2 - 2x - 2y - 7 = 0$

19. a

20. b

21. b

22. c

23. c

24. b

25. 10

List of Problem Codes for BCA Testing

Gustafson/Frisk - College Algebra 8E Chapter 2 Form G

1. gfca.02.01.4.33_NoAlgs
2. gfca.02.03.4.77_NoAlgs
3. gfca.02.01.4.35m_NoAlgs
4. gfca.02.01.4.47m_NoAlgs
5. gfca.02.01.4.68m_NoAlgs
6. gfca.02.02.4.17m_NoAlgs
7. gfca.02.02.4.29_NoAlgs
8. gfca.02.02.4.30_NoAlgs
9. gfca.02.02.4.39m_NoAlgs
10. gfca.02.02.4.47m_NoAlgs
11. gfca.02.03.4.07m_NoAlgs
12. gfca.02.03.4.15m_NoAlgs
13. gfca.02.03.4.30m_NoAlgs
14. gfca.02.03.4.59_NoAlgs
15. gfca.02.04.4.51m_NoAlgs
16. gfca.02.04.4.56m_NoAlgs
17. gfca.02.04.4.59_NoAlgs
18. gfca.02.04.4.65_NoAlgs
19. gfca.02.04.4.11m_NoAlgs
20. gfca.02.04.4.80m_NoAlgs
21. gfca.02.05.4.13m_NoAlgs
22. gfca.02.05.4.17m_NoAlgs
23. gfca.02.05.4.24m_NoAlgs
24. gfca.02.05.4.25m_NoAlgs
25. gfca.02.05.4.35_NoAlgs

1. Find the x-intercepts of the graph.

   $y = x^2 - 49$

   Select the correct answer.

   a. (7, 0), (−7, 0)
   b. (49, 0), (−49, 0)
   c. (49, 0)
   d. (7, 0)

2. Find the slope of the line passing through the pair of points:

   P (−19,−5); Q (9,6);

   Select the correct answer.

   a. $m = \dfrac{11}{28}$

   b. $m = \dfrac{11}{29}$

   c. $m = \dfrac{21}{28}$

   d. $m = -\dfrac{11}{28}$

3. Use point-slope form to write the equation of the line with the properties:

   $m = 5$, passing through $P(3, 3)$

   Select the correct answer.

   a. $5x + y = -12$
   b. $y = 5x - 12$
   c. $-y = 5x - 12$
   d. $y - 5x = 12$

4. Write the equation of the line that passes through the point P (0; 0), and is parallel to the line $y = 7x - 7$. Write the answer in slope-intercept form.

5. Graph the equation

   $2x^2 + 2y^2 + 2y = 16$

   Select the correct answer.

   a.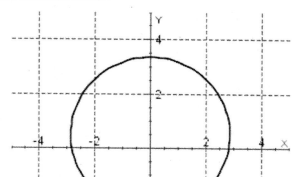

   b.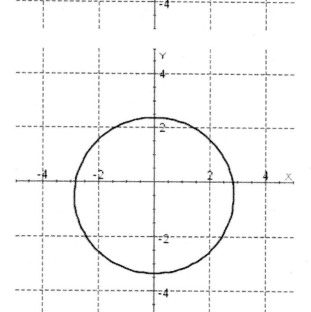

6. Graph the equation.

   $y = |x - 4|$

   Select the correct answer.

   a.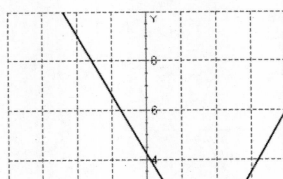

   b.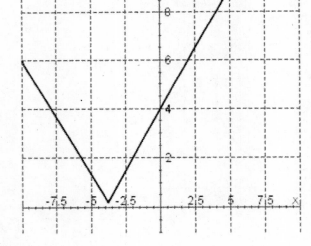

Gustafson/Frisk - College Algebra 8E Chapter 2 Form H

7. Determine whether the lines with the given slopes are parallel, perpendicular, or neither.

   $m_1 = \sqrt{200}$; $m_2 = 5\sqrt{8}$

   Select the correct answer.

   a. neither
   b. parallel
   c. perpendicular

8. Find the equation of the line parallel to the line $x = 4$ and passing through the midpoint of the segment joining (1, 2) and (5, -8).

9. Find the slope of the line.

   $20x - 18 = 40(y + x)$

10. Given that P varies jointly with r and s and $P = 6$ when $r = 9$ and $s = 2$, find P when $r = 54$ and $s = 18$.

    Select the correct answer.

    a. $P = 54$
    b. $P = 324$
    c. $P = 6$
    d. $P = 18$

11. Given that w is directly proportional to z and $w = 7$ when $z = 5$, find w when $z = 40$.

    Select the correct answer.

    a. $w = 56$
    b. $w = -56$
    c. $w = 8$
    d. $w = 40$

12. Find the distance between the point P ($\sqrt{5}$, $\sqrt{95}$) and Q (0, 0).

    Select the correct answer.

    a. $d(PQ) = 9$
    b. $d(PQ) = 12$
    c. $d(PQ) = 14$
    d. $d(PQ) = 10$

13. Graph the equation

$y^2 = 4x$

Select the correct answer.

a.
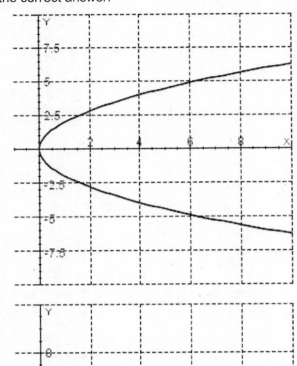

b.
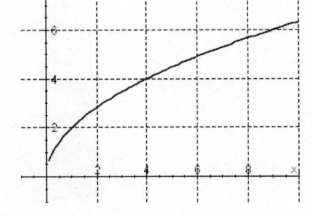

14. Find the constant of proportionality for the stated conditions: y is directly proportional to x, and $x = 7$ when $y = 35$.

Select the correct answer.

a. $k = 35$
b. $k = -5$
c. $k = 5$
d. $k = 7$

15. Find the y-intercept of the function.

    $3x + 4y = 4$

    Select the correct answer.

    a.   (0, 1)
    b.   (0, 4)
    c.   (0, - 4)
    d.   (0, - 1)

16. Solve the proportion.

    $$\frac{x}{21} = \frac{2}{x+1}$$

    Select the correct answer.

    a. $x_1 = 7, x_2 = -6$
    b. $x_1 = 7, x_2 = 6$
    c. $x_1 = -7, x_2 = 6$
    d.   no solutions

17. A line passes through the two points P(5,5), and Q(-3, -3). Write the equation in slope-intercept form.

    Select the correct answer.

    a. $y = 5$
    b. $x = -3$
    c. $y = 5x$
    d. $y = x$

18. Use the slope-intercept form to write the equation of the line passing through the point P(12, 0) and having the slope $m = -\frac{1}{4}$. Express the answer in general form.

    Select the correct answer.

    a. $x + 4y = 12$
    b. $x + 4y = -12$
    c. $x - 4y = 12$
    d. $x - 4y = -9$

19. Graph the equation.

    $2x - y = 4$

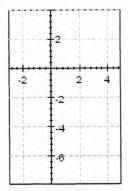

20. The power, in watts, dissipated as heat in a resistor varies directly with the square of the voltage and inversely with the resistance. If 80 volts are placed across a 80-ohm resistor, it will dissipate 80 watts. What voltage across a 10-ohm resistor will dissipate 10 watts?

21. Find the graph of the equation.

    $4(x - y) = 3x + 3$

a.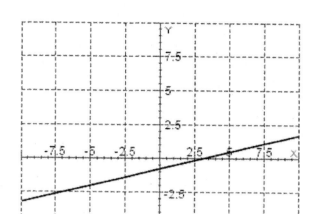

b.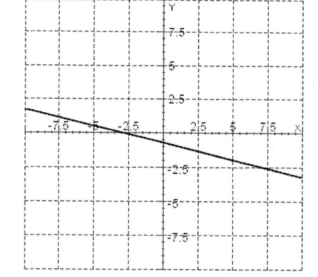

22. Graph the equation

    $y = \sqrt{x} - 1$

    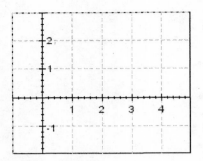

23. Determine whether the line through the points $P(8,10)$, $Q(10,8)$ and the line through $R(4,5)$, $S(5,4)$ are parallel, perpendicular, or neither.

    Select the correct answer.

    a. perpendicular
    b. parallel
    c. neither

24. Find the slope of the line.

    $5(y + x) = 5(x - 7)$

25. Find the equation in general form of the circle with center at $(1, 1)$ and $r = 3$.

ANSWER KEY

Gustafson/Frisk - College Algebra 8E Chapter 2 Form H

1. a
2. a
3. b
4. $y = 7x$
5. b
6. a
7. b
8. $x = 3$
9. $m = -\dfrac{20}{40}$
10. b
11. a
12. d
13. a
14. c
15. a
16. c
17. d
18. a
19.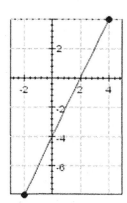

ANSWER KEY

Gustafson/Frisk - College Algebra 8E Chapter 2 Form H

20. 10

21. a

22.

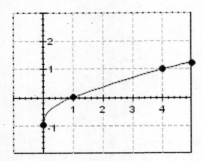

23. b

24. m = 0

25. $x^2 + y^2 - 2x - 2y - 7 = 0$

List of Problem Codes for BCA Testing

Gustafson/Frisk - College Algebra 8E Chapter 2 Form H

1. gfca.02.04.4.11m_NoAlgs
2. gfca.02.02.4.17m_NoAlgs
3. gfca.02.03.4.07m_NoAlgs
4. gfca.02.03.4.59_NoAlgs
5. gfca.02.04.4.80m_NoAlgs
6. gfca.02.04.4.51m_NoAlgs
7. gfca.02.02.4.39m_NoAlgs
8. gfca.02.03.4.77_NoAlgs
9. gfca.02.02.4.30_NoAlgs
10. gfca.02.05.4.25m_NoAlgs
11. gfca.02.05.4.24m_NoAlgs
12. gfca.02.01.4.68m_NoAlgs
13. gfca.02.04.4.56m_NoAlgs
14. gfca.02.05.4.17m_NoAlgs
15. gfca.02.01.4.35m_NoAlgs
16. gfca.02.05.4.13m_NoAlgs
17. gfca.02.03.4.15m_NoAlgs
18. gfca.02.03.4.30m_NoAlgs
19. gfca.02.01.4.33_NoAlgs
20. gfca.02.05.4.35_NoAlgs
21. gfca.02.01.4.47m_NoAlgs
22. gfca.02.04.4.59_NoAlgs
23. gfca.02.02.4.47m_NoAlgs
24. gfca.02.02.4.29_NoAlgs
25. gfca.02.04.4.65_NoAlgs

Gustafson/Frisk - College Algebra 8E Chapter 3 Form A

1. Tell where the function is increasing.

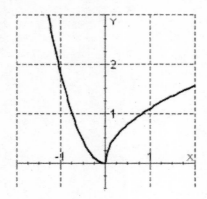

2. Let $f(x) = 3x$, $g(x) = x + 1$. Find the composite function.

   $(g \circ f)(x)$

3. Let the function f be defined by $y = f(x)$, where x and $f(x)$ are real numbers. Find $f(2)$.

   $f(x) = 93 - 2x^2$

4. Give the domain of the function.

   $f(x) = -\sqrt[3]{9x + 35}$

5. A farmer wants to partition a rectangular feed storage area in a corner of his barn. The barn walls form two sides of the stall, and the farmer has 58 feet of partition for the remaining two sides. What dimensions will maximize the area of the partition?

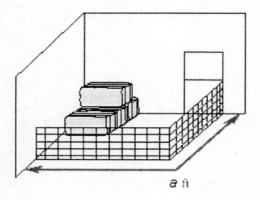

a ft

a = 58

6. Graph the function

$$f(x) = \left(\frac{1}{2}x\right)^3$$

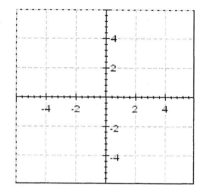

7. Graph the equation

$$h(x) = \sqrt{x-2} + 1, \quad x \geq 2$$

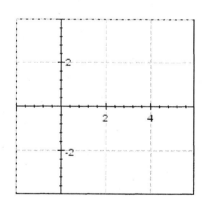

8. Find *y*-intercept of the function

$$f(x) = \frac{4x-6}{x-1}$$

9. Find the vertex of the parabola.

   $y = 4x^2 + 12x + 19$

10. Find the inverse of the one-to-one function.

    $y = 5x + 7$

11. Graph the piecewise-defined function.

    $y = f(x) = \begin{cases} x + 2 & \text{if } x < 0 \\ 2 & \text{if } x \geq 0 \end{cases}$

    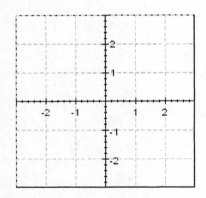

12. Graph the function

    $g(x) = (x - 2)^3$

    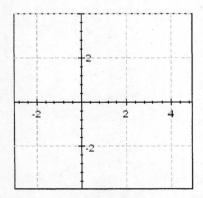

13. Let $f(x) = 2x - 1$, $g(x) = 3x - 2$. Find the domain of the function.

    $(f - g)(x)$

14. Let the function f be defined by the equation $y = f(x)$, where x and $f(x)$ are real numbers. Find the domain of the function

    $$f(x) = \frac{47x + 41}{x - 26}$$

15. Graph the function

    $$h(x) = \frac{x^2 - 2x - 8}{x - 1}$$

    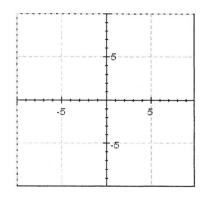

16. Graph the polynomial function

    $$f(x) = x^3 + x^2$$

    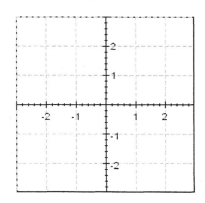

17. A pizzeria charges $9.50 plus $0.90 per topping for a large pizza. Find a linear function that expresses the cost y of a large pizza in terms of the number of toppings x.

18. Let $f(x) = \sqrt{x}$, $g(x) = x + 1$. Find the domain of the composite function. Please express the answer in interval notation.

$(g \circ f)(x)$

19. The function $f(x) = x^2 - 8$ is one-to-one on the domain $x \leq 0$. Find $f^{-1}(x)$.

20. Graph the quadratic function.

$f(x) = -x^2 - 4x + 1$

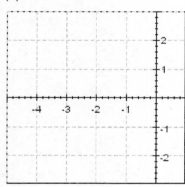

21. A plumber charges $40, plus $30 per hour (or fraction of an hour), to install a new bathtub. Graph the points (t, c), where t is the time it takes to do the job and c is the cost.

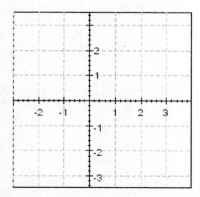

22. Let $f(x) = 2x - 5$, $g(x) = 5x - 2$. Find the value of the function.

    $(f \circ g)(-4)$

23. Graph the function

    $f(x) = \dfrac{x^2}{x}$

    Note that the numerator and denominator of the fraction share a common factor.

    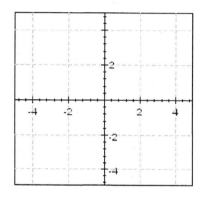

24. The function $f(x) = \dfrac{2}{x^2}$ is one-to-one on the domain $x > 0$. Find $f^{-1}(x)$.

25. Graph the quadratic function.

    $f(x) = x^2 + 2x$

    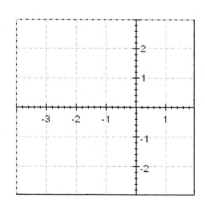

ANSWER KEY

Gustafson/Frisk - College Algebra 8E Chapter 3 Form A

1. $(0, \infty)$
2. $3x + 1$
3. 85
4. $(-\infty, \infty)$
5. 29, 29
6.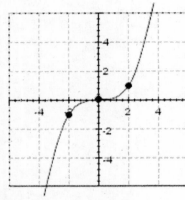
7.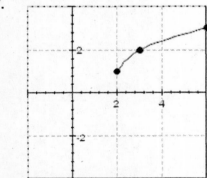
8. $(0, 6)$
9. $(-1, 5, 10)$
10. $\dfrac{(x-7)}{5}$

ANSWER KEY

Gustafson/Frisk - College Algebra 8E Chapter 3 Form A

11.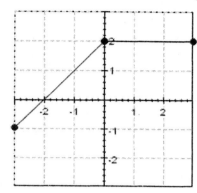

12.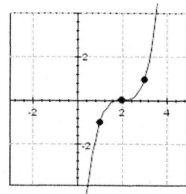

13. $(-\infty, \infty)$

14. $(-\infty, 26) \cup (26, \infty)$

15.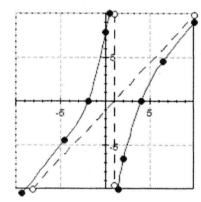

ANSWER KEY

Gustafson/Frisk - College Algebra 8E Chapter 3 Form A

16.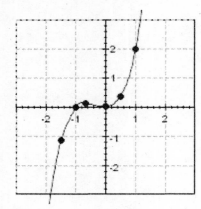

17. $y = 0.9x + 9.5$

18. $[0, \infty)$

19. $-\sqrt{x+8}$

20.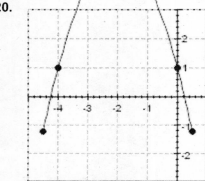

21.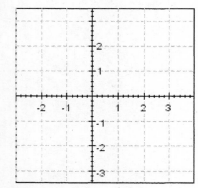

22. $-4, 9$

ANSWER KEY

Gustafson/Frisk - College Algebra 8E Chapter 3 Form A

23.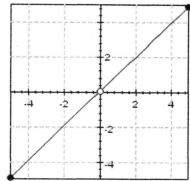

24. $\sqrt{\dfrac{2}{x}}$

25.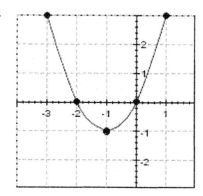

List of Problem Codes for BCA Testing

Gustafson/Frisk - College Algebra 8E Chapter 3 Form A

1. gfca.03.03.4.29_NoAlgs
2. gfca.03.06.4.43_NoAlgs
3. gfca.03.01.4.40_NoAlgs
4. gfca.03.01.4.82_NoAlgs
5. gfca.03.02.4.29_NoAlgs
6. gfca.03.04.4.35_NoAlgs
7. gfca.03.04.4.43_NoAlgs
8. gfca.03.05.4.33_NoAlgs
9. gfca.03.02.4.20_NoAlgs
10. gfca.03.07.4.27_NoAlgs
11. gfca.03.03.4.35_NoAlgs
12. gfca.03.04.4.21_NoAlgs
13. gfca.03.06.4.11_NoAlgs
14. gfca.03.01.4.34_NoAlgs
15. gfca.03.05.4.51_NoAlgs
16. gfca.03.03.4.12_NoAlgs
17. gfca.03.07.4.55a_NoAlgs
18. gfca.03.06.4.52_NoAlgs
19. gfca.03.07.4.45_NoAlgs
20. gfca.03.02.4.12_NoAlgs
21. gfca.03.03.4.50_NoAlgs
22. gfca.03.06.4.35_NoAlgs
23. gfca.03.05.4.55_NoAlgs
24. gfca.03.07.4.46_NoAlgs
25. gfca.03.02.4.06_NoAlgs

1. Find the inverse of the one-to-one function.

    $y = 5x + 7$

2. Graph the function.

    $f(x) = \dfrac{x^2}{x}$

    Note that the numerator and denominator of the fraction share a common factor.

    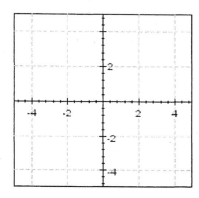

3. Let $f(x) = \sqrt{x}$, $g(x) = x + 1$. Find the domain of the composite function. Please express the answer in interval notation.

    $(g \circ f)(x)$

4. Graph the function.

$$h(x) = \frac{x^2 - 2x - 8}{x - 1}$$

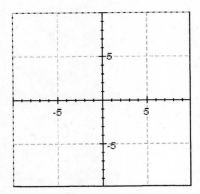

5. Graph the piecewise-defined function.

$$y = f(x) = \begin{cases} x + 2 & \text{if } x < 0 \\ 2 & \text{if } x \geq 0 \end{cases}$$

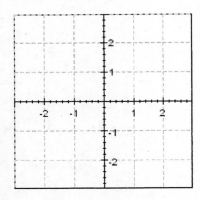

6. Graph the quadratic function.

$f(x) = -x^2 - 4x + 1$

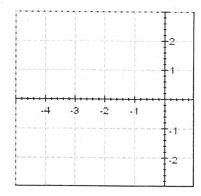

7. Find *y*-intercept of the function.

$f(x) = \dfrac{4x - 6}{x - 1}$

8. The function $f(x) = x^2 - 8$ is one-to-one on the domain $x \leq 0$. Find $f^{-1}(x)$.

9. Graph the function.

$g(x) = (x - 2)^3$

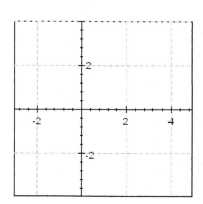

10. Give the domain of the function.

$$f(x) = -\sqrt[3]{9x + 35}$$

11. Graph the equation.

$$h(x) = \sqrt{x-2} + 1, \quad x \geq 2$$

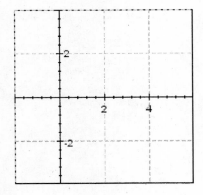

12. The function $f(x) = \dfrac{2}{x^2}$ is one-to-one on the domain $x > 0$. Find $f^{-1}(x)$.

13. Graph the polynomial function.

$$f(x) = x^3 + x^2$$

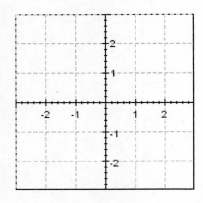

14. Find the vertex of the parabola.

    $y = 4x^2 + 12x + 19$

15. Tell where the function is increasing.

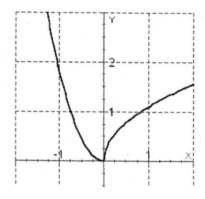

16. Let $f(x) = 3x$, $g(x) = x + 1$. Find the composite function.

    $(g \circ f)(x)$

17. A farmer wants to partition a rectangular feed storage area in a corner of his barn. The barn walls form two sides of the stall, and the farmer has 58 feet of partition for the remaining two sides. What dimensions will maximize the area of the partition?

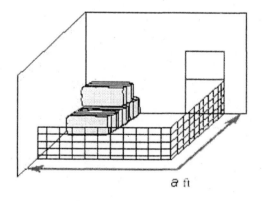

a ft

$a = 58$

18. Let $f(x) = 2x - 5$, $g(x) = 5x - 2$. Find the value of the function.

    $(f \circ g)(-4)$

19. Graph the quadratic function.

    $f(x) = x^2 + 2x$

    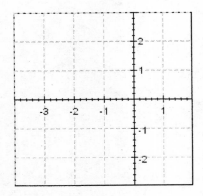

20. Let the function f be defined by the equation $y = f(x)$, where x and $f(x)$ are real numbers. Find the domain of the function.

    $f(x) = \dfrac{47x + 41}{x - 26}$

21. A pizzeria charges $9.50 plus $0.90 per topping for a large pizza. Find a linear function that expresses the cost y of a large pizza in terms of the number of toppings x.

22. Let the function f be defined by $y = f(x)$, where x and $f(x)$ are real numbers. Find $f(2)$.

    $f(x) = 93 - 2x^2$

23. Graph the function.

$$f(x) = \left(\frac{1}{2}x\right)^3$$

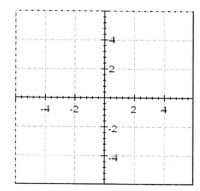

24. A plumber charges $40, plus $30 per hour (or fraction of an hour), to install a new bathtub. Graph the points (t, c), where t is the time it takes to do the job and c is the cost.

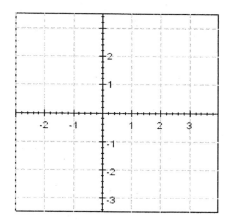

25. Let $f(x) = 2x - 1$, $g(x) = 3x - 2$. Find the domain of the function.

$(f - g)(x)$

ANSWER KEY

Gustafson/Frisk - College Algebra 8E Chapter 3 Form B

1. $\dfrac{(x-7)}{5}$

2.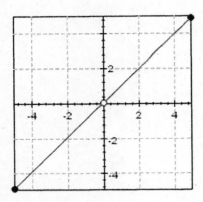

3. $[0, \infty)$

4.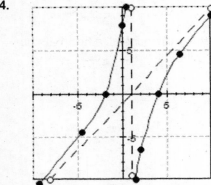

5.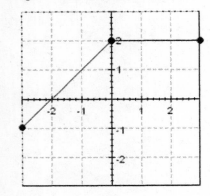

ANSWER KEY

Gustafson/Frisk - College Algebra 8E Chapter 3 Form B

6.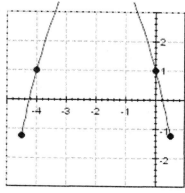

7. $(0, 6)$

8. $-\sqrt{x+8}$

9.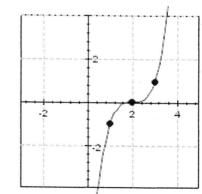

10. $(-\infty, \infty)$

11.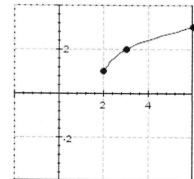

12. $\sqrt{\dfrac{2}{x}}$

13.
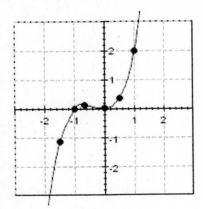

14. $(-1.5, 10)$

15. $(0, \infty)$

16. $3x + 1$

17. $29, 29$

18. $-4, 9$

19.
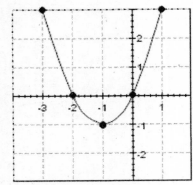

20. $(-\infty, 26) \cup (26, \infty)$

21. $y = 0.9x + 9.5$

22. 85

ANSWER KEY

Gustafson/Frisk - College Algebra 8E Chapter 3 Form B

23.

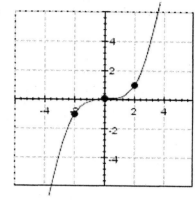

24.

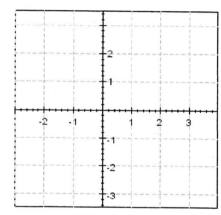

25. $(-\infty, \infty)$

List of Problem Codes for BCA Testing

Gustafson/Frisk - College Algebra 8E Chapter 3 Form B

1. gfca.03.07.4.27_NoAlgs
2. gfca.03.05.4.55_NoAlgs
3. gfca.03.06.4.52_NoAlgs
4. gfca.03.05.4.51_NoAlgs
5. gfca.03.03.4.35_NoAlgs
6. gfca.03.02.4.12_NoAlgs
7. gfca.03.05.4.33_NoAlgs
8. gfca.03.07.4.45_NoAlgs
9. gfca.03.04.4.21_NoAlgs
10. gfca.03.01.4.82_NoAlgs
11. gfca.03.04.4.43_NoAlgs
12. gfca.03.07.4.46_NoAlgs
13. gfca.03.03.4.12_NoAlgs
14. gfca.03.02.4.20_NoAlgs
15. gfca.03.03.4.29_NoAlgs
16. gfca.03.06.4.43_NoAlgs
17. gfca.03.02.4.29_NoAlgs
18. gfca.03.06.4.35_NoAlgs
19. gfca.03.02.4.06_NoAlgs
20. gfca.03.01.4.34_NoAlgs
21. gfca.03.07.4.55a_NoAlgs
22. gfca.03.01.4.40_NoAlgs
23. gfca.03.04.4.35_NoAlgs
24. gfca.03.03.4.50_NoAlgs
25. gfca.03.06.4.11_NoAlgs

Gustafson/Frisk - College Algebra 8E Chapter 3 Form C

1. Let the function f be defined by the equation $y = f(x)$, where x and $f(x)$ are real numbers. Find the domain of the function

   $$f(x) = \frac{43x + 11}{x - 22}$$

   Select the correct answer.

   a. domain: $(-\infty, -22] \cup [22, +\infty)$
   b. domain: $(-\infty, -43) \cup (43, +\infty)$
   c. domain: $(-\infty, 22) \cup (22, +\infty)$
   d. domain: $(-\infty, +\infty)$

2. Let the function f be defined by $y = f(x)$, where x and $f(x)$ are real numbers. Find $f(10)$.
   $$f(x) = 61 - 7x^2$$

   Select the correct answer.

   a. $f(10) = 179$
   b. $f(10) = -639$
   c. $f(10) = -662$

3. Give the domain of the function.

   $$f(x) = -\sqrt[3]{3x + 15}$$

   Select the correct answer.

   a. $(-\infty, 3)$
   b. $(-\infty, \infty)$
   c. $(-\infty, 3]$

4. Find y-intercept of the function.

   $$f(x) = \frac{3x - 3}{x - 3}$$

   Select the correct answer.

   a. $(0, -2)$
   b. $(0, 9)$
   c. $(0, 7)$
   d. $(0, 1)$

5. Graph the quadratic function.

$f(x) = x^2 + 2x$

Select the correct answer.

a.
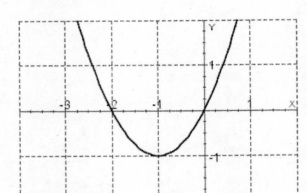

b.
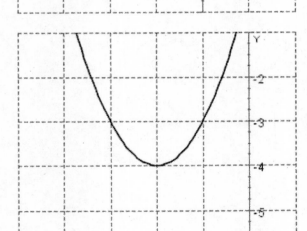

6. Graph the quadratic function.

   $f(x) = -x^2 - 3x + 1$

   Select the correct answer.

   a.
   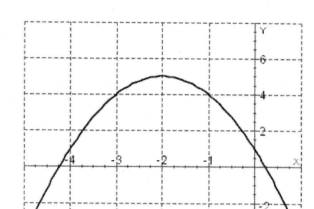

   b.
   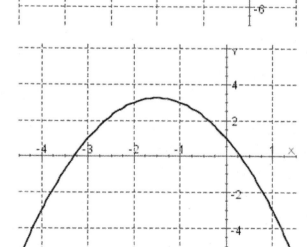

7. Find the vertex of the parabola.

    $y = 16x^2 + 40x + 32$

    Select the correct answer.

    a. $\left(-\dfrac{5}{4},\ 32\right)$

    b. $\left(\dfrac{5}{4},\ 7\right)$

    c. $\left(-\dfrac{5}{4},\ 25\right)$

    d. $\left(-\dfrac{5}{4},\ 7\right)$

8. A farmer wants to partition a rectangular feed storage area in a corner of his barn. The barn walls form two sides of the stall, and the farmer has 50 feet of partition for the remaining two sides. What dimensions will maximize the area of the partition?

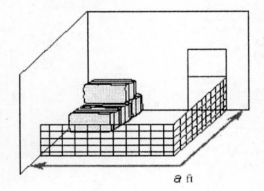

a ft

$a = 50$

Select the correct answer.

    a. 12.5 ft by 37.5 ft
    b. 20 ft by 30 ft
    c. 25 ft by 25 ft
    d. 10 ft by 40 ft

9. Graph the polynomial function

$$y = x^3 + 3x^2$$

Select the correct answer.

a.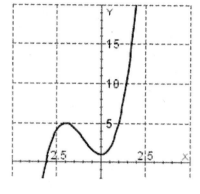

b.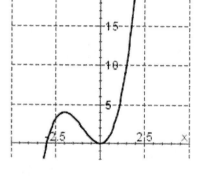

c.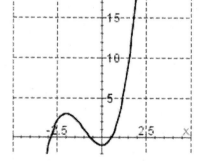

10. Tell where the function is increasing.

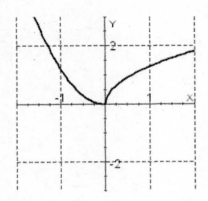

Select the correct answer.

a. $(0, \infty)$
b. $(-\infty, 0)$
c. always decreasing
d. always constant

11. Tell where the function is increasing.

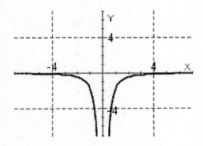

Select the correct answer.

a. always constant
b. always decreasing
c. $(-\infty, 0)$
d. $(0, \infty)$

12. A plumber charges $40, plus $50 per hour (or fraction of an hour), to install a new bathtub. Graph the points (t, c), where t is the time it takes to do the job and c is the cost.

   Select the correct answer.

   a.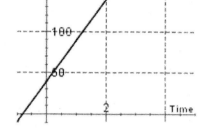

   b.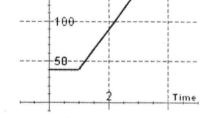

   c.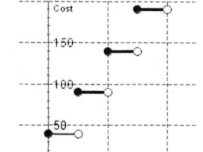

13. The graph of the function $g(x)$ is a translation of the graph of $f(x) = x^3$. Graph the function $g(x) = (x - 3)^3$

Select the correct answer.

a.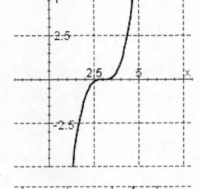

b.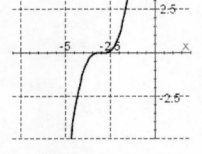

c.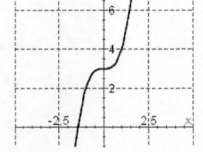

14. The graph of the function $f(x)$ is a stretching of the graph of $y = x^5$. Graph the function

$$f(x) = \left(\frac{1}{5}x\right)^5$$

Select the correct answer.

a.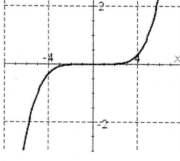

b.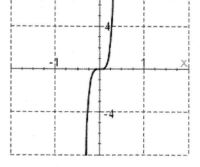

c.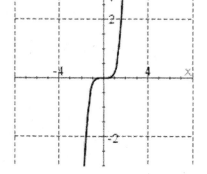

15. Use a translation to graph the equation.

$$h(x) = \sqrt{x-1} + 5, \quad x \geq 1$$

Select the correct answer.

a.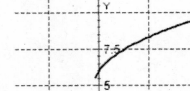

b.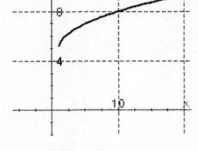

c.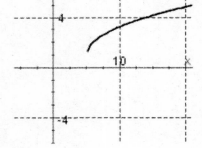

16. Graph the function.

$$h(x) = \frac{x^2 - 2x - 1}{x - 2}$$

Select the correct answer.

a.
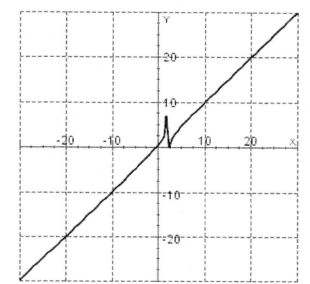

b.
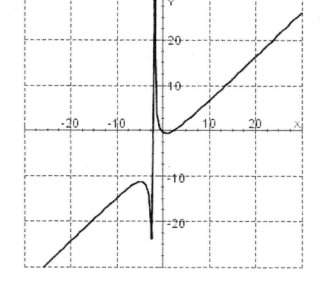

17. Graph the rational function.

$$f(x) = \frac{4x^2}{5x}$$

Note that the numerator and denominator of the fraction share a common factor.

Select the correct answer.

a.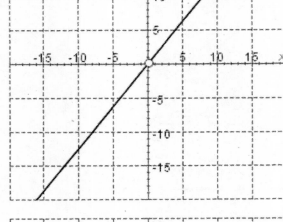

b.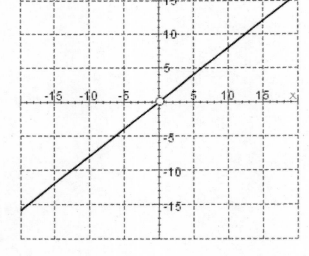

18. Let $f(x) = 2x - 1$, $g(x) = 3x - 2$. Find the domain of the function.

$(f \cdot g)(x)$

Select the correct answer.

a. $(-\infty, \infty)$
b. $(-\infty, 0]$
c. $[0, \infty)$
d. $(-\infty, 0)$
e. $(0, \infty)$

19. Let $f(x) = 2x - 5$, $g(x) = 5x - 2$. Find the value of the function.

$(f \circ f)(5)$

Select the correct answer.

a. $(f \circ f)(5) = 6$
b. $(f \circ f)(5) = 4$
c. $(f \circ f)(5) = 5$
d. $(f \circ f)(5) = 3$

20. Let $f(x) = 3x$, $g(x) = x + 1$. Find the composite function.

$(g \circ f)(x)$

Select the correct answer.

a. $(g \circ f)(x) = 3x + 3$
b. $(g \circ f)(x) = x + 2$
c. $(g \circ f)(x) = 3x + 1$
d. $(g \circ f)(x) = 9x$

21. Let $f(x) = \sqrt{x}$, $g(x) = x + 1$. Find the domain of the composite function.

    $(f \circ f)(x)$

    Select the correct answer.

    a. $[0, \infty)$
    b. $(0, \infty)$
    c. $[0, \infty]$
    d. $(-\infty, 0]$
    e. $(-\infty, 0)$

22. Find the inverse of the one-to-one function.

    $y = 6x + 3$

    Select the correct answer.

    a. $y = \dfrac{x + 3}{6}$
    b. $y = \dfrac{x - 3}{6}$
    c. $y = \dfrac{x - 6}{3}$
    d. $y = \dfrac{6}{x - 3}$

23. The function $f(x) = x^2 - 3$ is one-to-one on the domain $(x \leq 0)$. Find $f^{-1}(x)$.

    Select the correct answer.

    a. $f^{-1}(x) = \sqrt{x - 3}$
    b. $f^{-1}(x) = \sqrt{x + 3}$
    c. $f^{-1}(x) = -\sqrt{x + 3}$
    d. $f^{-1}(x) = x^2 + 3$

24. The function $f(x) = \dfrac{7}{x^2}$ is one-to-one on the domain $x > 0$. Find $f^{-1}(x)$.

Select the correct answer.

a. $f^{-1}(x) = \sqrt{\dfrac{x}{7}}$

b. $f^{-1}(x) = \dfrac{\sqrt{7}}{x}$

c. $f^{-1}(x) = \dfrac{x^2}{7}$

d. $f^{-1}(x) = \sqrt{\dfrac{7}{x}}$

25. A pizzeria charges $12.50 plus $0.75 per topping for a large pizza. Find the cost of a pizza that has 4 toppings.

Select the correct answer.

a. $15.50
b. $9.50
c. $9.25
d. $16.85

ANSWER KEY

Gustafson/Frisk - College Algebra 8E Chapter 3 Form C

1. c
2. b
3. b
4. d
5. a
6. b
7. d
8. c
9. b
10. a
11. d
12. c
13. a
14. a
15. b
16. a
17. b
18. a
19. c
20. c
21. a
22. b
23. c
24. d
25. a

List of Problem Codes for BCA Testing

Gustafson/Frisk - College Algebra 8E Chapter 3 Form C

1. gfca.03.01.4.34m_NoAlgs
2. gfca.03.01.4.40m_NoAlgs
3. gfca.03.01.4.82m_NoAlgs
4. gfca.03.05.4.33m_NoAlgs
5. gfca.03.02.4.06m_NoAlgs
6. gfca.03.02.4.12m_NoAlgs
7. gfca.03.02.4.20m_NoAlgs
8. gfca.03.02.4.29m_NoAlgs
9. gfca.03.03.4.12m_NoAlgs
10. gfca.03.03.4.29m_NoAlgs
11. gfca.03.03.4.31m_NoAlgs
12. gfca.03.03.4.50m_NoAlgs
13. gfca.03.04.4.21m_NoAlgs
14. gfca.03.04.4.35m_NoAlgs
15. gfca.03.04.4.43m_NoAlgs
16. gfca.03.05.4.51m_NoAlgs
17. gfca.03.05.4.55m_NoAlgs
18. gfca.03.06.4.11m_NoAlgs
19. gfca.03.06.4.35m_NoAlgs
20. gfca.03.06.4.43m_NoAlgs
21. gfca.03.06.4.52m_NoAlgs
22. gfca.03.07.4.27m_NoAlgs
23. gfca.03.07.4.45m_NoAlgs
24. gfca.03.07.4.46m_NoAlgs
25. gfca.03.07.4.55bm_NoAlgs

1. The graph of the function g(x) is a translation of the graph of $f(x) = x^3$. Graph the function $g(x) = (x - 3)^3$

   Select the correct answer.

   a.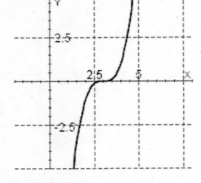

   b.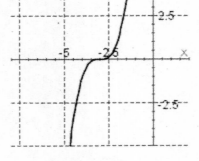

   c.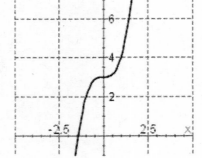

2. Graph the function.

$$h(x) = \frac{x^2 - 2x - 1}{x - 2}$$

Select the correct answer.

a.
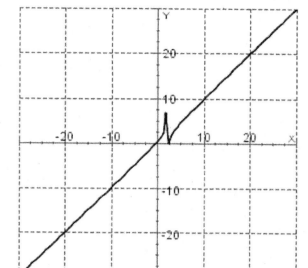

b.
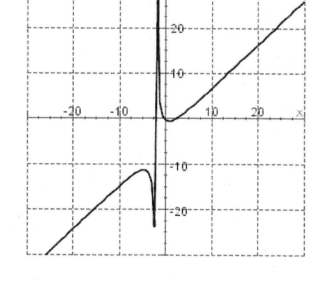

3. The function $f(x) = x^2 - 3$ is one-to-one on the domain $(x \leq 0)$. Find $f^{-1}(x)$.

Select the correct answer.

a. $f^{-1}(x) = \sqrt{x-3}$

b. $f^{-1}(x) = \sqrt{x+3}$

c. $f^{-1}(x) = -\sqrt{x+3}$

d. $f^{-1}(x) = x^2 + 3$

4. A pizzeria charges $12.50 plus $0.75 per topping for a large pizza. Find the cost of a pizza that has 4 toppings.

Select the correct answer.

a. $15.50
b. $9.50
c. $9.25
d. $16.85

5. Let the function f be defined by y = f(x), where x and f(x) are real numbers. Find f(10).

$$f(x) = 61 - 7x^2$$

Select the correct answer.

a. $f(10) = 179$
b. $f(10) = -639$
c. $f(10) = -662$

6. Let $f(x) = 3x$, $g(x) = x + 1$. Find the composite function.

$(g \circ f)(x)$

Select the correct answer.

a. $(g \circ f)(x) = 3x + 3$
b. $(g \circ f)(x) = x + 2$
c. $(g \circ f)(x) = 3x + 1$
d. $(g \circ f)(x) = 9x$

7. Graph the rational function.

$$f(x) = \frac{4x^2}{5x}$$

Note that the numerator and denominator of the fraction share a common factor.

Select the correct answer.

a.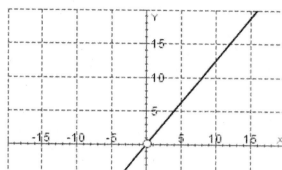

b.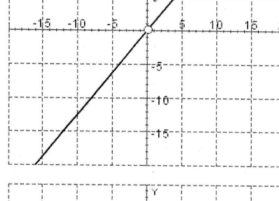

8. Graph the polynomial function.

$$y = x^3 + 3x^2$$

Select the correct answer.

a.
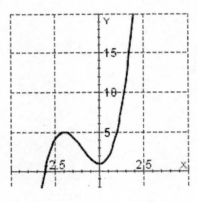

b.
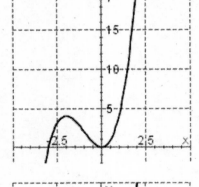

c.
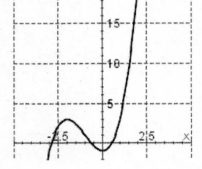

9. Use a translation to graph the equation.

$h(x) = \sqrt{x-1} + 5$, $x \geq 1$

Select the correct answer.

a.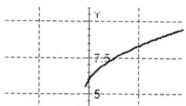

b.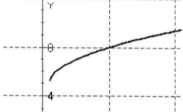

c.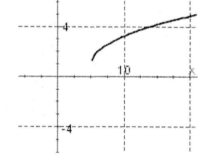

10. A plumber charges $40, plus $50 per hour (or fraction of an hour), to install a new bathtub. Graph the points (t, c), where t is the time it takes to do the job and c is the cost.

   Select the correct answer.

   a.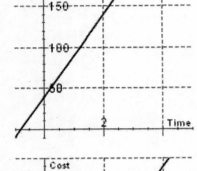

   b.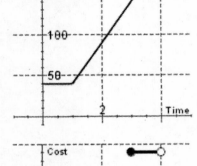

   c.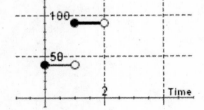

11. Let $f(x) = 2x - 1$, $g(x) = 3x - 2$. Find the domain of the function.

    $(f \cdot g)(x)$

    Select the correct answer.

    a. $(-\infty, \infty)$
    b. $(-\infty, 0]$
    c. $[0, \infty)$
    d. $(-\infty, 0)$
    e. $(0, \infty)$

12. A farmer wants to partition a rectangular feed storage area in a corner of his barn. The barn walls form two sides of the stall, and the farmer has 50 feet of partition for the remaining two sides.

    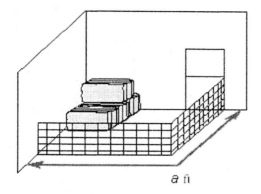

    a ft

    $a = 50$

    What dimensions will maximize the area of the partition?

    Select the correct answer.

    a. 12.5 ft by 37.5 ft
    b. 20 ft by 30 ft
    c. 25 ft by 25 ft
    d. 10 ft by 40 ft

13. Let $f(x) = \sqrt{x}$, $g(x) = x + 1$. Find the domain of the composite function.

    $(f \circ f)(x)$

    Select the correct answer.

    a. $[0, \infty)$
    b. $(0, \infty)$
    c. $[0, \infty]$
    d. $(-\infty, 0]$
    e. $(-\infty, 0)$

14. Find the vertex of the parabola.

    $y = 16x^2 + 40x + 32$

    Select the correct answer.

    a. $\left(-\dfrac{5}{4}, 32\right)$
    b. $\left(\dfrac{5}{4}, 7\right)$
    c. $\left(-\dfrac{5}{4}, 25\right)$
    d. $\left(-\dfrac{5}{4}, 7\right)$

15. The function $f(x) = \dfrac{7}{x^2}$ is one-to-one on the domain $x > 0$. Find $f^{-1}(x)$.

    Select the correct answer.

    a. $f^{-1}(x) = \sqrt{\dfrac{x}{7}}$
    b. $f^{-1}(x) = \dfrac{\sqrt{7}}{x}$
    c. $f^{-1}(x) = \dfrac{x^2}{7}$
    d. $f^{-1}(x) = \sqrt{\dfrac{7}{x}}$

16. Graph the quadratic function.

    $f(x) = -x^2 - 3x + 1$

    Select the correct answer.

a.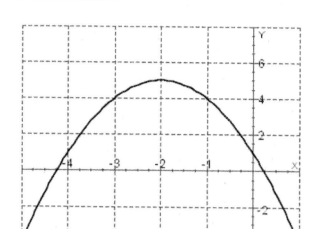

b.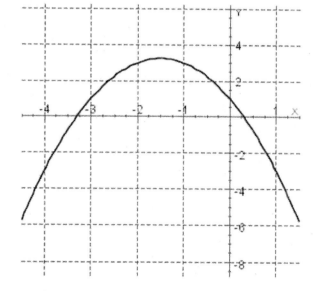

17. Give the domain of the function.

    $f(x) = -\sqrt[3]{3x + 15}$

    Select the correct answer.

    a. $(-\infty, 3)$
    b. $(-\infty, \infty)$
    c. $(-\infty, 3]$

18. Find the inverse of the one-to-one function.

    $y = 6x + 3$

    Select the correct answer.

    a. $y = \dfrac{x + 3}{6}$

    b. $y = \dfrac{x - 3}{6}$

    c. $y = \dfrac{x - 6}{3}$

    d. $y = \dfrac{6}{x - 3}$

19. Tell where the function is increasing.

    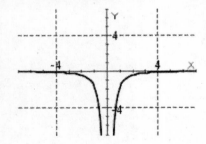

    Select the correct answer.

    a. always decreasing
    b. always constant
    c. $(0, \infty)$
    d. $(-\infty, 0)$

20. Graph the quadratic function.

    $f(x) = x^2 + 2x$

    Select the correct answer.

    a.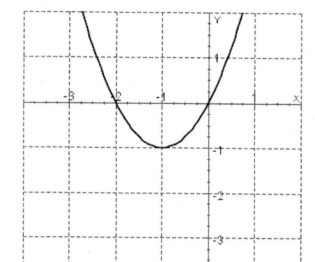

    b.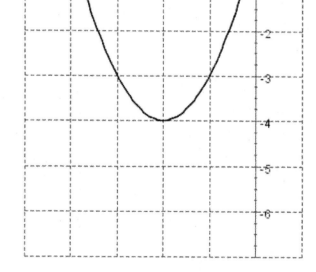

21. Find *y*-intercept of the function

$$f(x) = \frac{3x - 3}{x - 3}$$

Select the correct answer.

a. (0, -2)
b. (0, 9)
c. (0, 7)
d. (0, 1)

22. Let the function *f* be defined by the equation $y = f(x)$, where *x* and $f(x)$ are real numbers. Find the domain of the function

$$f(x) = \frac{43x + 11}{x - 22}$$

Select the correct answer.

a. domain: $(-\infty, -22] \cup [22, +\infty)$
b. domain: $(-\infty, 22) \cup (22, +\infty)$
c. domain: $(-\infty, -43) \cup (43, +\infty)$
d. domain: $(-\infty, +\infty)$

23. Let $f(x) = 2x - 5$, $g(x) = 5x - 2$. Find the value of the function.

$(f \circ f)(5)$

Select the correct answer.

a. $(f \circ f)(5) = 6$
b. $(f \circ f)(5) = 4$
c. $(f \circ f)(5) = 5$
d. $(f \circ f)(5) = 3$

24. The graph of the function $f(x)$ is a stretching of the graph of $y = x^5$. Graph the function
$$f(x) = \left(\frac{1}{5}x\right)^5$$

Select the correct answer.

a.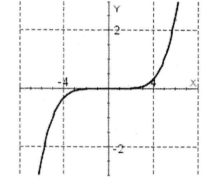

b.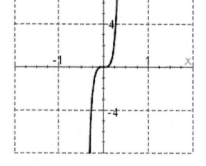

c.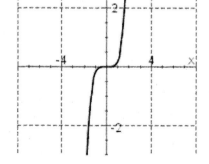

25. Tell where the function is increasing.

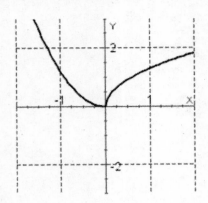

Select the correct answer.

a. $(0, \infty)$
b. $(-\infty, 0)$
c. always decreasing
d. always constant

ANSWER KEY

Gustafson/Frisk - College Algebra 8E Chapter 3 Form D

1. a
2. a
3. c
4. a
5. b
6. c
7. b
8. b
9. b
10. c
11. a
12. c
13. a
14. d
15. d
16. b
17. b
18. b
19. c
20. a
21. d
22. b
23. c
24. a
25. a

List of Problem Codes for BCA Testing

Gustafson/Frisk - College Algebra 8E Chapter 3 Form D

1. gfca.03.04.4.21m_NoAlgs
2. gfca.03.05.4.51m_NoAlgs
3. gfca.03.07.4.45m_NoAlgs
4. gfca.03.07.4.55bm_NoAlgs
5. gfca.03.01.4.40m_NoAlgs
6. gfca.03.06.4.43m_NoAlgs
7. gfca.03.05.4.55m_NoAlgs
8. gfca.03.03.4.12m_NoAlgs
9. gfca.03.04.4.43m_NoAlgs
10. gfca.03.03.4.50m_NoAlgs
11. gfca.03.06.4.11m_NoAlgs
12. gfca.03.02.4.29m_NoAlgs
13. gfca.03.06.4.52m_NoAlgs
14. gfca.03.02.4.20m_NoAlgs
15. gfca.03.07.4.46m_NoAlgs
16. gfca.03.02.4.12m_NoAlgs
17. gfca.03.01.4.82m_NoAlgs
18. gfca.03.07.4.27m_NoAlgs
19. gfca.03.03.4.31m_NoAlgs
20. gfca.03.02.4.06m_NoAlgs
21. gfca.03.05.4.33m_NoAlgs
22. gfca.03.01.4.34m_NoAlgs
23. gfca.03.06.4.35m_NoAlgs
24. gfca.03.04.4.35m_NoAlgs
25. gfca.03.03.4.29m_NoAlgs

1. Let the function f be defined by the equation y = f(x), where x and f(x) are real numbers. Find the domain of the function

$$f(x) = \frac{43x + 11}{x - 22}$$

Select the correct answer.

a. domain: $(-\infty, +\infty)$
b. domain: $(-\infty, -22] \cup [22, +\infty)$
c. domain: $(-\infty, -43) \cup (43, +\infty)$
d. domain: $(-\infty, 22) \cup (22, +\infty)$

2. Graph the quadratic function.

$f(x) = x^2 + 2x$

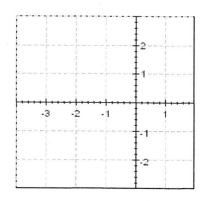

3. Let the function f be defined by y = f(x), where x and f(x) are real numbers. Find f(2).

$f(x) = 93 - 2x^2$

4. Give the domain of the function.

$f(x) = -\sqrt[3]{9x + 35}$

5. Find the vertex of the parabola.

   $y = x^2 - 8x + 16$

6. A farmer wants to partition a rectangular feed storage area in a corner of his barn. The barn walls form two sides of the stall, and the farmer has 50 feet of partition for the remaining two sides. What dimensions will maximize the area of the partition?

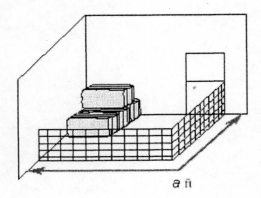

$a = 50$

Select the correct answer.

a. 25 ft by 25 ft
b. 20 ft by 30 ft
c. 12.5 ft by 37.5 ft
d. 10 ft by 40 ft

7. Tell where the function is decreasing.

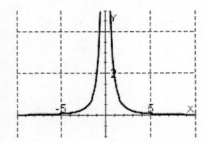

8. Find the vertex of the parabola.

   $y = 16x^2 + 40x + 32$

   Select the correct answer.

   a. $\left(-\dfrac{5}{4}, 7\right)$

   b. $\left(-\dfrac{5}{4}, 32\right)$

   c. $\left(\dfrac{5}{4}, 7\right)$

   d. $\left(-\dfrac{5}{4}, 25\right)$

9. Graph the polynomial function

   $f(x) = x^3 + x^2$

   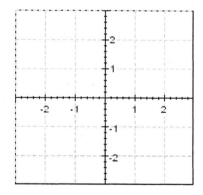

10. Determine whether the function is one-to-one.

    $y = 8x^3 - 4x$

    Select the correct answer.

    a. No, it isn't one-to-one.
    b. Yes, it is one-to-one.

11. The function $f(x) = x^2 - 8$ is one-to-one on the domain $x \leq 0$. Find $f^{-1}(x)$.

12. The function $f(x) = \dfrac{2}{x^2}$ is one-to-one on the domain $x > 0$. Find $f^{-1}(x)$.

13. A pizzeria charges $12.50 plus $0.75 per topping for a large pizza. Find the cost of a pizza that has 4 toppings.

    Select the correct answer.

    a. $9.25
    b. $16.85
    c. $15.50
    d. $9.50

14. Graph the function

    $g(x) = (x + 3)^2$

    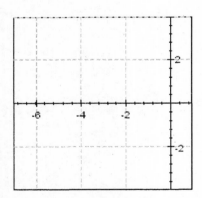

15. A plumber charges $40, plus $50 per hour (or fraction of an hour), to install a new bathtub. Graph the points (*t*, *c*), where *t* is the time it takes to do the job and *c* is the cost.

Select the correct answer.

a.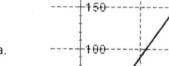

b.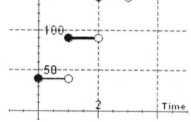

c.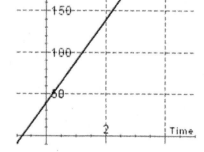

16. The graph of the function g(x) is a translation of the graph of $f(x) = x^3$. Graph the function

$g(x) = (x-3)^3$

Select the correct answer.

a.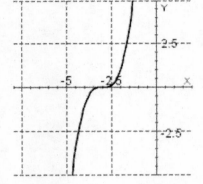

b.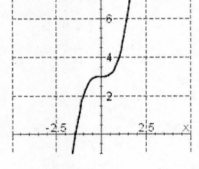

c.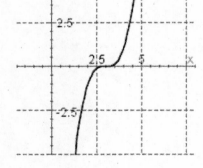

17. Graph the equation

$$h(x) = \sqrt{x-2} + 1, \quad x \geq 2$$

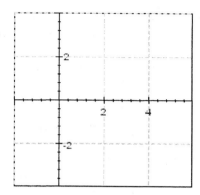

18. Find the domain of the rational function.

$$y = \frac{x - 1}{x^3 - 9x}$$

Select the correct answer.

a. $(-\infty, -3) \cup (-3, 0) \cup (0, 3) \cup (3, \infty)$
b. $(-\infty, -8) \cup (-8, 0) \cup (0, 8) \cup (8, \infty)$
c. $(-\infty, -5) \cup (-5, 0) \cup (0, 5) \cup (5, \infty)$
d. $(-\infty, -6) \cup (-6, 0) \cup (0, 6) \cup (6, \infty)$

19. Find vertical and horizontal asymptotes of the function

$$f(x) = \frac{1}{x - 8}$$

Select the correct answer.

a. Horizontal: $y = -1$, Vertical: $x = 8$
b. Horizontal: $y = 1$, Vertical: $x = 8$
c. Horizontal: $y = 0$, Vertical: $x = -8$
d. Horizontal: $y = 0$, Vertical: $x = 8$

20. Graph the function.

$$h(x) = \frac{x^2 - 2x - 8}{x - 1}$$

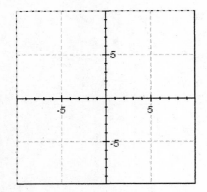

21. Let $f(x) = x^2 + x$, $g(x) = x^2 - 1$. Find $(f \cdot g)(x)$.

Select the correct answer.

a. $(f \cdot g)(x) = x + 1$

b. $(f \cdot g)(x) = x^4 + x^3 - x^2 - x$

c. $(f \cdot g)(x) = \dfrac{x}{x - 1}$

d. $(f \cdot g)(x) = 2x^2 + x - 1$

22. Let $f(x) = 2x - 5$, $g(x) = 5x - 2$. Find the value of the function.

$(f \circ f)(5)$

Select the correct answer.

a. $(f \circ f)(5) = 3$

b. $(f \circ f)(5) = 5$

c. $(f \circ f)(5) = 4$

d. $(f \circ f)(5) = 6$

23. Tell where the function is increasing.

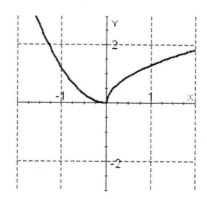

Select the correct answer.

a. $(-\infty, 0)$
b. $(0, \infty)$
c. always constant
d. always decreasing

24. Let $f(x) = 3x$, $g(x) = x + 1$. Find the composite function.

$(g \circ f)(x)$

25. Let $f(x) = \sqrt{x}$, $g(x) = x + 1$. Find the domain of the composite function. Please express the answer in interval notation.

$(g \circ f)(x)$

ANSWER KEY

Gustafson/Frisk - College Algebra 8E Chapter 3 Form E

1. d

2.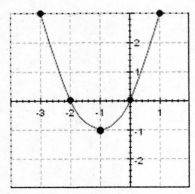

3. 85

4. $(-\infty, \infty)$

5. $(4, 0)$

6. a

7. $(0, \infty)$

8. a

9.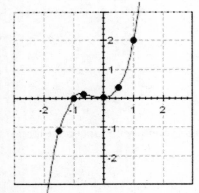

10. a

11. $-\sqrt{x+8}$

12. $\sqrt{\dfrac{2}{x}}$

13. c

ANSWER KEY

Gustafson/Frisk - College Algebra 8E Chapter 3 Form E

14.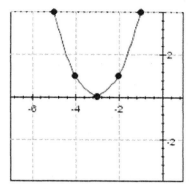

15. b
16. c
17.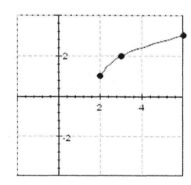

18. a

19. d

20.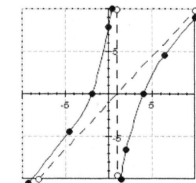

21. b

22. b

23. b

24. $3x + 1$

25. $[0, \infty)$

List of Problem Codes for BCA Testing

Gustafson/Frisk - College Algebra 8E Chapter 3 Form E

1. gfca.03.01.4.34m_NoAlgs
2. gfca.03.02.4.06_NoAlgs
3. gfca.03.01.4.40_NoAlgs
4. gfca.03.01.4.82_NoAlgs
5. gfca.03.02.4.16_NoAlgs
6. gfca.03.02.4.29m_NoAlgs
7. gfca.03.03.4.31_NoAlgs
8. gfca.03.02.4.20m_NoAlgs
9. gfca.03.03.4.12_NoAlgs
10. gfca.03.07.4.09m_NoAlgs
11. gfca.03.07.4.45_NoAlgs
12. gfca.03.07.4.46_NoAlgs
13. gfca.03.07.4.55bm_NoAlgs
14. gfca.03.04.4.13_NoAlgs
15. gfca.03.03.4.50m_NoAlgs
16. gfca.03.04.4.21m_NoAlgs
17. gfca.03.04.4.43_NoAlgs
18. gfca.03.05.4.23m_NoAlgs
19. gfca.03.05.4.27m_NoAlgs
20. gfca.03.05.4.51_NoAlgs
21. gfca.03.06.4.17m_NoAlgs
22. gfca.03.06.4.35m_NoAlgs
23. gfca.03.03.4.29m_NoAlgs
24. gfca.03.06.4.43_NoAlgs
25. gfca.03.06.4.52_NoAlgs

1. A plumber charges $40, plus $50 per hour (or fraction of an hour), to install a new bathtub. Graph the points (t, c), where t is the time it takes to do the job and c is the cost.

   Select the correct answer.

   a.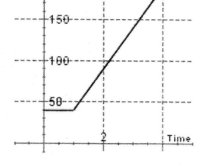

   b.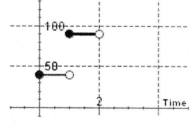

   c.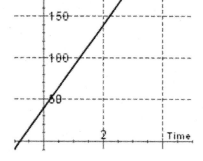

2. Find the vertex of the parabola.

   $y = x^2 - 8x + 16$

3. Graph the equation

$$h(x) = \sqrt{x-2} + 1, \quad x \geq 2$$

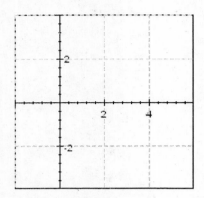

4. Let $f(x) = x^2 + x$, $g(x) = x^2 - 1$. Find $(f \cdot g)(x)$.

Select the correct answer.

a. $(f \cdot g)(x) = x^4 + x^3 - x^2 - x$

b. $(f \cdot g)(x) = \dfrac{x}{x-1}$

c. $(f \cdot g)(x) = 2x^2 + x - 1$

d. $(f \cdot g)(x) = x + 1$

5. Graph the function

$$h(x) = \dfrac{x^2 - 2x - 8}{x - 1}$$

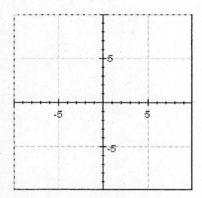

6. The graph of the function g(x) is a translation of the graph of $f(x) = x^3$. Graph the function $g(x) = (x - 3)^3$

   Select the correct answer.

   a.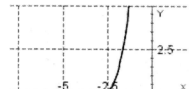

   b.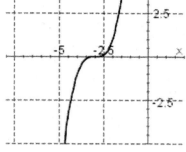

   c.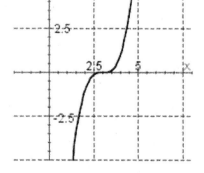

7. Give the domain of the function.

   $f(x) = -\sqrt[3]{9x + 35}$

8. The function $f(x) = x^2 - 8$ is one-to-one on the domain $x \leq 0$. Find $f^{-1}(x)$.

9. Graph the polynomial function

$$f(x) = x^3 + x^2$$

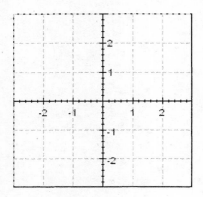

10. Find the domain of the rational function.

$$y = \frac{x-1}{x^3 - 9x}$$

Select the correct answer.

a. $(-\infty, -3) \cup (-3, 0) \cup (0, 3) \cup (3, \infty)$
b. $(-\infty, -8) \cup (-8, 0) \cup (0, 8) \cup (8, \infty)$
c. $(-\infty, -5) \cup (-5, 0) \cup (0, 5) \cup (5, \infty)$
d. $(-\infty, -6) \cup (-6, 0) \cup (0, 6) \cup (6, \infty)$

11. Graph the function.

    $g(x) = (x + 3)^2$

    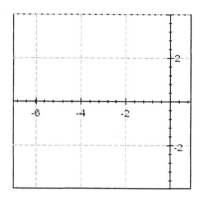

12. Graph the quadratic function.

    $f(x) = x^2 + 2x$

    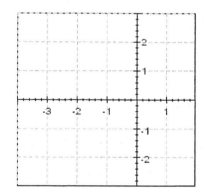

13. A pizzeria charges $12.50 plus $0.75 per topping for a large pizza. Find the cost of a pizza that has 4 toppings.

    Select the correct answer.

    a. $9.25
    b. $16.85
    c. $15.50
    d. $9.50

14. The function

    $f(x) = \dfrac{2}{x^2}$ is one-to-one on the domain $x > 0$. Find $f^{-1}(x)$.

15. Let the function *f* be defined by the equation $y = f(x)$, where x and $f(x)$ are real numbers. Find the domain of the function

$$f(x) = \frac{43x + 11}{x - 22}$$

Select the correct answer.

a.   domain: $(-\infty, +\infty)$

b.   domain: $(-\infty, -22] \cup [22, +\infty)$

c.   domain: $(-\infty, -43) \cup (43, +\infty)$

d.   domain: $(-\infty, 22) \cup (22, +\infty)$

16. Find the vertex of the parabola.

$y = 16x^2 + 40x + 32$

Select the correct answer.

a. $\left(-\frac{5}{4}, 7\right)$

b. $\left(-\frac{5}{4}, 32\right)$

c. $\left(\frac{5}{4}, 7\right)$

d. $\left(-\frac{5}{4}, 25\right)$

17. Tell where the function is decreasing.

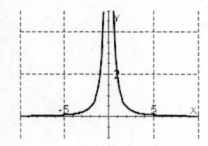

18. Tell where the function is increasing.

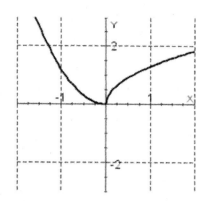

Select the correct answer.

a. $(-\infty, 0)$
b. $(0, \infty)$
c. always constant
d. always decreasing

19. Let $f(x) = 2x - 5$, $g(x) = 5x - 2$. Find the value of the function.

$(f \circ f)(5)$

Select the correct answer.

a. $(f \circ f)(5) = 3$
b. $(f \circ f)(5) = 5$
c. $(f \circ f)(5) = 4$
d. $(f \circ f)(5) = 6$

20. Let the function f be defined by $y = f(x)$, where x and $f(x)$ are real numbers. Find $f(2)$.

$f(x) = 93 - 2x^2$

21. Let $f(x) = 3x$, $g(x) = x + 1$. Find the composite function.

$(g \circ f)(x)$

22. Let $f(x) = \sqrt{x}$, $g(x) = x + 1$. Find the domain of the composite function. Please express the answer in interval notation.

$(g \circ f)(x)$

23. A farmer wants to partition a rectangular feed storage area in a corner of his barn. The barn walls form two sides of the stall, and the farmer has 50 feet of partition for the remaining two sides. What dimensions will maximize the area of the partition?

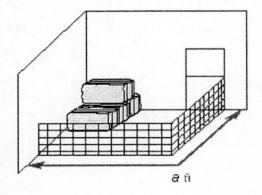

a ft

a = 50

Select the correct answer.

a. 25 ft by 25 ft
b. 20 ft by 30 ft
c. 12.5 ft by 37.5 ft
d. 10 ft by 40 ft

24. Find vertical and horizontal asymptotes of the function
$$f(x) = \frac{1}{x-8}$$

Select the correct answer.

a. Horizontal: $y = -1$, Vertical: $x = 8$
b. Horizontal: $y = 1$, Vertical: $x = 8$
c. Horizontal: $y = 0$, Vertical: $x = -8$
d. Horizontal: $y = 0$, Vertical: $x = 8$

25. Determine whether the function is one-to-one.

$$y = 8x^3 - 4x$$

Select the correct answer.

a. No, it isn't one-to-one.
b. Yes, it is one-to-one.

ANSWER KEY

Gustafson/Frisk - College Algebra 8E Chapter 3 Form F

1. b

2. $(4, 0)$

3.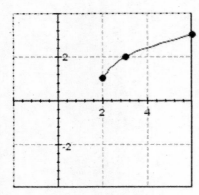

4. a

5.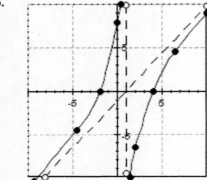

6. c

7. $(-\infty, \infty)$

8. $-\sqrt{x+8}$

9.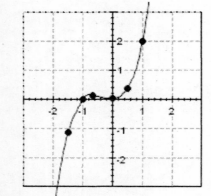

10. a

ANSWER KEY

Gustafson/Frisk - College Algebra 8E Chapter 3 Form F

11.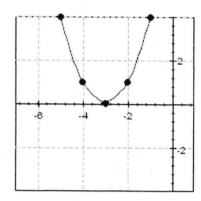

12.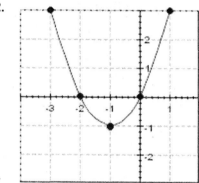

13. c

14. $\sqrt{\dfrac{2}{x}}$

15. d

16. a

17. $(0, \infty)$

18. b

19. b

20. 85

21. $3x + 1$

22. $[0, \infty)$

23. a

24. d

25. a

List of Problem Codes for BCA Testing

Gustafson/Frisk - College Algebra 8E Chapter 3 Form F

1. gfca.03.03.4.50m_NoAlgs
2. gfca.03.02.4.16_NoAlgs
3. gfca.03.04.4.43_NoAlgs
4. gfca.03.06.4.17m_NoAlgs
5. gfca.03.05.4.51_NoAlgs
6. gfca.03.04.4.21m_NoAlgs
7. gfca.03.01.4.82_NoAlgs
8. gfca.03.07.4.45_NoAlgs
9. gfca.03.03.4.12_NoAlgs
10. gfca.03.05.4.23m_NoAlgs
11. gfca.03.04.4.13_NoAlgs
12. gfca.03.02.4.06_NoAlgs
13. gfca.03.07.4.55bm_NoAlgs
14. gfca.03.07.4.46_NoAlgs
15. gfca.03.01.4.34m_NoAlgs
16. gfca.03.02.4.20m_NoAlgs
17. gfca.03.03.4.31_NoAlgs
18. gfca.03.03.4.29m_NoAlgs
19. gfca.03.06.4.35m_NoAlgs
20. gfca.03.01.4.40_NoAlgs
21. gfca.03.06.4.43_NoAlgs
22. gfca.03.06.4.52_NoAlgs
23. gfca.03.02.4.29m_NoAlgs
24. gfca.03.05.4.27m_NoAlgs
25. gfca.03.07.4.09m_NoAlgs

1. Graph the piecewise-defined function.

$$y = f(x) = \begin{cases} x + 6 & \text{if } x < 0 \\ 6 & \text{if } x \geq 0 \end{cases}$$

Select the correct answer.

a.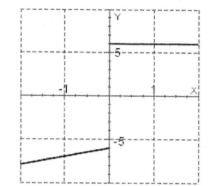

b.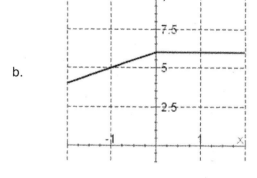

c.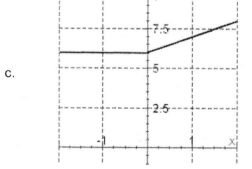

2. Graph the rational function

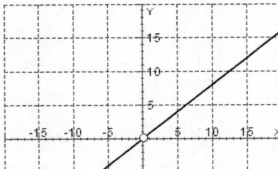

$$f(x) = \frac{4x^2}{5x}$$

Note that the numerator and denominator of the fraction share a common factor.

Select the correct answer.

a.
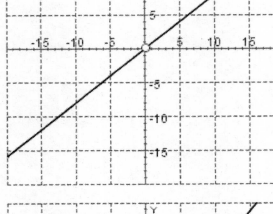

b.
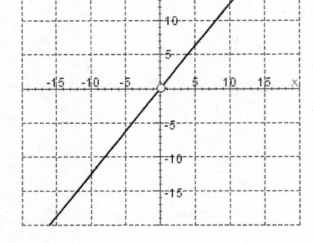

3. Graph the quadratic function.

$f(x) = -x^2 - 3x + 1$

Select the correct answer.

a.
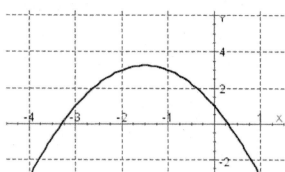

b.
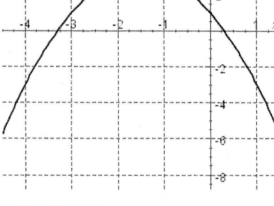

4. Let the function f be defined by $y = f(x)$, where x and $f(x)$ are real numbers. Find $f(2)$.

$f(x) = 93 - 2x^2$

5. Let $f(x) = 2x - 1$, $g(x) = 3x - 2$. Find the domain of the function.

   $(f - g)(x)$

6. The function $f(x) = x^2 - 3$ is one-to-one on the domain $(x \leq 0)$. Find $f^{-1}(x)$.

   Select the correct answer.

   a. $f^{-1}(x) = \sqrt{x - 3}$

   b. $f^{-1}(x) = -\sqrt{x + 3}$

   c. $f^{-1}(x) = \sqrt{x + 3}$

   d. $f^{-1}(x) = x^2 + 3$

7. Let the function f be defined by the equation $y = f(x)$, where x and $f(x)$ are real numbers. Find the domain of the function

   $f(x) = \dfrac{4\,3x + 11}{x - 22}$

   Select the correct answer.

   a. domain: $(-\infty, 22) \cup (22, +\infty)$
   b. domain: $(-\infty, -43) \cup (43, +\infty)$
   c. domain: $(-\infty, -22] \cup [22, +\infty)$
   d. domain: $(-\infty, +\infty)$

8. Let $f(x) = 3x$, $g(x) = x + 1$. Find the composite function.

   $(g \circ f)(x)$

   Select the correct answer.

   a. $(g \circ f)(x) = 3x + 3$
   b. $(g \circ f)(x) = 3x + 1$
   c. $(g \circ f)(x) = x + 2$
   d. $(g \circ f)(x) = 9x$

9. A plumber charges $40, plus $50 per hour (or fraction of an hour), to install a new bathtub. Graph the points (t, c), where t is the time it takes to do the job and c is the cost.

Select the correct answer.

a.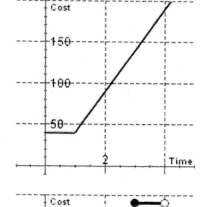

b.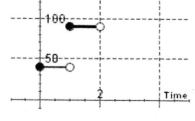

c.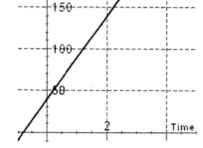

10. Let $f(x) = x^2 + x$, $g(x) = x^2 - 1$. Find $(f \cdot g)(x)$.

   Select the correct answer.

   a. $(f \cdot g)(x) = x + 1$

   b. $(f \cdot g)(x) = \dfrac{x}{x - 1}$

   c. $(f \cdot g)(x) = 2x^2 + x - 1$

   d. $(f \cdot g)(x) = x^4 + x^3 - x^2 - x$

11. Graph the equation

    $h(x) = \sqrt{x - 2} + 1$, $x \geq 2$

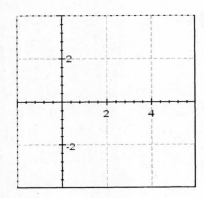

12. Find vertical and horizontal asymptotes of the function

    $f(x) = \dfrac{1}{x - 8}$

    Select the correct answer.

    a. Horizontal: $y = -1$, Vertical: $x = 8$

    b. Horizontal: $y = 0$, Vertical: $x = 8$

    c. Horizontal: $y = 1$, Vertical: $x = 8$

    d. Horizontal: $y = 0$, Vertical: $x = -8$

13. Give the domain of the function.

    $f(x) = -\sqrt[3]{3x + 15}$

    Select the correct answer.

    a. $(-\infty, 3]$
    b. $(-\infty, 3)$
    c. $(-\infty, \infty)$

14. A farmer wants to partition a rectangular feed storage area in a corner of his barn. The barn walls form two sides of the stall, and the farmer has 58 feet of partition for the remaining two sides. What dimensions will maximize the area of the partition?

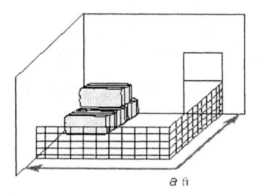

a ft

$a = 58$

15. Tell where the function is increasing.

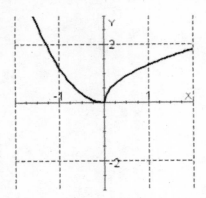

Select the correct answer.

a. always constant
b. (0, ∞)
c. always decreasing
d. (−∞, 0)

16. Graph the polynomial function.

$$y = x^3 + 3x^2$$

Select the correct answer.

a.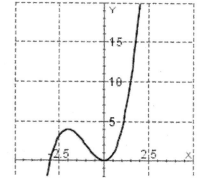

b.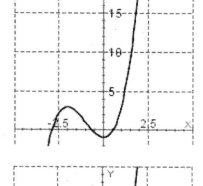

c.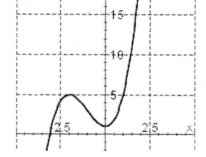

17. Let the function *f* be defined by $y = f(x)$, where x and $f(x)$ are real numbers. Find $f(10)$.

$f(x) = 61 - 7x^2$

Select the correct answer.

a. $f(10) = -639$
b. $f(10) = 179$
c. $f(10) = -662$

18. Graph the function

$h(x) = \dfrac{x^2 - 2x - 8}{x - 1}$

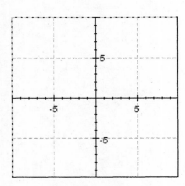

19. The graph of the function g(x) is a translation of the graph of $f(x) = x^3$. Graph the function $g(x) = (x - 3)^3$

Select the correct answer.

a.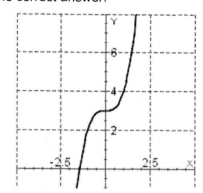

b.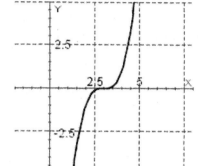

c.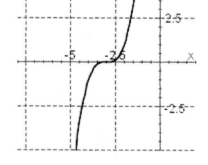

20. Graph the function

 $g(x) = (x + 3)^2$

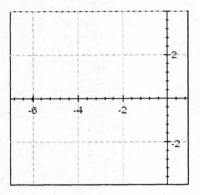

21. Find the inverse of the one-to-one function.

 $y = 6x + 3$

 Select the correct answer.

 a. $y = \dfrac{x - 6}{3}$

 b. $y = \dfrac{x - 3}{6}$

 c. $y = \dfrac{x + 3}{6}$

 d. $y = \dfrac{6}{x - 3}$

22. The graph of the function $f(x)$ is a stretching of the graph of $y = x^5$. Graph the function

$$f(x) = \left(\frac{1}{5}x\right)^5$$

Select the correct answer.

a.
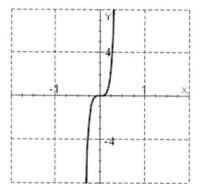

b.
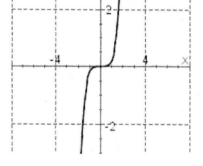

c.
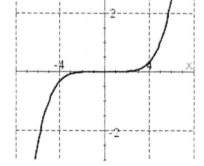

23. The function $f(x) = \dfrac{2}{x^2}$ is one-to-one on the domain $x > 0$. Find $f^{-1}(x)$.

24. Find the vertex of the parabola.

$y = 16x^2 + 40x + 32$

Select the correct answer.

a. $\left(\dfrac{5}{4}, 7\right)$

b. $\left(-\dfrac{5}{4}, 7\right)$

c. $\left(-\dfrac{5}{4}, 25\right)$

d. $\left(-\dfrac{5}{4}, 32\right)$

25. Graph the quadratic function.

$f(x) = -x^2 - 4x + 1$

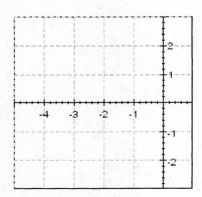

ANSWER KEY

Gustafson/Frisk - College Algebra 8E Chapter 3 Form G

1. b
2. a
3. a
4. 85
5. $(-\infty, \infty)$
6. b
7. a
8. b
9. b
10. d
11.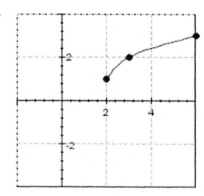
12. b
13. c
14. 29, 29
15. b
16. a
17. a

ANSWER KEY

Gustafson/Frisk - College Algebra 8E Chapter 3 Form G

18.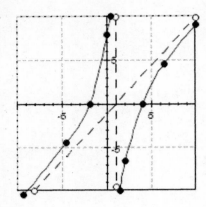

19. b

20.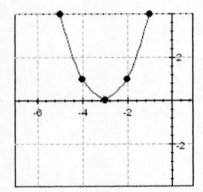

21. b

22. c

23. $\sqrt{\dfrac{2}{x}}$

24. b

25.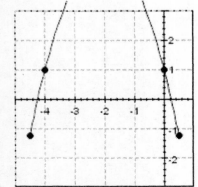

List of Problem Codes for BCA Testing

Gustafson/Frisk - College Algebra 8E Chapter 3 Form G

1. gfca.03.03.4.35m_NoAlgs
2. gfca.03.05.4.55m_NoAlgs
3. gfca.03.02.4.12m_NoAlgs
4. gfca.03.01.4.40_NoAlgs
5. gfca.03.06.4.11_NoAlgs
6. gfca.03.07.4.45m_NoAlgs
7. gfca.03.01.4.34m_NoAlgs
8. gfca.03.06.4.43m_NoAlgs
9. gfca.03.03.4.50m_NoAlgs
10. gfca.03.06.4.17m_NoAlgs
11. gfca.03.04.4.43_NoAlgs
12. gfca.03.05.4.27m_NoAlgs
13. gfca.03.01.4.82m_NoAlgs
14. gfca.03.02.4.29_NoAlgs
15. gfca.03.03.4.29m_NoAlgs
16. gfca.03.03.4.12m_NoAlgs
17. gfca.03.01.4.40m_NoAlgs
18. gfca.03.05.4.51_NoAlgs
19. gfca.03.04.4.21m_NoAlgs
20. gfca.03.04.4.13_NoAlgs
21. gfca.03.07.4.27m_NoAlgs
22. gfca.03.04.4.35m_NoAlgs
23. gfca.03.07.4.46_NoAlgs
24. gfca.03.02.4.20m_NoAlgs
25. gfca.03.02.4.12_NoAlgs

1. Graph the function.

   $g(x) = (x + 3)^2$

   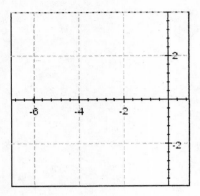

2. Let $f(x) = 3x$, $g(x) = x + 1$. Find the composite function.

   $(g \circ f)(x)$

   Select the correct answer.

   a. $(g \circ f)(x) = 3x + 3$
   b. $(g \circ f)(x) = 3x + 1$
   c. $(g \circ f)(x) = x + 2$
   d. $(g \circ f)(x) = 9x$

3. The graph of the function $f(x)$ is a stretching of the graph of $y = x^5$. Graph the function

$$f(x) = \left(\frac{1}{5}x\right)^5$$

Select the correct answer.

a.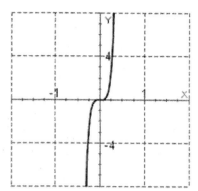

b.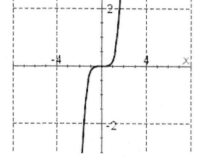

c.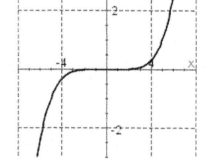

4. Let $f(x) = 2x - 1$, $g(x) = 3x - 2$. Find the domain of the function.

$(f - g)(x)$

5. Find vertical and horizontal asymptotes of the function.

$$f(x) = \frac{1}{x-8}$$

Select the correct answer.

a. Horizontal: $y = -1$, Vertical: $x = 8$
b. Horizontal: $y = 0$, Vertical: $x = 8$
c. Horizontal: $y = 1$, Vertical: $x = 8$
d. Horizontal: $y = 0$, Vertical: $x = -8$

6. Graph the quadratic function.

$f(x) = -x^2 - 4x + 1$

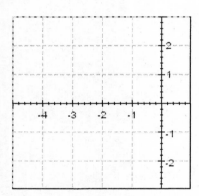

7. The graph of the function g(x) is a translation of the graph of $f(x) = x^3$. Graph the function $g(x) = (x - 3)^3$

Select the correct answer.

a.
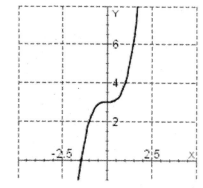

b.
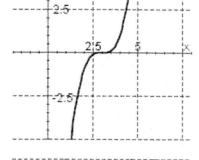

c.
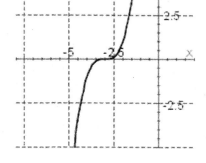

8. Find the inverse of the one-to-one function.

y = 6x + 3

Select the correct answer.

a. $y = \dfrac{x - 6}{3}$

b. $y = \dfrac{x - 3}{6}$

c. $y = \dfrac{x + 3}{6}$

d. $y = \dfrac{6}{x - 3}$

9. Tell where the function is increasing.

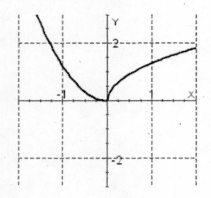

Select the correct answer.

a. always constant
b. $(0, \infty)$
c. always decreasing
d. $(-\infty, 0)$

10. Graph the quadratic function.

    $f(x) = -x^2 - 3x + 1$

    Select the correct answer.

    a.
    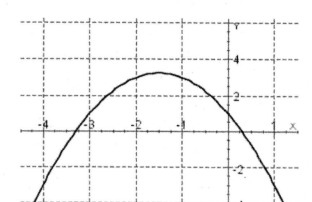

    b.
    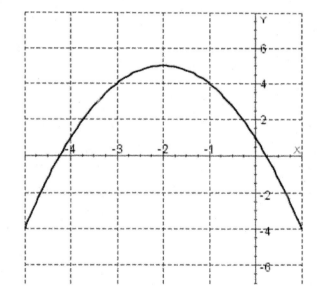

11. Graph the function

$$h(x) = \frac{x^2 - 2x - 8}{x - 1}$$

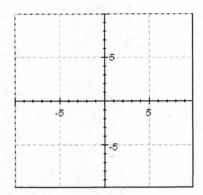

12. The function $f(x) = x^2 - 3$ is one-to-one on the domain $(x \leq 0)$. Find $f^{-1}(x)$.

Select the correct answer.

a. $f^{-1}(x) = \sqrt{x - 3}$

b. $f^{-1}(x) = -\sqrt{x + 3}$

c. $f^{-1}(x) = \sqrt{x + 3}$

d. $f^{-1}(x) = x^2 + 3$

13. Graph the piecewise-defined function.

$$y = f(x) = \begin{cases} x + 6 & \text{if } x < 0 \\ 6 & \text{if } x \geq 0 \end{cases}$$

Select the correct answer.

a.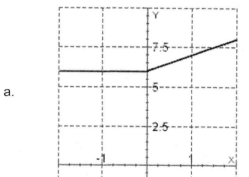

b.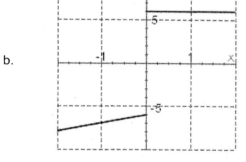

c.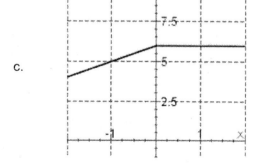

14. A plumber charges $40, plus $50 per hour (or fraction of an hour), to install a new bathtub. Graph the points (t, c), where t is the time it takes to do the job and c is the cost.

Select the correct answer.

a.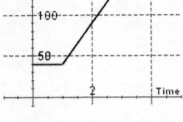

b.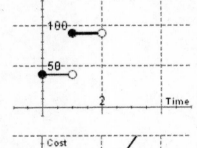

c.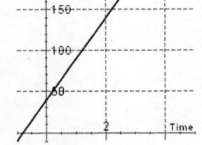

15. Let the function f be defined by the equation $y = f(x)$, where x and $f(x)$ are real numbers. Find the domain of the function

$$f(x) = \frac{43x + 11}{x - 22}$$

Select the correct answer.

a.   domain: $(-\infty, 22) \cup (22, +\infty)$
b.   domain: $(-\infty, -43) \cup (43, +\infty)$
c.   domain: $(-\infty, -22] \cup [22, +\infty)$
d.   domain: $(-\infty, +\infty)$

16. Let the function f be defined by $y = f(x)$, where x and $f(x)$ are real numbers. Find $f(10)$.

$$f(x) = 61 - 7x^2$$

Select the correct answer.

a. $f(10) = -639$
b. $f(10) = 179$
c. $f(10) = -662$

17. Let $f(x) = x^2 + x$, $g(x) = x^2 - 1$. Find $(f \cdot g)(x)$.

Select the correct answer.

a. $(f \cdot g)(x) = x + 1$
b. $(f \cdot g)(x) = \dfrac{x}{x - 1}$
c. $(f \cdot g)(x) = 2x^2 + x - 1$
d. $(f \cdot g)(x) = x^4 + x^3 - x^2 - x$

18. Graph the equation.

$$h(x) = \sqrt{x-2} + 1, \quad x \geq 2$$

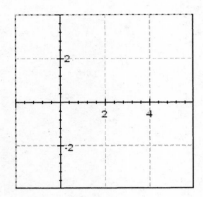

19. The function $f(x) = \dfrac{2}{x^2}$ is one-to-one on the domain $x > 0$. Find $f^{-1}(x)$.

20. Let the function f be defined by $y = f(x)$, where x and $f(x)$ are real numbers. Find $f(2)$.

$$f(x) = 93 - 2x^2$$

21. Find the vertex of the parabola.

$y = 16x^2 + 40x + 32$

Select the correct answer.

a. $\left(\dfrac{5}{4}, 7\right)$

b. $\left(-\dfrac{5}{4}, 7\right)$

c. $\left(-\dfrac{5}{4}, 25\right)$

d. $\left(-\dfrac{5}{4}, 32\right)$

22. A farmer wants to partition a rectangular feed storage area in a corner of his barn. The barn walls form two sides of the stall, and the farmer has 58 feet of partition for the remaining two sides. What dimensions will maximize the area of the partition?

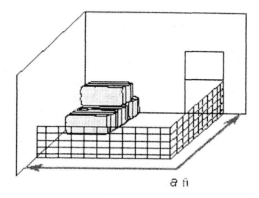

$a = 58$

23. Give the domain of the function.

$$f(x) = -\sqrt[3]{3x + 15}$$

Select the correct answer.

a. $(-\infty, 3]$

b. $(-\infty, 3)$

c. $(-\infty, \infty)$

24. Graph the polynomial function

$$y = x^3 + 3x^2$$

Select the correct answer.

a.
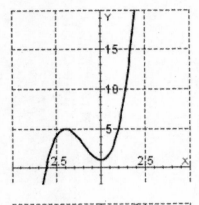

b.
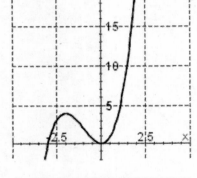

c.
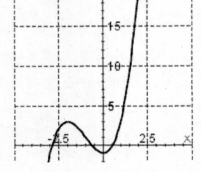

25. Graph the rational function

$$f(x) = \frac{4x^2}{5x}$$

Note that the numerator and denominator of the fraction share a common factor.

Select the correct answer.

a.
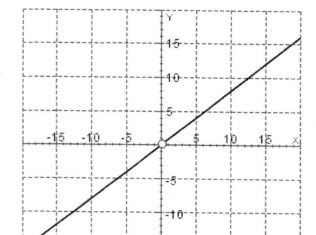

b.
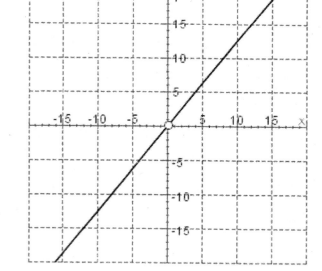

ANSWER KEY

Gustafson/Frisk - College Algebra 8E Chapter 3 Form H

1.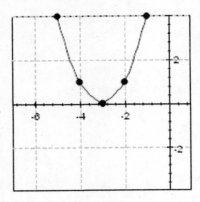

2. b

3. c

4. $(-\infty, \infty)$

5. b

6.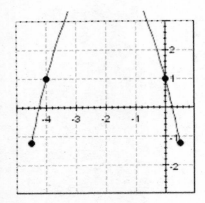

7. b

8. b

9. b

10. a

ANSWER KEY

Gustafson/Frisk - College Algebra 8E Chapter 3 Form H

11.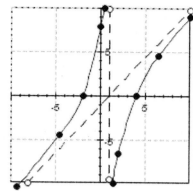

12. b

13. c

14. b

15. a

16. a

17. d

18.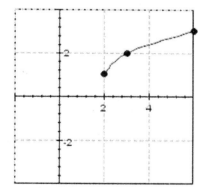

19. $\sqrt{\dfrac{2}{x}}$

20. 85

21. b

22. 29, 29

23. c

24. b

25. a

List of Problem Codes for BCA Testing

Gustafson/Frisk - College Algebra 8E Chapter 3 Form H

1. gfca.03.04.4.13_NoAlgs
2. gfca.03.06.4.43m_NoAlgs
3. gfca.03.04.4.35m_NoAlgs
4. gfca.03.06.4.11_NoAlgs
5. gfca.03.05.4.27m_NoAlgs
6. gfca.03.02.4.12_NoAlgs
7. gfca.03.04.4.21m_NoAlgs
8. gfca.03.07.4.27m_NoAlgs
9. gfca.03.03.4.29m_NoAlgs
10. gfca.03.02.4.12m_NoAlgs
11. gfca.03.05.4.51_NoAlgs
12. gfca.03.07.4.45m_NoAlgs
13. gfca.03.03.4.35m_NoAlgs
14. gfca.03.03.4.50m_NoAlgs
15. gfca.03.01.4.34m_NoAlgs
16. gfca.03.01.4.40m_NoAlgs
17. gfca.03.06.4.17m_NoAlgs
18. gfca.03.04.4.43_NoAlgs
19. gfca.03.07.4.46_NoAlgs
20. gfca.03.01.4.40_NoAlgs
21. gfca.03.02.4.20m_NoAlgs
22. gfca.03.02.4.29_NoAlgs
23. gfca.03.01.4.82m_NoAlgs
24. gfca.03.03.4.12m_NoAlgs
25. gfca.03.05.4.55m_NoAlgs

Gustafson/Frisk - College Algebra 8E Chapter 4 Form A

1. Solve the equation.

   $\log(x-44) - \log 9 = \log(x-40) - \log x$

2. Find the value of x.

   $\log_6 x = -2$

3. Use a calculator to find y from $\log y = 1.0638$ to four decimal places.

4. An amplifier produces an output of 13 volts when the input signal is 0.72 volt. Find the decibel voltage gain.

5. Before the parachute opens, the velocity v (in meters per second) of a skydiver is given by $v = 45(1 - e^{-0.2t})$. After t seconds, a certain falling object has a velocity v given by $v = 46(1 - e^{-0.3t})$. Which is falling faster after 46 seconds, this object or the skydiver?

          The falling object   faster

          The skydiver   slower

6. Tritium, a radioactive isotope of hydrogen, has a half-life of 12.4 years. Of an initial sample of 42 grams, how much will remain after 33 years?

7. Solve the equation. If an answer is not exact, give the answer to four decimal places.

   $6^{x^2} = 9$

8. The intensity I of light (in lumens) at a distance x meters below the surface is given by $I = I_0 k^x$, where I_0 is the intensity at the surface and k depends on the clarity of the water. At one location in the Arctic Ocean, $I_0 = 9$ and $k = 0.8$. Find the intensity at a depth of 6 meters.

9. Find the value of x.

$$2^{\log_2 3} = x$$

10. The concentration x of a certain drug in an organ after t minutes is given by $x = 0.08(1 - e^{-0.1t})$. Find the concentration of the drug at 40 minutes.

11. Simplify the expression.

$$5^{\log_5 2}$$

12. Graph the function using translations

$f(x) = e^{x+3}$

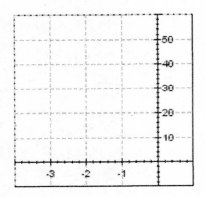

13. Assume that x, y, z and b are positive numbers. Use the properties of logarithms to write the expression in terms of the logarithms of x, y, and z.

$$\log_b \sqrt[16]{\frac{x^7 y^8}{z^{16}}}$$

14. Solve the equation. If an answer is not exact, give the answer to four decimal places.

    $6(4^{x+1}) = 4(6^{x-1})$

15. Graph the function.

    $y = \log_{1/3} x$

    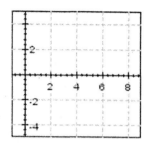

16. Use a calculator and the change-of-base formula to find the logarithm.

    $\log_{.7} e^4$

17. A population growing at an annual rate r will triple in a time t given by the formula $t = \dfrac{\ln 3}{r}$. If the growth rate remains constant and equals 14% per year, how long will it take the population of the town to triple?

18. Assume that x, y, and z are positive numbers. Use the properties of logarithms to write the expression as the logarithm of one quantity.

    $-3 \log_b x - 9 \log_b y + \dfrac{1}{6} \log_b z$

19. Simplify the expression.

    $(10^{\sqrt{2}})^{\sqrt{2}}$

20. An initial deposit of $200 earns 2% interest, compounded quarterly. How much will be in the account in 4 years?

21. Solve the equation. If an answer is not exact, give the answer to four decimal places.

    $\log(9x - 2) = \log(7x + 14)$

22. If $k = 0.15$, how long will it take a battery to reach a 25% charge? Assume that the battery was fully discharged when it began charging.

23. An earthquake has an amplitude of 5,500 micrometers and a period of 0.5 second. Find its measure on the Richter scale.

24. Graph the function using translations

    $f(x) = 3^{x+2} - 1$

    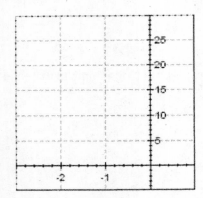

25. Solve the equation. If an answer is not exact, give the answer to four decimal places.

    $2\log_2 x = 1 + \log_2(x + 24)$

ANSWER KEY

Gustafson/Frisk - College Algebra 8E Chapter 4 Form A

1. 45

2. $\dfrac{1}{36}$

3. 11.5824

4. 25

5. The skydiver → slower,
   The falling object → faster

6. 6.6393

7. 1.1074, −1.1074

8. 2.36

9. 3

10. 0.0785

11. 2

12.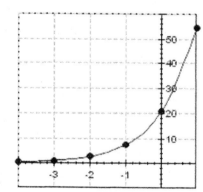

13. $\dfrac{7}{16} \cdot \log_b(x) + \dfrac{1}{2} \cdot \log_b(y) - \log_b(z)$

14. 8.8380

15.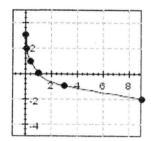

ANSWER KEY

Gustafson/Frisk - College Algebra 8E Chapter 4 Form A

16. 3.4943

17. 7.8

18. $\log_b \left(\dfrac{z^{\frac{1}{6}}}{\left(x^3 \cdot y^9\right)} \right)$

19. 100

20. 216.61

21. $x = 8$

22. 1.9

23. 4.0

24.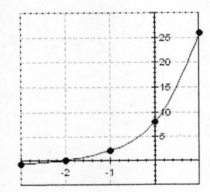

25. $x = 8$

List of Problem Codes for BCA Testing

Gustafson/Frisk - College Algebra 8E Chapter 4 Form A

1. gfca.04.06.4.41_NoAlgs
2. gfca.04.03.4.50_NoAlgs
3. gfca.04.03.4.81_NoAlgs
4. gfca.04.04.4.07_NoAlgs
5. gfca.04.02.4.34_NoAlgs
6. gfca.04.02.4.03_NoAlgs
7. gfca.04.06.4.13_NoAlgs
8. gfca.04.02.4.13_NoAlgs
9. gfca.04.03.4.61_NoAlgs
10. gfca.04.02.4.26_NoAlgs
11. gfca.04.05.4.15_NoAlgs
12. gfca.04.01.4.54_NoAlgs
13. gfca.04.05.4.36_NoAlgs
14. gfca.04.06.4.24_NoAlgs
15. gfca.04.03.4.99_NoAlgs
16. gfca.04.05.4.84_NoAlgs
17. gfca.04.04.4.19_NoAlgs
18. gfca.04.05.4.40_NoAlgs
19. gfca.04.01.4.20_NoAlgs
20. gfca.04.01.4.67_NoAlgs
21. gfca.04.06.4.29_NoAlgs
22. gfca.04.04.4.16_NoAlgs
23. gfca.04.04.4.11_NoAlgs
24. gfca.04.01.4.58_NoAlgs
25. gfca.04.06.4.51_NoAlgs

1. An earthquake has an amplitude of 5,500 micrometers and a period of 0.5 second. Find its measure on the Richter scale.

2. Solve the equation. If an answer is not exact, give the answer to four decimal places.

   log (9x - 2) = log (7x + 14)

3. Graph the function using translations

   $f(x) = 3^{x+2} - 1$

   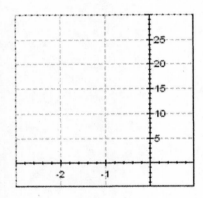

4. Assume that x, y, and z are positive numbers. Use the properties of logarithms to write the expression as the logarithm of one quantity.

   $-3 \log_b x - 9 \log_b y + \frac{1}{6} \log_b z$

5. An initial deposit of $ 200 earns 2% interest, compounded quarterly. How much will be in the account in 4 years?

6. Solve the equation. If an answer is not exact, give the answer to four decimal places.

   $6^{x^2} = 9$

7. Simplify the expression.

    $5^{\log_5 2}$

8. Find the value of x.

    $2^{\log_2 3} = x$

9. Solve the equation. If an answer is not exact, give the answer to four decimal places.

    $6(4^{x+1}) = 4(6^{x-1})$

10. Before the parachute opens, the velocity v (in meters per second) of a skydiver is given by $v = 45(1 - e^{-0.2t})$. After t seconds, a certain falling object has a velocity v given by $v = 46(1 - e^{-0.3t})$. Which is falling faster after 46 seconds, this object or the skydiver?

    The falling object   faster

    The skydiver   slower

11. Use a calculator and the change-of-base formula to find the logarithm.

    $\log_7 e^4$

12. The intensity I of light (in lumens) at a distance x meters below the surface is given by $I = I_0 k^x$, where I_0 is the intensity at the surface and k depends on the clarity of the water. At one location in the Arctic Ocean, $I_0 = 9$ and $k = 0.8$. Find the intensity at a depth of 6 meters.

13. Solve the equation. If an answer is not exact, give the answer to four decimal places.

    $2 \log_2 x = 1 + \log_2 (x + 24)$

14. A population growing at an annual rate *r* will triple in a time *t* given by the formula $t = \dfrac{\ln 3}{r}$. If the growth rate remains constant and equals 14% per year, how long will it take the population of the town to triple?

15. Solve the equation.

    $\log(x - 44) - \log 9 = \log(x - 40) - \log x$

16. The concentration *x* of a certain drug in an organ after *t* minutes is given by $x = 0.08(1 - e^{-0.1t})$. Find the concentration of the drug at 40 minutes.

17. Simplify the expression.

    $(10^{\sqrt{2}})^{\sqrt{2}}$

18. An amplifier produces an output of 13 volts when the input signal is 0.72 volt. Find the decibel voltage gain.

19. Use a calculator to find *y* from $\log y = 1.0638$ to four decimal places.

20. Graph the function.
    $y = \log_{1/3} x$

    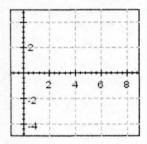

21. Find the value of *x*.

$\log_6 x = -2$

22. If $k = 0.15$, how long will it take a battery to reach a 25% charge? Assume that the battery was fully discharged when it began charging.

23. Assume that x, y, z and b are positive numbers. Use the properties of logarithms to write the expression in terms of the logarithms of x, y, and z.

$$\log_b \sqrt[16]{\frac{x^7 y^8}{z^{16}}}$$

24. Tritium, a radioactive isotope of hydrogen, has a half-life of 12.4 years. Of an initial sample of 42 grams, how much will remain after 33 years?

25. Graph the function using translations.

$f(x) = e^{x+3}$

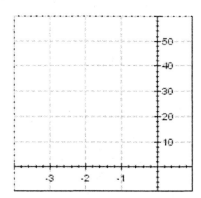

ANSWER KEY

Gustafson/Frisk - College Algebra 8E Chapter 4 Form B

1. 4.0

2. $x = 8$

3.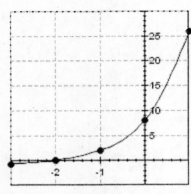

4.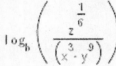

5. 216.61

6. 1.1074, -1.1074

7. 2

8. 3

9. 8.8380

10. The skydiver → slower,
    The falling object → faster

11. 3.4943

12. 2.36

13. $x = 8$

14. 7.8

15. 45

16. 0.0785

17. 100

18. 25

ANSWER KEY

Gustafson/Frisk - College Algebra 8E Chapter 4 Form B

19. 11.5824

20.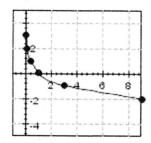

21. $\dfrac{1}{36}$

22. 1.9

23. $\dfrac{7}{16} \cdot \log_b(x) + \dfrac{1}{2} \cdot \log_b(y) - \log_b(z)$

24. 6.6393

25.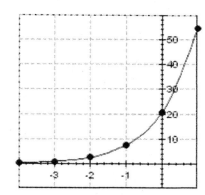

List of Problem Codes for BCA Testing

Gustafson/Frisk - College Algebra 8E Chapter 4 Form B

1. gfca.04.04.4.11_NoAlgs
2. gfca.04.06.4.29_NoAlgs
3. gfca.04.01.4.58_NoAlgs
4. gfca.04.05.4.40_NoAlgs
5. gfca.04.01.4.67_NoAlgs
6. gfca.04.06.4.13_NoAlgs
7. gfca.04.05.4.15_NoAlgs
8. gfca.04.03.4.61_NoAlgs
9. gfca.04.06.4.24_NoAlgs
10. gfca.04.02.4.34_NoAlgs
11. gfca.04.05.4.84_NoAlgs
12. gfca.04.02.4.13_NoAlgs
13. gfca.04.06.4.51_NoAlgs
14. gfca.04.04.4.19_NoAlgs
15. gfca.04.06.4.41_NoAlgs
16. gfca.04.02.4.26_NoAlgs
17. gfca.04.01.4.20_NoAlgs
18. gfca.04.04.4.07_NoAlgs
19. gfca.04.03.4.81_NoAlgs
20. gfca.04.03.4.99_NoAlgs
21. gfca.04.03.4.50_NoAlgs
22. gfca.04.04.4.16_NoAlgs
23. gfca.04.05.4.36_NoAlgs
24. gfca.04.02.4.03_NoAlgs
25. gfca.04.01.4.54_NoAlgs

Gustafson/Frisk - College Algebra 8E Chapter 4 Form C

1. Simplify the expression.

   $14^{\sqrt{2} \cdot \sqrt{2}}$

   Select the correct answer.

   a. 16
   b. 15
   c. 32
   d. 10
   e. 8

2. Graph the function using translations.

   $f(x) = e^{x+1} + 2$

   Select the correct answer.

   a.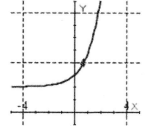

   b.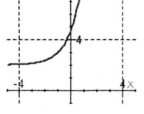

   c.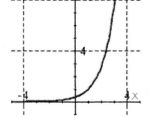

3. Graph the function using translations.

   $f(x) = 2^{x+1} + 1$

   Select the correct answer.

   a.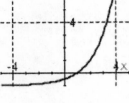

   b.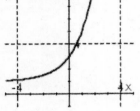

   c.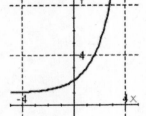

4. An initial deposit of $ 800 earns 9% interest, compounded quarterly. How much will be in the account in 14 years?

   Select the correct answer.

   a. $8,933.71
   b. $2,781.22
   c. $2,768.56
   d. $99,764.00
   e. $1,092.39

5. Tritium, a radioactive isotope of hydrogen, has a half-life of 12.4 years. Of an initial sample of 44 grams, how much will remain after 62 years?

   Select the correct answer.

   a. 7.8184 grams
   b. 0.6875 grams
   c. 1.3750 grams

6. The population of the Earth is approximately 6 billion people and is growing at an annual rate of 1.4%. Assuming a Malthusian growth model, find the world population in 20 years.

   Select the correct answer.

   a. 8.74 billion
   b. 9.24 billion
   c. 7.94 billion
   d. 8.64 billion

7. The concentration x of a certain drug in an organ after t minutes is given by $x = 0.08 (1 - e^{-0.1t})$. Find the concentration of the drug at 15 minutes.

   Select the correct answer.

   a. 0.0808
   b. 0.0621
   c. 0.0392
   d. 0.0559

8. Before the parachute opens, the velocity v (in meters per second) of a skydiver is given by $v = 44(1 - e^{-0.2t})$. After t seconds, a certain falling object has a velocity v given by $v = 50(1 - e^{-0.3t})$. Which is falling faster after 8 seconds, this object or the skydiver?

   Select the correct answer.

   a. The falling object
   b. The skydiver
   c. Their speeds are equal.

Gustafson/Frisk - College Algebra 8E Chapter 4 Form C

9. Find the value of x.

$$\log_6 x = -2$$

Select the correct answer.

a. $x = \dfrac{1}{36}$
b. $x = 6$
c. $x = 36$
d. $x = 2$
e. no solutions

10. Find the value of x.

$$3^{\log_3 6} = x$$

Select the correct answer.

a. $x = 3$
b. $x = 0$
c. $x = 6$
d. no solutions

11. Use a calculator to find y from $\log y = 1.0974$ to four decimal places.

Select the correct answer.

a. $y = 12.5137$
b. $y = 12.5138$
c. $y = 12.5134$
d. $y = 12.5141$

12. Find the graph of the function.

$$y = \log_{1/3} x$$

Select the correct answer.

a.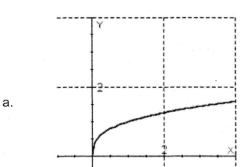

b.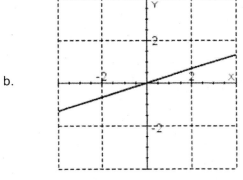

c.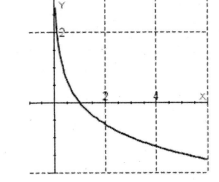

13. An amplifier produces an output of 7 volts when the input signal is 0.18 volt. Find the decibel voltage gain.

    Select the correct answer.

    a. 16 db
    b. 73 db
    c. 48 db
    d. 32 db

14. An earthquake has an amplitude of 4,500 micrometers and a period of 0.3 second. Find its measure on the Richter scale.

    Select the correct answer.

    a. 6.3
    b. 4.8
    c. 12.2
    d. 4.2

15. If $k = 0.15$, how long will it take a battery to reach a 40% charge? Assume that the battery was fully discharged when it began charging.

    Select the correct answer.

    a. 2.2 minutes
    b. 1.0 minutes
    c. 1.5 minutes
    d. 3.4 minutes

16. A population growing at an annual rate r will triple in a time t given by the formula $t = \dfrac{\ln 3}{r}$ If the growth rate remains constant and equals 14% per year, how long will it take the population of the town to triple?

    Select the correct answer.

    a. 5 years
    b. 7.8 years
    c. 3.4 years
    d. 2.2 years

17. Simplify the expression.

$$3^{\log_3 4}$$

Select the correct answer.

a. 16
b. 4
c. 12
d. 3

18. Assume that *x*, *y*, *z* and *b* are positive numbers. Use the properties of logarithms to write the expression in terms of the logarithms of *x*, *y*, and *z*.

$$\log_b \sqrt[8]{\frac{x^7 y^4}{z^8}}$$

Select the correct answer.

a. $\log_b x + \frac{1}{2} \log_b y - \log_b z$

b. $\frac{7}{8} \log_b x + \frac{1}{2} \log_b y - \log_b z$

c. $56 \log_b x + 32 \log_b y - 64 \log_b z$

d. $\frac{7}{8} \log_b x + \log_b y - \log_b z$

19. Tell whether the statement is true or false.

$$\log_a 7 = \log_7 a$$

Select the correct answer.
a. false
b. true

20. Use a calculator and the change-of-base formula to find the logarithm.

$$\log_\pi e^2$$

Select the correct answer.

a. 1.7471
b. 1.7473
c. 1.7468
d. 1.7469

21. Solve the equation.

$$5^{x^2} = 6$$

Select the correct answer(s).

a. $x = -1.1801$
b. $x = -1.0551$
c. $x = -1.5551$
d. $x = 1.0551$
e. $x = 2.5551$
f. $x = 1.5551$

22. Solve the equation. If an answer is not exact, give the answer to four decimal places.

$$4(2^{x+1}) = 2(4^{x-1})$$

Select the correct answer.

a. $x = 4.0112$
b. $x = 4.0053$
c. $x = 4.4011$
d. $x = 4.0000$

23. Solve the equation.

    $\log(8x - 8) = \log(3x + 37)$

    Select the correct answer.

    a. $x = 9$
    b. $x = 8$
    c. $x = 12$
    d. $x = 7$

24. Solve the equation.

    $\log(x - 14) - \log 5 = \log(x - 12) - \log x$

    Select the correct answer(s).

    a. $x = 4$
    b. $x = 17$
    c. $x = 15$
    d. $x = 19$

25. Solve the equation.

    $2\log_2 x = 1 + \log_2(x + 40)$

    Select the correct answer.

    a. $x = -8$
    b. $x = 12$
    c. $x = 9$
    d. $x = 10$

ANSWER KEY

Gustafson/Frisk - College Algebra 8E Chapter 4 Form C

1. a
2. b
3. b
4. b
5. c
6. c
7. b
8. a
9. a
10. c
11. d
12. c
13. d
14. d
15. d
16. b
17. b
18. b
19. a
20. a
21. b,d
22. d
23. a
24. c
25. d

List of Problem Codes for BCA Testing

Gustafson/Frisk - College Algebra 8E Chapter 4 Form C

1. gfca.04.01.4.20m_NoAlgs
2. gfca.04.01.4.54m_NoAlgs
3. gfca.04.01.4.58m_NoAlgs
4. gfca.04.01.4.67m_NoAlgs
5. gfca.04.02.4.03m_NoAlgs
6. gfca.04.02.4.21m_NoAlgs
7. gfca.04.02.4.26m_NoAlgs
8. gfca.04.02.4.34m_NoAlgs
9. gfca.04.03.4.50m_NoAlgs
10. gfca.04.03.4.61m_NoAlgs
11. gfca.04.03.4.81m_NoAlgs
12. gfca.04.03.4.99m_NoAlgs
13. gfca.04.04.4.07m_NoAlgs
14. gfca.04.04.4.11m_NoAlgs
15. gfca.04.04.4.16m_NoAlgs
16. gfca.04.04.4.19m_NoAlgs
17. gfca.04.05.4.15m_NoAlgs
18. gfca.04.05.4.36m_NoAlgs
19. gfca.04.05.4.48m_NoAlgs
20. gfca.04.05.4.84m_NoAlgs
21. gfca.04.06.4.13m_NoAlgs
22. gfca.04.06.4.24m_NoAlgs
23. gfca.04.06.4.29m_NoAlgs
24. gfca.04.06.4.41m_NoAlgs
25. gfca.04.06.4.51m_NoAlgs

1. Graph the function using translations.

   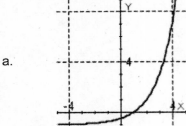

   Select the correct answer.

   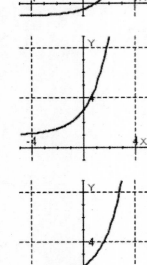

   a.

   b.

   c.

2. An initial deposit of $ 800 earns 9% interest, compounded quarterly. How much will be in the account in 14 years?

   Select the correct answer.

   a. $8,933.71
   b. $2,781.22
   c. $2,768.56
   d. $99,764.00
   e. $1,092.39

3. Simplify the expression.

$$\left(4^{\sqrt{2}}\right)^{\sqrt{2}}$$

Select the correct answer.

a. 8
b. 15
c. 10
d. 16
e. 32

4. Before the parachute opens, the velocity v (in meters per second) of a skydiver is given by $v = 44(1 - e^{-0.2t})$. After t seconds, a certain falling object has a velocity v given by $v = 50(1 - e^{-0.3t})$. Which is falling faster after 8 seconds, this object or the skydiver?

Select the correct answer.

a. The falling object
b. The skydiver
c. Their speeds are equal.

5. A population growing at an annual rate r will triple in a time t given by the formula $t = \dfrac{\ln 3}{r}$.

If the growth rate remains constant and equals 14% per year, how long will it take the population of the town to triple?

Select the correct answer.

a. 3.4 years
b. 2.2 years
c. 7.8 years
d. 5 years

6. Solve the equation

$2\log_2 x = 1 + \log_2 (x + 40)$

Select the correct answer.

a. $x = -8$
b. $x = 10$
c. $x = 12$
d. $x = 9$

7. Solve the equation.

   $\log(x - 14) - \log 5 = \log(x - 12) - \log x$

   Select the correct answer(s).

   a. $x = 4$
   b. $x = 17$
   c. $x = 15$
   d. $x = 19$

8. The population of the Earth is approximately 6 billion people and is growing at an annual rate of 1.4%. Assuming a Malthusian growth model, find the world population in 20 years.

   Select the correct answer.

   a. 8.74 billion
   b. 9.24 billion
   c. 7.94 billion
   d. 8.64 billion

9. An amplifier produces an output of 7 volts when the input signal is 0.18 volt. Find the decibel voltage gain.

   Select the correct answer.

   a. 73 db
   b. 48 db
   c. 32 db
   d. 16 db

10. Find the value of x.

    $$3^{\log_3 6} = x$$

    Select the correct answer.

    a. $x = 3$
    b. $x = 0$
    c. $x = 6$
    d. no solutions

11. Find the graph of the function.

    $y = \log_{1/3} x$

    Select the correct answer.

    a.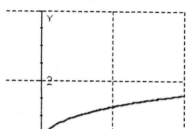

    b.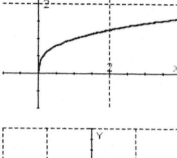

    c.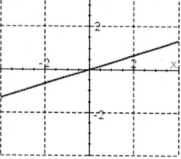

12. Tritium, a radioactive isotope of hydrogen, has a half-life of 12.4 years. Of an initial sample of 44 grams, how much will remain after 62 years?

    Select the correct answer.

    a.   1.3750 grams
    b.   7.8184 grams
    c.   0.6875 grams

13. Tell whether the statement is true or false.

    $\log_a 7 = \log_7 a$

    Select the correct answer.

    a.   true
    b.   false

14. Assume that x, y, z and b are positive numbers. Use the properties of logarithms to write the expression in terms of the logarithms of x, y, and z.

    $$\log_b \sqrt[8]{\frac{x^7 y^4}{z^8}}$$

    Select the correct answer.

    a. $\log_b x + \frac{1}{2} \log_b y - \log_b z$

    b. $\frac{7}{8} \log_b x + \frac{1}{2} \log_b y - \log_b z$

    c. $56 \log_b x + 32 \log_b y - 64 \log_b z$

    d. $\frac{7}{8} \log_b x + \log_b y - \log_b z$

15. Graph the function using translations.

$$f(x) = e^{x+1} + 2$$

Select the correct answer.

a.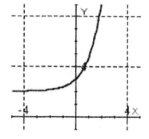

b.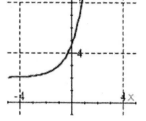

c.

16. An earthquake has an amplitude of 4,500 micrometers and a period of 0.3 second. Find its measure on the Richter scale.

Select the correct answer.

a. 6.3
b. 4.8
c. 12.2
d. 4.2

17. The concentration x of a certain drug in an organ after t minutes is given by $x = 0.08(1 - e^{-0.1t})$. Find the concentration of the drug at 15 minutes.

Select the correct answer.

a. 0.0808
b. 0.0621
c. 0.0392
d. 0.0559

18. Solve the equation. If an answer is not exact, give the answer to four decimal places.

$4(2^{x+1}) = 2(4^{x-1})$

Select the correct answer.

a. $x = 4.0112$
b. $x = 4.0053$
c. $x = 4.0000$
d. $x = 4.4011$

19. Use a calculator and the change-of-base formula to find the logarithm.

$\log_\pi e^2$

Select the correct answer.

a. 1.7471
b. 1.7473
c. 1.7468
d. 1.7469

20. If $k = 0.15$, how long will it take a battery to reach a 40% charge? Assume that the battery was fully discharged when it began charging.

Select the correct answer.

a. 3.4 minutes
b. 1.5 minutes
c. 2.2 minutes
d. 1.0 minutes

21. Solve the equation.

$$5^{x^2} = 6$$

Select the correct answer(s).

a. $x = 2.5551$
b. $x = 1.5551$
c. $x = -1.5551$
d. $x = -1.1801$
e. $x = -1.0551$
f. $x = 1.0551$

22. Find the value of x.

$$\log_6 x = -2$$

Select the correct answer.

a. $x = \dfrac{1}{36}$
b. $x = 6$
c. $x = 2$
d. $x = 36$
e. no solutions

23. Simplify the expression.

$$3^{\log_3 4}$$

Select the correct answer.

a. 4
b. 16
c. 3
d. 12

24. Use a calculator to find y from $\log y = 1.0974$ to four decimal places.

Select the correct answer.

a. $y = 12.5134$
b. $y = 12.5141$
c. $y = 12.5138$
d. $y = 12.5137$

25. Solve the equation.

$\log(8x - 8) = \log(3x + 37)$

Select the correct answer.

a. $x = 9$
b. $x = 8$
c. $x = 12$
d. $x = 7$

ANSWER KEY

Gustafson/Frisk - College Algebra 8E Chapter 4 Form D

1. b
2. b
3. d
4. a
5. c
6. b
7. c
8. c
9. c
10. c
11. c
12. a
13. b
14. b
15. b
16. d
17. b
18. c
19. a
20. a
21. e,f
22. a
23. a
24. b
25. a

List of Problem Codes for BCA Testing

Gustafson/Frisk - College Algebra 8E Chapter 4 Form D

1. gfca.04.01.4.58m_NoAlgs
2. gfca.04.01.4.67m_NoAlgs
3. gfca.04.01.4.20m_NoAlgs
4. gfca.04.02.4.34m_NoAlgs
5. gfca.04.04.4.19m_NoAlgs
6. gfca.04.06.4.51m_NoAlgs
7. gfca.04.06.4.41m_NoAlgs
8. gfca.04.02.4.21m_NoAlgs
9. gfca.04.04.4.07m_NoAlgs
10. gfca.04.03.4.61m_NoAlgs
11. gfca.04.03.4.99m_NoAlgs
12. gfca.04.02.4.03m_NoAlgs
13. gfca.04.05.4.48m_NoAlgs
14. gfca.04.05.4.36m_NoAlgs
15. gfca.04.01.4.54m_NoAlgs
16. gfca.04.04.4.11m_NoAlgs
17. gfca.04.02.4.26m_NoAlgs
18. gfca.04.06.4.24m_NoAlgs
19. gfca.04.05.4.84m_NoAlgs
20. gfca.04.04.4.16m_NoAlgs
21. gfca.04.06.4.13m_NoAlgs
22. gfca.04.03.4.50m_NoAlgs
23. gfca.04.05.4.15m_NoAlgs
24. gfca.04.03.4.81m_NoAlgs
25. gfca.04.06.4.29m_NoAlgs

1. Simplify the expression.

   $(10^{\sqrt{2}})^{\sqrt{2}}$

2. Find the value of b that would cause the graph of $y = b^x$ to look like the graph indicated.

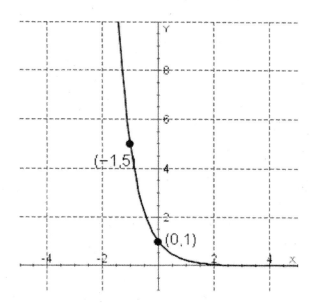

   Select the correct answer.

   a. $b = \dfrac{1}{5}$

   b. $b = \dfrac{3}{5}$

   c. $b = 5$

3. Solve the equation. If an answer is not exact, give the answer to four decimal places.

   $2 \log_2 x = 1 + \log_2 (x + 24)$

4. Graph the function using translations.

   $f(x) = e^{x+3}$

   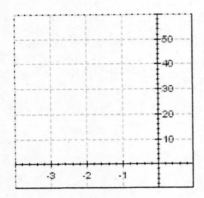

5. An initial investment of $1,700 earns 27% interest, compounded continuously. What will the investment be worth in 10 years?

   Select the correct answer.

   a. $25,316.54
   b. $25,262.54
   c. $25,288.54
   d. $25,295.54
   e. $25,274.54

6. Tritium, a radioactive isotope of hydrogen, has a half-life of 12.4 years. Of an initial sample of 44 grams, how much will remain after 62 years?

   Select the correct answer.

   a. 7.8184 grams
   b. 1.3750 grams
   c. 0.6875 grams

7. The population of the Earth is approximately 6 billion people and is growing at an annual rate of 1.4%. Assuming a Malthusian growth model, find the world population in 20 years.

   Select the correct answer.

   a. 8.74 billion
   b. 7.94 billion
   c. 8.64 billion
   d. 9.24 billion

8. The concentration x of a certain drug in an organ after t minutes is given by $x = 0.08(1 - e^{-0.1t})$. Find the concentration of the drug at 40 minutes.

9. Before the parachute opens, the velocity v (in meters per second) of a skydiver is given by $v = 44(1 - e^{-0.2t})$. After t seconds, a certain falling object has a velocity v given by $v = 50(1 - e^{-0.3t})$. Which is falling faster after 8 seconds, this object or the skydiver?

   Select the correct answer.

   a. The falling object
   b. The skydiver
   c. Their speeds are equal.

10. Find the value of x.

    $$\log_6 x = -2$$

11. Find the value of x.

    $$3^{\log_3 6} = x$$

    Select the correct answer.

    a. $x = 3$
    b. $x = 0$
    c. $x = 6$
    d. no solutions

12. Use a calculator to find y from $\log y = 1.0974$ to four decimal places.

Select the correct answer.

a. $y = 12.5138$
b. $y = 12.5141$
c. $y = 12.5137$
d. $y = 12.5134$

13. Graph the function.

$$y = \log_{1/3} x$$

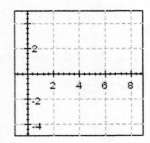

14. Solve the equation. If an answer is not exact, give the answer to four decimal places.

$$3^{2x+1} - 9(3^x) + 3 = 0$$

15. An amplifier produces an output of 7 volts when the input signal is 0.18 volt. Find the decibel voltage gain.

Select the correct answer.

a. 32 db
b. 48 db
c. 73 db
d. 16 db

16. An earthquake has an amplitude of 4,500 micrometers and a period of 0.3 second. Find its measure on the Richter scale.

Select the correct answer.

a. 6.3
b. 4.8
c. 12.2
d. 4.2

17. If $k = 0.15$, how long will it take a battery to reach a 25% charge? Assume that the battery was fully discharged when it began charging.

18. A population growing at an annual rate r will triple in a time t given by the formula $t = \dfrac{\ln 3}{r}$. If the growth rate remains constant and equals 14% per year, how long will it take the population of the town to triple?

19. Simplify the expression.

$$3^{\log_3 4}$$

Select the correct answer.

a. 16
b. 4
c. 3
d. 12

20. Assume that x, y, z and b are positive numbers. Use the properties of logarithms to write the expression in terms of the logarithms of x, y, and z.

$$\log_b \sqrt[16]{\dfrac{x^7 y^8}{z^{16}}}$$

21. Assume that x, y, and z are positive numbers. Use the properties of logarithms to write the expression as the logarithm of one quantity.

$$-3\log_b x - 9\log_b y + \frac{1}{6}\log_b z$$

22. Use a calculator and the change-of-base formula to find the logarithm.

$$\log_\pi e^2$$

Select the correct answer.

   a.   1.7469
   b.   1.7473
   c.   1.7471
   d.   1.7468

23. Solve the equation.

$$5^{x^2} = 6$$

Select the correct answer(s).

   a.   x = 1.0551
   b.   x = -1.1801
   c.   x = -1.0551
   d.   x = 1.5551
   e.   x = -1.5551

24. Solve the equation. If an answer is not exact, give the answer to four decimal places.

   log (9x - 2) = log (7x + 14)

25. Solve the equation.

   log (x - 14) - log 5 = log (x - 12) - log x

Select the correct answer(s).

   a.   x = 17
   b.   x = 19
   c.   x = 4
   d.   x = 15

ANSWER KEY

Gustafson/Frisk - College Algebra 8E Chapter 4 Form E

1. 100

2. a

3. $x = 8$

4.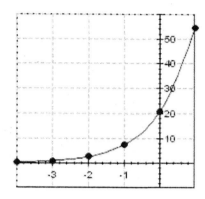

5. d

6. b

7. b

8. 0.0785

9. a

10. $\dfrac{1}{36}$

11. c

12. b

13.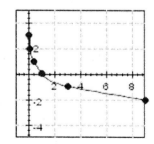

14. 0.88, -0.88

ANSWER KEY

Gustafson/Frisk - College Algebra 8E Chapter 4 Form E

15. a

16. d

17. 1.9

18. 7.8

19. b

20. $\dfrac{7}{16} \cdot \log_b(x) + \dfrac{1}{2} \cdot \log_b(y) - \log_b(z)$

21. $\log_b\left(\dfrac{z^{\frac{1}{6}}}{\left(x^3 \cdot y^9\right)}\right)$

22. c

23. a,c

24. $x = 8$

25. d

List of Problem Codes for BCA Testing

Gustafson/Frisk - College Algebra 8E Chapter 4 Form E

1. gfca.04.01.4.20_NoAlgs
2. gfca.04.01.4.35m_NoAlgs
3. gfca.04.06.4.51_NoAlgs
4. gfca.04.01.4.54_NoAlgs
5. gfca.04.01.4.75m_NoAlgs
6. gfca.04.02.4.03m_NoAlgs
7. gfca.04.02.4.21m_NoAlgs
8. gfca.04.02.4.26_NoAlgs
9. gfca.04.02.4.34m_NoAlgs
10. gfca.04.03.4.50_NoAlgs
11. gfca.04.03.4.61m_NoAlgs
12. gfca.04.03.4.81m_NoAlgs
13. gfca.04.03.4.99_NoAlgs
14. gfca.04.06.4.28_NoAlgs
15. gfca.04.04.4.07m_NoAlgs
16. gfca.04.04.4.11m_NoAlgs
17. gfca.04.04.4.16_NoAlgs
18. gfca.04.04.4.19_NoAlgs
19. gfca.04.05.4.15m_NoAlgs
20. gfca.04.05.4.36_NoAlgs
21. gfca.04.05.4.40_NoAlgs
22. gfca.04.05.4.84m_NoAlgs
23. gfca.04.06.4.13m_NoAlgs
24. gfca.04.06.4.29_NoAlgs
25. gfca.04.06.4.41m_NoAlgs

1. Graph the function.

   $y = \log_{1/3} x$

   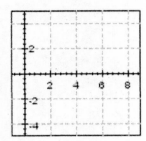

2. Solve the equation. If an answer is not exact, give the answer to four decimal places.

   $2 \log_2 x = 1 + \log_2 (x + 24)$

3. An initial investment of $1,700 earns 27% interest, compounded continuously. What will the investment be worth in 10 years?

   Select the correct answer.

   a. $25,316.54
   b. $25,262.54
   c. $25,288.54
   d. $25,295.54
   e. $25,274.54

4. Simplify the expression.

   $(10^{\sqrt{2}})^{\sqrt{2}}$

5. The concentration x of a certain drug in an organ after t minutes is given by $x = 0.08 (1 - e^{-0.1t})$. Find the concentration of the drug at 40 minutes.

6. Graph the function using translations.

   $f(x) = e^{x+3}$

   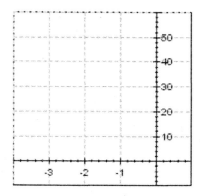

7. Use a calculator and the change-of-base formula to find the logarithm.

   $\log_{\frac{2}{7}} e^2$

   Select the correct answer.

   a.   1.7469
   b.   1.7473
   c.   1.7471
   d.   1.7468

8. Assume that x, y, and z are positive numbers. Use the properties of logarithms to write the expression as the logarithm of one quantity.

   $-3 \log_b x - 9 \log_b y + \dfrac{1}{6} \log_b z$

9. Tritium, a radioactive isotope of hydrogen, has a half-life of 12.4 years. Of an initial sample of 44 grams, how much will remain after 62 years?

   Select the correct answer.

   a. 0.6875 grams
   b. 7.8184 grams
   c. 1.3750 grams

10. Solve the equation. If an answer is not exact, give the answer to four decimal places.

    $$3^{2x+1} - 9(3^x) + 3 = 0$$

11. Find the value of x.

    $$3^{\log_3 6} = x$$

    Select the correct answer.

    a. $x = 3$
    b. $x = 0$
    c. $x = 6$
    d. no solutions

12. Simplify the expression.

    $$3^{\log_3 4}$$

    Select the correct answer.

    a. 16
    b. 4
    c. 3
    d. 12

13. An amplifier produces an output of 7 volts when the input signal is 0.18 volt. Find the decibel voltage gain.

    Select the correct answer.

    a. 32 db
    b. 48 db
    c. 73 db
    d. 16 db

14. Solve the equation. If an answer is not exact, give the answer to four decimal places.

    log (9x - 2) = log (7x + 14)

15. If $k = 0.15$, how long will it take a battery to reach a 25% charge? Assume that the battery was fully discharged when it began charging.

16. Solve the equation.

    $$5^{x^2} = 6$$

    Select the correct answer(s).
    a. $x = 1.0551$
    b. $x = -1.1801$
    c. $x = -1.0551$
    d. $x = 1.5551$
    e. $x = -1.5551$
    f. $x = 2.5551$

17. A population growing at an annual rate r will triple in a time t given by the formula $t = \dfrac{\ln 3}{r}$. If the growth rate remains constant and equals 14% per year, how long will it take the population of the town to triple?

18. Solve the equation.

log (x - 14) - log 5 = log (x - 12) - log x

Select the correct answer(s).

a. $x = 17$
b. $x = 19$
c. $x = 4$
d. $x = 15$

19. Find the value of b that would cause the graph of $y = b^x$ to look like the graph indicated.

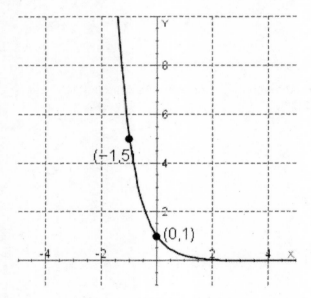

Select the correct answer.

a. $b = \dfrac{1}{5}$

b. $b = \dfrac{3}{5}$

c. $b = 5$

20. The population of the Earth is approximately 6 billion people and is growing at an annual rate of 1.4%. Assuming a Malthusian growth model, find the world population in 20 years.

Select the correct answer.

a. 8.74 billion
b. 7.94 billion
c. 8.64 billion
d. 9.24 billion

21. An earthquake has an amplitude of 4,500 micrometers and a period of 0.3 second. Find its measure on the Richter scale.

    Select the correct answer.

    a. 6.3
    b. 4.8
    c. 12.2
    d. 4.2

22. Before the parachute opens, the velocity v (in meters per second) of a skydiver is given by $v = 44(1 - e^{-0.2t})$. After t seconds, a certain falling object has a velocity v given by $v = 50(1 - e^{-0.3t})$. Which is falling faster after 8 seconds, this object or the skydiver?

    Select the correct answer.

    a. The falling object
    b. The skydiver
    c. Their speeds are equal.

23. Find the value of x.

    $$\log_6 x = -2$$

24. Assume that x, y, z and b are positive numbers. Use the properties of logarithms to write the expression in terms of the logarithms of x, y, and z.

    $$\log_b \sqrt[16]{\frac{x^7 y^8}{z^{16}}}$$

25. Use a calculator to find y from $\log y = 1.0974$ to four decimal places.

Select the correct answer.

a. $y = 12.5138$
b. $y = 12.5141$
c. $y = 12.5137$
d. $y = 12.5134$

ANSWER KEY

Gustafson/Frisk - College Algebra 8E Chapter 4 Form F

1.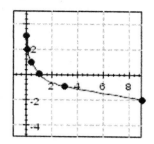

2. $x = 8$

3. d

4. 100

5. 0.0785

6.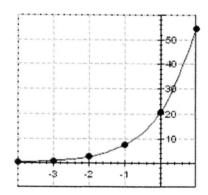

7. c

8. $\log_b\left(\dfrac{z^{\frac{1}{6}}}{\left(x^3 \cdot y^9\right)}\right)$

9. c

10. 0.88, -0.88

11. c

12. b

13. a

ANSWER KEY

Gustafson/Frisk - College Algebra 8E Chapter 4 Form F

14. $x = 3$

15. 1.9

16. a,c

17. 7.8

18. d

19. a

20. b

21. d

22. a

23. $\dfrac{1}{36}$

24. $\dfrac{7}{16} \cdot \log_b(x) + \dfrac{1}{2} \cdot \log_b(y) - \log_b(z)$

25. b

List of Problem Codes for BCA Testing

Gustafson/Frisk - College Algebra 8E Chapter 4 Form F

1. gfca.04.03.4.99_NoAlgs
2. gfca.04.06.4.51_NoAlgs
3. gfca.04.01.4.75m_NoAlgs
4. gfca.04.01.4.20_NoAlgs
5. gfca.04.02.4.26_NoAlgs
6. gfca.04.01.4.54_NoAlgs
7. gfca.04.05.4.84m_NoAlgs
8. gfca.04.05.4.40_NoAlgs
9. gfca.04.02.4.03m_NoAlgs
10. gfca.04.06.4.28_NoAlgs
11. gfca.04.03.4.61m_NoAlgs
12. gfca.04.05.4.15m_NoAlgs
13. gfca.04.04.4.07m_NoAlgs
14. gfca.04.06.4.29_NoAlgs
15. gfca.04.04.4.16_NoAlgs
16. gfca.04.06.4.13m_NoAlgs
17. gfca.04.04.4.19_NoAlgs
18. gfca.04.06.4.41m_NoAlgs
19. gfca.04.01.4.35m_NoAlgs
20. gfca.04.02.4.21m_NoAlgs
21. gfca.04.04.4.11m_NoAlgs
22. gfca.04.02.4.34m_NoAlgs
23. gfca.04.03.4.50_NoAlgs
24. gfca.04.05.4.36_NoAlgs
25. gfca.04.03.4.81m_NoAlgs

1. Find the value of *b* that would cause the graph of $y = b^x$ to look like the graph indicated.

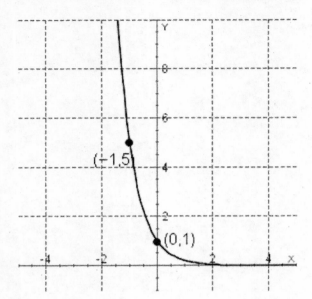

Select the correct answer.

a. $b = \dfrac{3}{5}$

b. $b = \dfrac{1}{5}$

c. $b = 5$

2. Solve the equation. If an answer is not exact, give the answer to four decimal places.

$$3^{2x+1} - 9(3^x) + 3 = 0$$

3. An initial investment of $1,700 earns 27% interest, compounded continuously. What will the investment be worth in 10 years?

   Select the correct answer.

   a. $25,288.54
   b. $25,295.54
   c. $25,262.54
   d. $25,274.54
   e. $25,316.54

4. Simplify the expression.

   $(4^{\sqrt{2}})^{\sqrt{2}}$

   Select the correct answer.

   a. 32
   b. 10
   c. 8
   d. 16
   e. 15

5. An amplifier produces an output of 13 volts when the input signal is 0.72 volt. Find the decibel voltage gain.

6. Solve the equation. If an answer is not exact, give the answer to four decimal places.

   $2 \log_2 x = 1 + \log_2 (x + 24)$

7. Use a calculator and the change-of-base formula to find the logarithm.

   $\log_7 e^4$

8. An initial deposit of $200 earns 2% interest, compounded quarterly. How much will be in the account in 4 years?

9. The concentration x of a certain drug in an organ after t minutes is given by $x = 0.08(1 - e^{-0.1t})$. Find the concentration of the drug at 15 minutes.

   Select the correct answer.

   a. 0.0559
   b. 0.0808
   c. 0.0392
   d. 0.0621

10. If $k = 0.15$, how long will it take a battery to reach a 40% charge? Assume that the battery was fully discharged when it began charging.

    Select the correct answer.

    a. 1.0 minutes
    b. 1.5 minutes
    c. 3.4 minutes
    d. 2.2 minutes

11. Before the parachute opens, the velocity v (in meters per second) of a skydiver is given by $v = 50(1 - e^{-0.2t})$. Find the velocity after 19 seconds.

    Select the correct answer.

    a. 71.37 m/sec
    b. 97.76 m/sec
    c. 195.53 m/sec
    d. 48.88 m/sec

12. Assume that x, y, z and b are positive numbers. Use the properties of logarithms to write the expression in terms of the logarithms of x, y, and z.

    $$\log_b \sqrt[8]{\frac{x^7 y^4}{z^8}}$$

    Select the correct answer.

    a. $56 \log_b x + 32 \log_b y - 64 \log_b z$

    b. $\log_b x + \frac{1}{2} \log_b y - \log_b z$

    c. $\frac{7}{8} \log_b x + \log_b y - \log_b z$

    d. $\frac{7}{8} \log_b x + \frac{1}{2} \log_b y - \log_b z$

13. Use a calculator to find y from $\log y = 1.0974$ to four decimal places.

    Select the correct answer.

    a. $y = 12.5141$
    b. $y = 12.5137$
    c. $y = 12.5138$
    d. $y = 12.5134$

14. Solve the equation.

    $$5^{x^2} = 6$$

    Select the correct answer(s).

    a. $x = -1.0551$
    b. $x = 2.5551$
    c. $x = -1.1801$
    d. $x = -1.5551$
    e. $x = 1.0551$
    f. $x = 1.5551$

15. A population growing at an annual rate r will triple in a time t given by the formula $t = \dfrac{\ln 3}{r}$. If the growth rate remains constant and equals 14% per year, how long will it take the population of the town to triple?

    Select the correct answer.

    a. 2.2 years
    b. 3.4 years
    c. 5 years
    d. 7.8 years

16. Tritium, a radioactive isotope of hydrogen, has a half-life of 12.4 years. Of an initial sample of 42 grams, how much will remain after 33 years?

17. Simplify the expression.

    $3^{\log_3 4}$

    Select the correct answer.

    a. 12
    b. 3
    c. 16
    d. 4

18. The population of the Earth is approximately 6 billion people and is growing at an annual rate of 1.4%. Assuming a Malthusian growth model, find the world population in 20 years.

    Select the correct answer.

    a. 7.94 billion
    b. 8.64 billion
    c. 8.74 billion
    d. 9.24 billion

19. Graph the function.

    $y = \log_{1/3} x$

    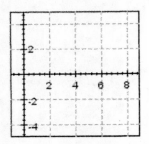

20. Solve the equation.

    $\log(x - 14) - \log 5 = \log(x - 12) - \log x$

    Select the correct answer(s).

    a. $x = 17$
    b. $x = 19$
    c. $x = 4$
    d. $x = 15$

21. Solve the equation. If an answer is not exact, give the answer to four decimal places.

$4(2^{x+1}) = 2(4^{x-1})$

Select the correct answer.

a. $x = 4.0000$
b. $x = 4.4011$
c. $x = 4.0053$
d. $x = 4.0112$

22. Find the value of x.

$2^{\log_2 3} = x$

23. Find the value of x.

$\log_6 x = -2$

Select the correct answer.

a. $x = 6$
b. $x = 36$
c. $x = \dfrac{1}{36}$
d. no solutions

24. Solve the equation.

$\log(8x - 8) = \log(3x + 37)$

Select the correct answer.

a. $x = 7$
b. $x = 9$
c. $x = 8$
d. $x = 12$

25. Graph the function using translations.

$$f(x) = e^{x+1} + 2$$

Select the correct answer.

a.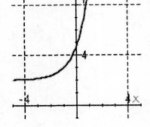

b.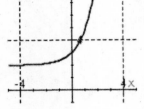

c.

ANSWER KEY

Gustafson/Frisk - College Algebra 8E Chapter 4 Form G

1. b
2. 0.88, -0.88
3. b
4. d
5. 25
6. x = 8
7. 3.4943
8. 216.61
9. d
10. c
11. d
12. d
13. a
14. a,e
15. d
16. 6.6393
17. d
18. a
19.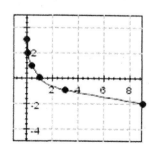
20. d
21. a
22. 3
23. c
24. b
25. a

List of Problem Codes for BCA Testing

Gustafson/Frisk - College Algebra 8E Chapter 4 Form G

1. gfca.04.01.4.35m_NoAlgs
2. gfca.04.06.4.28_NoAlgs
3. gfca.04.01.4.75m_NoAlgs
4. gfca.04.01.4.20m_NoAlgs
5. gfca.04.04.4.07_NoAlgs
6. gfca.04.06.4.51_NoAlgs
7. gfca.04.05.4.84_NoAlgs
8. gfca.04.01.4.67_NoAlgs
9. gfca.04.02.4.26m_NoAlgs
10. gfca.04.04.4.16m_NoAlgs
11. gfca.04.02.4.33m_NoAlgs
12. gfca.04.05.4.36m_NoAlgs
13. gfca.04.03.4.81m_NoAlgs
14. gfca.04.06.4.13m_NoAlgs
15. gfca.04.04.4.19m_NoAlgs
16. gfca.04.02.4.03_NoAlgs
17. gfca.04.05.4.15m_NoAlgs
18. gfca.04.02.4.21m_NoAlgs
19. gfca.04.03.4.99_NoAlgs
20. gfca.04.06.4.41m_NoAlgs
21. gfca.04.06.4.24m_NoAlgs
22. gfca.04.03.4.61_NoAlgs
23. gfca.04.03.4.50m_NoAlgs
24. gfca.04.06.4.29m_NoAlgs
25. gfca.04.01.4.54m_NoAlgs

1. Solve the equation.

   $\log(x - 14) - \log 5 = \log(x - 12) - \log x$

   Select the correct answer(s).

   a. $x = 15$
   b. $x = 4$
   c. $x = 19$
   d. $x = 17$

2. Graph the function using translations.

   $f(x) = e^{x+1} + 2$

   Select the correct answer.

   a.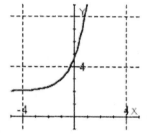

   b.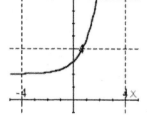

   c.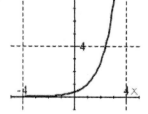

3. Find the value of x.

$$\log_6 x = -2$$

Select the correct answer.

a. $x = 6$
b. $x = 2$
c. $x = 36$
d. $x = \dfrac{1}{36}$
e. no solutions

4. Use a calculator to find y from $\log y = 1.0974$ to four decimal places.

Select the correct answer.

a. $y = 12.5141$
b. $y = 12.5137$
c. $y = 12.5138$
d. $y = 12.5134$

5. Tritium, a radioactive isotope of hydrogen, has a half-life of 12.4 years. Of an initial sample of 42 grams, how much will remain after 33 years?

6. Solve the equation. If an answer is not exact, give the answer to four decimal places.

$$3^{2x+1} - 9(3^x) + 3 = 0$$

7. Before the parachute opens, the velocity v (in meters per second) of a skydiver is given by $v = 50(1 - e^{-0.2t})$. Find the velocity after 19 seconds.

Select the correct answer.

a. 71.37 m/sec
b. 97.76 m/sec
c. 195.53 m/sec
d. 48.88 m/sec

8. Assume that x, y, z and b are positive numbers. Use the properties of logarithms to write the expression in terms of the logarithms of x, y, and z.

$$\log_b \sqrt[8]{\frac{x^7 y^4}{z^8}}$$

Select the correct answer.

a. $56 \log_b x + 32 \log_b y - 64 \log_b z$

b. $\log_b x + \frac{1}{2} \log_b y - \log_b z$

c. $\frac{7}{8} \log_b x + \log_b y - \log_b z$

d. $\frac{7}{8} \log_b x + \frac{1}{2} \log_b y - \log_b z$

9. Simplify the expression.

$$14^{\sqrt{2} \cdot \sqrt{2}}$$

Select the correct answer.

a. 32
b. 10
c. 8
d. 16
e. 15

10. An amplifier produces an output of 13 volts when the input signal is 0.72 volt. Find the decibel voltage gain.

11. A population growing at an annual rate *r* will triple in a time *t* given by the formula $t = \dfrac{\ln 3}{r}$. If the growth rate remains constant and equals 14% per year, how long will it take the population of the town to triple?

Select the correct answer.

    a.   2.2 years
    b.   3.4 years
    c.   5 years
    d.   7.8 years

12. The population of the Earth is approximately 6 billion people and is growing at an annual rate of 1.4%. Assuming a Malthusian growth model, find the world population in 20 years.

Select the correct answer.

    a.   7.94 billion
    b.   8.64 billion
    c.   8.74 billion
    d.   9.24 billion

13. Solve the equation. If an answer is not exact, give the answer to four decimal places.

$$4(2^{x+1}) = 2(4^{x-1})$$

Select the correct answer.

    a. $x = 4.4011$
    b. $x = 4.0000$
    c. $x = 4.0053$
    d. $x = 4.0112$

14. If $k = 0.15$, how long will it take a battery to reach a 40% charge? Assume that the battery was fully discharged when it began charging.

Select the correct answer.

    a.   1.0 minutes
    b.   3.4 minutes
    c.   1.5 minutes
    d.   2.2 minutes

15. The concentration x of a certain drug in an organ after t minutes is given by $x = 0.08(1 - e^{-0.1t})$. Find the concentration of the drug at 15 minutes.

Select the correct answer.

a. 0.0559
b. 0.0808
c. 0.0392
d. 0.0621

16. Solve the equation.

$$5^{x^2} = 6$$

Select the correct answer(s).

a. $x = -1.0551$
b. $x = 2.5551$
c. $x = -1.1801$
d. $x = -1.5551$
e. $x = 1.0551$
f. $x = 1.5551$

17. Graph the function.

$$y = \log_{1/3} x$$

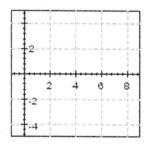

18. Use a calculator and the change-of-base formula to find the logarithm.

$$\log_\pi e^4$$

19. Simplify the expression.

$$3^{\log_3 4}$$

Select the correct answer.

a. 12
b. 3
c. 16
d. 4

20. An initial investment of $1,700 earns 27% interest, compounded continuously. What will the investment be worth in 10 years?

Select the correct answer.

a. $25,288.54
b. $25,295.54
c. $25,262.54
d. $25,274.54
e. $25,316.54

21. Find the value of x.

$$2^{\log_2 3} = x$$

22. An initial deposit of $200 earns 2% interest, compounded quarterly. How much will be in the account in 4 years?

23. Solve the equation. If an answer is not exact, give the answer to four decimal places.

$2 \log_2 x = 1 + \log_2 (x + 24)$

24. Solve the equation.

log (8x - 8) = log (3x + 37)

Select the correct answer.

a. x = 7
b. x = 9
c. x = 8
d. x = 12

25. Find the value of *b* that would cause the graph of $y = b^x$ to look like the graph indicated.

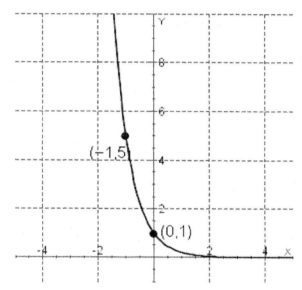

Select the correct answer.

a. $b = \dfrac{3}{5}$

b. $b = \dfrac{1}{5}$

c. $b = 5$

Page 7

ANSWER KEY

Gustafson/Frisk - College Algebra 8E Chapter 4 Form H

1. a
2. a
3. d
4. a
5. 6.6393
6. 0.88, -0.88
7. d
8. d
9. d
10. 25
11. d
12. a
13. b
14. b
15. d
16. a,e
17.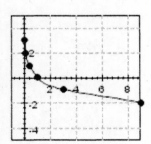
18. 3.4943
19. d
20. b
21. 3
22. 216.61
23. x = 8
24. b
25. b

List of Problem Codes for BCA Testing

Gustafson/Frisk - College Algebra 8E Chapter 4 Form H

1. gfca.04.06.4.41m_NoAlgs
2. gfca.04.01.4.54m_NoAlgs
3. gfca.04.03.4.50m_NoAlgs
4. gfca.04.03.4.81m_NoAlgs
5. gfca.04.02.4.03_NoAlgs
6. gfca.04.06.4.28_NoAlgs
7. gfca.04.02.4.33m_NoAlgs
8. gfca.04.05.4.36m_NoAlgs
9. gfca.04.01.4.20m_NoAlgs
10. gfca.04.04.4.07_NoAlgs
11. gfca.04.04.4.19m_NoAlgs
12. gfca.04.02.4.21m_NoAlgs
13. gfca.04.06.4.24m_NoAlgs
14. gfca.04.04.4.16m_NoAlgs
15. gfca.04.02.4.26m_NoAlgs
16. gfca.04.06.4.13m_NoAlgs
17. gfca.04.03.4.99_NoAlgs
18. gfca.04.05.4.84_NoAlgs
19. gfca.04.05.4.15m_NoAlgs
20. gfca.04.01.4.75m_NoAlgs
21. gfca.04.03.4.61_NoAlgs
22. gfca.04.01.4.67_NoAlgs
23. gfca.04.06.4.51_NoAlgs
24. gfca.04.06.4.29m_NoAlgs
25. gfca.04.01.4.35m_NoAlgs

Gustafson/Frisk - College Algebra 8E Chapter 5 Form A

1. Use a graphing calculator to find the real solution of the equation.

   $x^2 - 10x + 25 = 0$

2. Let $P(x) = 3x^3 + 3x^2 - 5x + 7$. Use synthetic division to find the value $P(8)$.

3. Write a fourth-degree polynomial equation with real coefficients and the roots $3, 1, 2 + i$.

4. Find all rational roots of the equation.

   $x^5 - 5x^4 - 5x^3 + 25x^2 + 4x - 20 = 0$

5. Use synthetic division to express $P(x) = 7x^3 + 30x^2 - 20x + 41$ in the form $(divisor)(quotient) + remainder$ for the divisor $x + 5$.

6. Find a polynomial with the zeros

   $(4, 6, 7)$.

7. The length of the shipping crate is to be 2 feet greater than its height and 5 feet greater than its width, and its volume is to be 32 cubic feet. Use the bisection method or a graphing calculator to find the height of the crate to the nearest tenth of a foot.

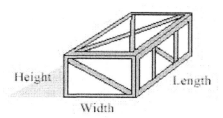

8. Write a third-degree polynomial equation with real coefficients and the roots 9 - *i*, -4.

9. Write a third-degree polynomial equation with real coefficients and the roots 2, -*i*.

10. Find the negative root of the equation using the bisection method.

    $3x^4 + 3x^3 - x^2 - 4x - 4 = 0$.

11. Tell how many roots the equation has.

    $x^7 = 5$

12. Find the negative root of the equation using the bisection method.

    $x^2 - 8 = 0$

13. Find all rational roots of the equation.

    $x^3 + 5x^2 - x - 5 = 0$

14. Find all rational roots of the equation.

    $x^3 - 6x^2 - x + 6 = 0$

15. Find all rational roots of the equation.

    $x^4 + 8x^3 + 24x^2 + 32x + 16 = 0$

16. The number $(1 + i)$ is a root of the equation. Find the other roots of the equation.

    $x^4 - 2x^3 - 7x^2 + 18x - 18 = 0$

17. Use synthetic division to do the division.

    $$\frac{3x^3 - 14x^2 - 62x + 40}{x + 3}$$

18. Solve the equation.

    $x^3 - \frac{44}{3}x^2 + \frac{100}{3}x - 16 = 0$

19. Find integer bounds for the roots of the equation.

    $5x^3 - 11x^2 - 82x = 0$

20. One root of the equation $x(4x^4 - 3) = 5x$ is 0. How many other roots are there?

21. Find the positive root of the equation using the bisection method.

    $x^2 - 7 = 0$

22. A partial solution set $(6, 6)$ is given for the equation. Find the complete solution set.

$$x^4 - 12x^3 + 37x^2 - 12x + 36 = 0$$

23. The number $(1 + i)$ is a root of the equation. Find the other roots of the equation.

$$x^3 - 11x^2 + 20x - 18 = 0$$

24. Use a graphing calculator to find the real solutions of the equation.

$$x^3 - 3x^2 - 2x + 6 = 0$$

25. A partial solution set (3) is given for the equation. Find the complete solution set.

$$x^3 + 2x^2 - 9x - 18 = 0$$

ANSWER KEY

Gustafson/Frisk - College Algebra 8E Chapter 5 Form A

1. 5

2. 1, 695

3. $x^4 - 8x^3 + 24x^2 - 32x + 15 = 0$

4. 1, 2, -1, -2, 5

5. $(x + 5)(7x^2 - 5x + 5) + 16$

6. $x^3 - 17x^2 + 94x - 168$

7. 4, 2

8. $x^3 - 14x^2 + 10x + 328 = 0$

9. $x^3 - (2x)^2 + x - 2 = 0$

10. -1, 2

11. 7

12. -2, 8

13. 1, -1, -5

14. 1, -1, 6

15. -2, -2, -2, -2

16. 3, -3, 1 - i

17. $3x^2 - 23x + 7 + \dfrac{19}{x + 3}$

ANSWER KEY

Gustafson/Frisk - College Algebra 8E Chapter 5 Form A

18. $\frac{2}{3}$, 1, 2, 2

19. −4, 6

20. 4

21. 2, 6

22. 6, 6, −i, i

23. 9, 1 − i

24. 1, 4, −1, 4, 3

25. 3, −2, −3

List of Problem Codes for BCA Testing

Gustafson/Frisk - College Algebra 8E Chapter 5 Form A

1. gfca.05.04.4.24_NoAlgs
2. gfca.05.01.4.57_NoAlgs
3. gfca.05.02.4.22_NoAlgs
4. gfca.05.03.4.16_NoAlgs
5. gfca.05.01.4.45_NoAlgs
6. gfca.05.01.4.33_NoAlgs
7. gfca.05.04.4.28_NoAlgs
8. gfca.05.02.4.20_NoAlgs
9. gfca.05.02.4.17_NoAlgs
10. gfca.05.04.4.21_NoAlgs
11. gfca.05.02.4.11_NoAlgs
12. gfca.05.04.4.16_NoAlgs
13. gfca.05.03.4.09_NoAlgs
14. gfca.05.03.4.05_NoAlgs
15. gfca.05.03.4.12_NoAlgs
16. gfca.05.03.4.36_NoAlgs
17. gfca.05.01.4.51_NoAlgs
18. gfca.05.03.4.37_NoAlgs
19. gfca.05.02.4.43_NoAlgs
20. gfca.05.02.4.15_NoAlgs
21. gfca.05.04.4.15_NoAlgs
22. gfca.05.01.4.27_NoAlgs
23. gfca.05.03.4.33_NoAlgs
24. gfca.05.04.4.26_NoAlgs
25. gfca.05.01.4.23_NoAlgs

Gustafson/Frisk - College Algebra 8E Chapter 5 Form B

1. Write a fourth-degree polynomial equation with real coefficients and the roots $3, 1, 2 + i$.

2. Find a polynomial with the zeros $(4, 6, 7)$.

3. Find all rational roots of the equation.
   $x^4 + 8x^3 + 24x^2 + 32x + 16 = 0$

4. Find the positive root of the equation using the bisection method.
   $x^2 - 7 = 0$

5. Find all rational roots of the equation.
   $x^3 - 6x^2 - x + 6 = 0$

6. Write a third-degree polynomial equation with real coefficients and the roots $9 - i, -4$.

7. One root of the equation $x(4x^4 - 3) = 5x$ is 0. How many other roots are there?

8. The number $(1 + i)$ is a root of the equation. Find the other roots of the equation.
   $x^3 - 11x^2 + 20x - 18 = 0$

9. Find all rational roots of the equation.
   $x^5 - 5x^4 - 5x^3 + 25x^2 + 4x - 20 = 0$

10. A partial solution set $(6, 6)$ is given for the equation. Find the complete solution set.
    $x^4 - 12x^3 + 37x^2 - 12x + 36 = 0$

11. Use a graphing calculator to find the real solution of the equation.

    $x^2 - 10x + 25 = 0$

12. Solve the equation.

    $x^3 - \dfrac{44}{3}x^2 + \dfrac{100}{3}x - 16 = 0$

13. Find all rational roots of the equation.

    $x^3 + 5x^2 - x - 5 = 0$

14. Find the negative root of the equation using the bisection method.

    $x^2 - 8 = 0$

15. Write a third-degree polynomial equation with real coefficients and the roots 2, -i.

16. Let $P(x) = 3x^3 + 3x^2 - 5x + 7$. Use synthetic division to find the value $P(8)$.

17. Use synthetic division to do the division.

    $$\dfrac{3x^3 - 14x^2 - 62x + 40}{x + 3}$$

18. The number (1 + i) is a root of the equation. Find the other roots of the equation.

    $x^4 - 2x^3 - 7x^2 + 18x - 18 = 0$

19. Find integer bounds for the roots of the equation.

    $5x^3 - 11x^2 - 82x = 0$

20. Use synthetic division to express $P(x) = 7x^3 + 30x^2 - 20x + 41$ in the form $(divisor)(quotient) + remainder$ for the divisor $x + 5$.

21. The length of the shipping crate is to be 2 feet greater than its height and 5 feet greater than its width, and its volume is to be 32 cubic feet. Use the bisection method or a graphing calculator to find the height of the crate to the nearest tenth of a foot.

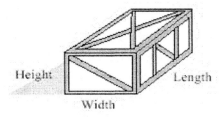

22. Use a graphing calculator to find the real solutions of the equation.

$x^3 - 3x^2 - 2x + 6 = 0$

23. Find the negative root of the equation using the bisection method.

$3x^4 + 3x^3 - x^2 - 4x - 4 = 0$.

24. A partial solution set (3) is given for the equation. Find the complete solution set.

$x^3 + 2x^2 - 9x - 18 = 0$

25. Tell how many roots the equation has.

$x^7 = 5$

ANSWER KEY

Gustafson/Frisk - College Algebra 8E Chapter 5 Form B

1. $x^4 - 8x^3 + 24x^2 - 32x + 15 = 0$
2. $x^3 - 17x^2 + 94x - 168$
3. $-2, -2, -2, -2$
4. $2, 6$
5. $1, -1, 6$
6. $x^3 - 14x^2 + 10x + 328 = 0$
7. 4
8. $9, 1 - i$
9. $1, 2, -1, -2, 5$
10. $6, 6, -i, i$
11. 5
12. $\frac{2}{3}, 1, 2, 2$
13. $1, -1, -5$
14. $-2, 8$
15. $x^3 - (2x)^2 + x - 2 = 0$
16. $1, 695$
17. $3x^2 - 23x + 7 + \dfrac{19}{x + 3}$
18. $3, -3, 1 - i$
19. $-4, 6$
20. $(x + 5) \cdot (7x^2 - 5x + 5) + 16$
21. $4, 2$
22. $1, 4, -1, 4, 3$
23. $-1, 2$
24. $3, -2, -3$
25. 7

List of Problem Codes for BCA Testing

Gustafson/Frisk - College Algebra 8E Chapter 5 Form B

1. gfca.05.02.4.22_NoAlgs
2. gfca.05.01.4.33_NoAlgs
3. gfca.05.03.4.12_NoAlgs
4. gfca.05.04.4.15_NoAlgs
5. gfca.05.03.4.05_NoAlgs
6. gfca.05.02.4.20_NoAlgs
7. gfca.05.02.4.15_NoAlgs
8. gfca.05.03.4.33_NoAlgs
9. gfca.05.03.4.16_NoAlgs
10. gfca.05.01.4.27_NoAlgs
11. gfca.05.04.4.24_NoAlgs
12. gfca.05.03.4.37_NoAlgs
13. gfca.05.03.4.09_NoAlgs
14. gfca.05.04.4.16_NoAlgs
15. gfca.05.02.4.17_NoAlgs
16. gfca.05.01.4.57_NoAlgs
17. gfca.05.01.4.51_NoAlgs
18. gfca.05.03.4.36_NoAlgs
19. gfca.05.02.4.43_NoAlgs
20. gfca.05.01.4.45_NoAlgs
21. gfca.05.04.4.28_NoAlgs
22. gfca.05.04.4.26_NoAlgs
23. gfca.05.04.4.21_NoAlgs
24. gfca.05.01.4.23_NoAlgs
25. gfca.05.02.4.11_NoAlgs

Gustafson/Frisk - College Algebra 8E Chapter 5 Form C

1. Find the negative root of the equation using the bisection method.

    $x^3 - x + 1 = 0.$

    Select the correct answer.

    a. -1.5
    b. -1.3
    c. -1.2
    d. -1.1
    e. -1.6

2. Solve the equation.

    $x^3 - \frac{53}{3}x^2 + \frac{124}{3}x - 20 = 0$

    Select the correct answer(s).

    a. $x = \frac{2}{3}$
    b. $x = -\frac{2}{3}$
    c. $x = -15$
    d. $x = 2$
    e. $x = -1$
    f. $x = 15$
    g. $x = 1$

3. Find all rational roots of the equation.

    $x^4 - 15x^3 + 80x^2 - 180x + 144 = 0$

    Select the correct answer(s).

    a. $x = 3$
    b. $x = 5$
    c. $x = 6$
    d. $x = -3$
    e. $x = 2$
    f. $x = -6$
    g. $x = 4$

4. Use synthetic division to express $P(x) = 2x^3 + 3x^2 - 8x + 29$ in the form $(divisor)(quotient) + remainder$ for the divisor $x + 3$.

Select the correct answer.

a. $(x + 3)(26x^2 - x + 1) + 26$

b. $(x - 3)(2x^2 - 3x + 1) - 26$

c. $(x + 3)(2x^2 - 3x + 26) + 1$

d. $(x + 3)(2x^2 - 3x + 1) + 26$

5. Use the bisection method or a graphing calculator to find the coordinates of the two points on the graph of $y = x^3$ that lie $r = 1$ unit from the origin. Give the result to the nearest hundredth.

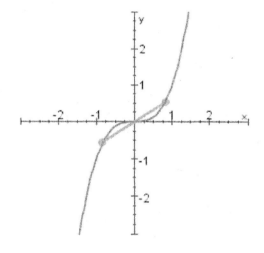

Select the correct answer.

a. $(1, 0.68), (-1, -0.68)$

b. $(0.76, 0.51), (-0.76, -0.51)$

c. $(0.83, 0.56), (-0.83, -0.56)$

d. $(1.04, 0.7), (-1.04, -0.7)$

e. $(0.62, 0.42), (-0.62, -0.42)$

6. Write a third-degree polynomial equation with real coefficients and the roots 7, -i.

   Select the correct answer.

   a. $x^3 - 7x^2 + x - 7 = 0$
   b. $x^3 - 8x^2 - x + 49 = 0$
   c. $x^3 - 7 = 0$
   d. $x^3 - 49 = 0$

7. Find a polynomial with the given zeros

   (5, 5, 2).

   Select the correct answer.

   a. $x^3 + 12x^2 - 45x + 50$
   b. $12x^2 + 45x - 50$
   c. $x^3 - 12x^2 + 45x - 50$
   d. $45x^3 - 12x^2 + 45x - 50$

8. Find integer bounds for the roots of the equation.

   $4x^3 - 9x^2 - 58x = 0$

   Select the correct answer.

   a. -3, 6
   b. -4, 3
   c. -4, 5
   d. -3, 4

9. Find all rational roots of the equation.

   $x^5 - 4x^4 - 10x^3 + 40x^2 + 9x - 36 = 0$

   Select the correct answer(s).

   a. $x = -3$
   b. $x = 2$
   c. $x = -2$
   d. $x = -1$
   e. $x = 4$
   f. $x = 3$
   g. $x = -4$
   h. $x = 1$

10. Use synthetic division to do the division.

$$\frac{2x^3 - 14x^2 - 33x + 21}{x + 2}$$

Select the correct answer.

a. $x^2 - 18x + 2 - \frac{15}{x+2}$

b. $2x^2 - 18x + 15 + \frac{21}{x+2}$

c. $2x^2 - 18x + 3$

d. $2x^2 - 18x + 3 + \frac{15}{x+2}$

11. Find the interval in which the equation has at least one real root.

$$x^4 - 17x^2 + 60 = 0$$

a. -2 and -1
b. 1 and 2
c. 0 and 1
d. 3 and 4
e. -1 and 0

12. Find the positive root of the equation using the bisection method.

$$x^2 - 15 = 0$$

Select the correct answer.

a. 4.3
b. 3.9
c. 3.8
d. 3.6
e. 3.7

13. Let $P(x) = 7x^3 + 3x^2 - 7x + 10$. Use synthetic division to find the value $P(5)$.

   Select the correct answer.

   a. 925
   b. 1,850
   c. 926
   d. 924

14. Find all rational roots of the equation.

   $x^7 - 12x^5 + 48x^3 - 64x = 0$

   Select the correct answer.

   a. $x = 0, x = 5, x = 5, x = 5, x = -2, x = -2, x = -2$
   b. $x = 0, x = 5, x = 5, x = 2, x = 2, x = -2$
   c. $x = 0, x = 2, x = 2, x = 2, x = 5, x = 5, x = 5$
   d. $x = 0, x = 2, x = 2, x = 2, x = -2, x = -2, x = -2$

15. A partial solution set $(7, 7)$ is given for the equation. Find the complete solution set.

   $x^4 - 14x^3 + 50x^2 - 14x + 49 = 0$

   Select the correct answer.

   a. $7, 7, -i, i$
   b. $7, 7, 7, i$
   c. $7, 7, 0, 0$
   d. $7, 7, -7i, 7i$

16. The length of the shipping crate is to be 2 feet greater than its height and 5 feet greater than its width, and its volume is to be 170 cubic feet. Use the bisection method or a graphing calculator to find the height of the crate to the nearest tenth of a foot.

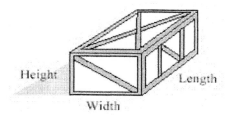

Select the correct answer.

a. 7.3 ft
b. 11.6 ft
c. 6.3 ft
d. 5.3 ft
e. 12.6 ft

17. Use a graphing calculator to find the solution of the equation.

$$x^2 - 8x + 16 = 0$$

Select the correct answer.

a. $x = -8$
b. $x = -4$
c. $x = 8$
d. $x = -16$
e. $x = 4$

18. Write a third-degree polynomial equation with real coefficients and the roots $8 - i$, -3.

Select the correct answer.

a. $x^3 + 15x^2 - 195 = 0$
b. $x^3 - 13x^2 + 17x + 195 = 0$
c. $x^3 + 17x^2 + 15x - 13 = 0$
d. $x^3 + 195 = 0$

19. The number (1 + i) is a root of the equation. Find the other roots of the equation.

    $x^4 - 2x^3 - 34x^2 + 72x - 72 = 0$

    Select the correct answer(s).

    a. $x = -6$
    b. $x = 1 - i$
    c. $x = -7$
    d. $x = 7$
    e. $x = 6$

20. A partial solution set (3) is given for the equation. Find the complete solution set.

    $x^3 + 4x^2 - 9x - 36 = 0$

    Select the correct answer.

    a. $3, -1, -3$
    b. $3, 4, -3$
    c. $1, -4, 2$
    d. $3, -4, -3$

21. The number (1 + i) is a root of the equation. Find the other roots of the equation.

    $x^3 - 10x^2 + 18x - 16 = 0$

    Select the correct answer(s).

    a. $x = 1 - i$
    b. $x = 1 + i$
    c. $x = 8$
    d. $x = 6$

22. Tell how many roots the equation has.

    $x^6 = 5$

    Select the correct answer.

    a. 5
    b. 9
    c. 7
    d. 6

Page 7

23. Write a fourth-degree polynomial equation with real coefficients and the roots 2, 3, 4+*i*.

    Select the correct answer.

    a. $x^4 + 102 = 0$
    b. $x^4 - 13x^3 + 63x^2 - 133x + 102 = 0$
    c. $x^4 - 13x^3 + 61x^2 - 133x + 96 = 0$
    d. $x^4 + 96x^3 - 102x^2 + 13 = 0$

24. Find all rational roots of the equation.

    $x^3 - 5x^2 - x + 5 = 0$

    Select the correct answer(s).

    a. $x = 5$
    b. $x = 1$
    c. $x = 8$
    d. $x = -1$
    e. $x = 0$
    f. $x = 2$

25. One root of the equation $x(8x^5 - 7) = 9x$ is 0. How many other roots are there?

    Select the correct answer.

    a. 4
    b. 6
    c. 0
    d. 5

ANSWER KEY

Gustafson/Frisk - College Algebra 8E Chapter 5 Form C

1. b
2. a,d,f
3. a,c,e,g
4. d
5. c
6. a
7. c
8. a
9. a,d,e,f,h
10. d
11. d
12. b
13. a
14. d
15. a
16. c
17. e
18. b
19. a,b,e
20. d
21. a,c
22. d
23. b
24. a,b,d
25. d

List of Problem Codes for BCA Testing

Gustafson/Frisk - College Algebra 8E Chapter 5 Form C

1. gfca.05.04.4.20m_NoAlgs
2. gfca.05.03.4.37m_NoAlgs
3. gfca.05.03.4.11m_NoAlgs
4. gfca.05.01.4.45m_NoAlgs
5. gfca.05.04.4.27m_NoAlgs
6. gfca.05.02.4.17m_NoAlgs
7. gfca.05.01.4.33m_NoAlgs
8. gfca.05.02.4.43m_NoAlgs
9. gfca.05.03.4.16m_NoAlgs
10. gfca.05.01.4.51m_NoAlgs
11. gfca.05.04.4.09m_NoAlgs
12. gfca.05.04.4.17m_NoAlgs
13. gfca.05.01.4.57m_NoAlgs
14. gfca.05.03.4.17m_NoAlgs
15. gfca.05.01.4.27m_NoAlgs
16. gfca.05.04.4.28m_NoAlgs
17. gfca.05.04.4.24m_NoAlgs
18. gfca.05.02.4.20m_NoAlgs
19. gfca.05.03.4.35m_NoAlgs
20. gfca.05.01.4.23m_NoAlgs
21. gfca.05.03.4.33m_NoAlgs
22. gfca.05.02.4.11m_NoAlgs
23. gfca.05.02.4.22m_NoAlgs
24. gfca.05.03.4.05m_NoAlgs
25. gfca.05.02.4.15m_NoAlgs

Gustafson/Frisk - College Algebra 8E Chapter 5 Form D

1. Find all rational roots of the equation.

   $x^7 - 12x^5 + 48x^3 - 64x = 0$

   Select the correct answer.

   a. $x = 0$, $x = 5$, $x = 5$, $x = 5$, $x = -2$, $x = -2$, $x = -2$
   b. $x = 0$, $x = 5$, $x = 5$, $x = 2$, $x = 2$, $x = -2$
   c. $x = 0$, $x = 2$, $x = 2$, $x = 2$, $x = 5$, $x = 5$, $x = 5$
   d. $x = 0$, $x = 2$, $x = 2$, $x = 2$, $x = -2$, $x = -2$, $x = -2$

2. Use a graphing calculator to find the solution of the equation.

   $x^2 - 8x + 16 = 0$

   Select the correct answer.

   a. $x = -8$
   b. $x = -4$
   c. $x = 8$
   d. $x = -16$
   e. $x = 4$

3. A partial solution set (3) is given for the equation. Find the complete solution set.

   $x^3 + 4x^2 - 9x - 36 = 0$

   Select the correct answer.

   a. $3, -1, -3$
   b. $3, 4, -3$
   c. $1, -4, 2$
   d. $3, -4, -3$

4. Write a fourth-degree polynomial equation with real coefficients and the roots 2, 3, 4+i.

   Select the correct answer.

   a. $x^4 + 96x^3 - 102x^2 + 13 = 0$
   b. $x^4 - 13x^3 + 61x^2 - 133x + 96 = 0$
   c. $x^4 - 13x^3 + 63x^2 - 133x + 102 = 0$
   d. $x^4 + 102 = 0$

5. One root of the equation $x(8x^5 - 7) = 9x$ is 0. How many other roots are there?

   Select the correct answer.

   a. 4
   b. 6
   c. 0
   d. 5

6. Write a third-degree polynomial equation with real coefficients and the roots 7, -i.

   Select the correct answer.

   a. $x^3 - 7x^2 + x - 7 = 0$
   b. $x^3 - 8x^2 - x + 49 = 0$
   c. $x^3 - 7 = 0$
   d. $x^3 - 49 = 0$

7. Find a polynomial with the given zeros

   (5, 5, 2).

   Select the correct answer.

   a. $x^3 + 12x^2 - 45x + 50$

   b. $12x^2 + 45x - 50$

   c. $x^3 - 12x^2 + 45x - 50$

   d. $45x^3 - 12x^2 + 45x - 50$

8. Find all rational roots of the equation.

   $x^4 - 15x^3 + 80x^2 - 180x + 144 = 0$

   Select the correct answer(s).

   a. $x = 3$
   b. $x = 5$
   c. $x = 6$
   d. $x = -3$
   e. $x = 2$
   f. $x = -6$
   g. $x = 4$

9. Use the bisection method or a graphing calculator to find the coordinates of the two points on the graph of $y = x^3$ that lie $r = 1$ unit from the origin. Give the result to the nearest hundredth.

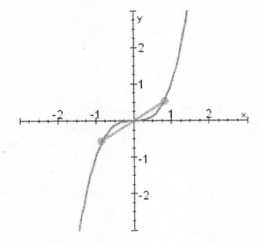

   Select the correct answer.

   a. $(1, 0.68), (-1, -0.68)$
   b. $(0.76, 0.51), (-0.76, -0.51)$
   c. $(0.83, 0.56), (-0.83, -0.56)$
   d. $(1.04, 0.7), (-1.04, -0.7)$
   e. $(0.62, 0.42), (-0.62, -0.42)$

10. Find the interval in which the equation has at least one real root.

    $x^4 - 17x^2 + 60 = 0$

    Select the correct answer.

    a. -2 and -1
    b. 1 and 2
    c. 0 and 1
    d. 3 and 4
    e. -1 and 0

11. Find the negative root of the equation using the bisection method.

    $x^3 - x + 1 = 0$.

    Select the correct answer.

    a. -1.5
    b. -1.3
    c. -1.2
    d. -1.1
    e. -1.6

12. Let $P(x) = 7x^3 + 3x^2 - 7x + 10$. Use synthetic division to find the value $P(5)$.

    Select the correct answer.

    a. 925
    b. $1,850$
    c. 926
    d. 924

Gustafson/Frisk - College Algebra 8E Chapter 5 Form D

13. The number $(1 + i)$ is a root of the equation. Find the other roots of the equation.

    $x^4 - 2x^3 - 34x^2 + 72x - 72 = 0$

    Select the correct answer(s).

    a. $x = -6$
    b. $x = 1 - i$
    c. $x = -7$
    d. $x = 7$
    e. $x = 6$

14. The number $(1 + i)$ is a root of the equation. Find the other roots of the equation.

    $x^3 - 10x^2 + 18x - 16 = 0$

    Select the correct answer(s).

    a. $x = 1 - i$
    b. $x = 1 + i$
    c. $x = 8$
    d. $x = 6$

15. Use synthetic division to do the division.

    $$\frac{2x^3 - 14x^2 - 33x + 21}{x + 2}$$

    Select the correct answer.

    a. $x^2 - 18x + 2 - \dfrac{15}{x + 2}$

    b. $2x^2 - 18x + 3$

    c. $2x^2 - 18x + 15 + \dfrac{21}{x + 2}$

    d. $2x^2 - 18x + 3 + \dfrac{15}{x + 2}$

16. Write a third-degree polynomial equation with real coefficients and the roots $8 - i$, -3.

Select the correct answer.

a. $x^3 + 15x^2 - 195 = 0$
b. $x^3 - 13x^2 + 17x + 195 = 0$
c. $x^3 + 17x^2 + 15x - 13 = 0$
d. $x^3 + 195 = 0$

17. Find integer bounds for the roots of the equation.

$4x^3 - 9x^2 - 58x = 0$

Select the correct answer.

a. $-3, 6$
b. $-4, 3$
c. $-4, 5$
d. $-3, 4$

18. Find all rational roots of the equation.

$x^3 - 5x^2 - x + 5 = 0$

Select the correct answer(s).

a. $x = 5$
b. $x = 1$
c. $x = 8$
d. $x = -1$
e. $x = 0$
f. $x = 2$

19. Tell how many roots the equation has.

$x^6 = 5$

Select the correct answer.

a. 5
b. 9
c. 7
d. 6

20. Use synthetic division to express $P(x) = 2x^3 + 3x^2 - 8x + 29$ in the form $(divisor)(quotient) + remainder$ for the divisor $x + 3$.

Select the correct answer.

a. $(x + 3)(2x^2 - 3x + 1) + 26$

b. $(x - 3)(2x^2 - 3x + 1) - 26$

c. $(x + 3)(26x^2 - x + 1) + 26$

d. $(x + 3)(2x^2 - 3x + 26) + 1$

21. Find all rational roots of the equation.

$x^5 - 4x^4 - 10x^3 + 40x^2 + 9x - 36 = 0$

Select the correct answer(s).

a. $x = -3$
b. $x = 2$
c. $x = -2$
d. $x = -1$
e. $x = 4$
f. $x = 3$
g. $x = -4$
h. $x = 1$

22. The length of the shipping crate is to be 2 feet greater than its height and 5 feet greater than its width, and its volume is to be 170 cubic feet. Use the bisection method or a graphing calculator to find the height of the crate to the nearest tenth of a foot.

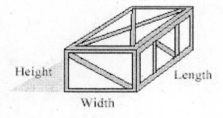

Select the correct answer.

a. 7.3 ft
b. 11.6 ft
c. 6.3 ft
d. 5.3 ft
e. 12.6 ft

23. Solve the equation.

$$x^3 - \frac{53}{3}x^2 + \frac{124}{3}x - 20 = 0$$

Select the correct answer(s).

a. $x = \frac{2}{3}$

b. $x = -\frac{2}{3}$

c. $x = -15$
d. $x = 2$
e. $x = -1$
f. $x = 15$
g. $x = 1$

24. Find the positive root of the equation using the bisection method.

$$x^2 - 15 = 0$$

Select the correct answer.

a. 3.6
b. 3.9
c. 3.7
d. 4.3
e. 3.8

25. A partial solution set $(7, 7)$ is given for the equation. Find the complete solution set.

$$x^4 - 14x^3 + 50x^2 - 14x + 49 = 0$$

Select the correct answer.

a. 7, 7, -i, i
b. 7, 7, 7, i
c. 7, 7, 0, 0
d. 7, 7, -7i, 7i

ANSWER KEY

Gustafson/Frisk - College Algebra 8E Chapter 5 Form D

1. d
2. e
3. d
4. c
5. d
6. a
7. c
8. a,c,e,g
9. c
10. d
11. b
12. a
13. a,b,e
14. a,c
15. d
16. b
17. a
18. a,b,d
19. d
20. a
21. a,d,e,f,h
22. c
23. a,d,f
24. b
25. a

List of Problem Codes for BCA Testing

Gustafson/Frisk - College Algebra 8E Chapter 5 Form D

1. gfca.05.03.4.17m_NoAlgs
2. gfca.05.04.4.24m_NoAlgs
3. gfca.05.01.4.23m_NoAlgs
4. gfca.05.02.4.22m_NoAlgs
5. gfca.05.02.4.15m_NoAlgs
6. gfca.05.02.4.17m_NoAlgs
7. gfca.05.01.4.33m_NoAlgs
8. gfca.05.03.4.11m_NoAlgs
9. gfca.05.04.4.27m_NoAlgs
10. gfca.05.04.4.09m_NoAlgs
11. gfca.05.04.4.20m_NoAlgs
12. gfca.05.01.4.57m_NoAlgs
13. gfca.05.03.4.35m_NoAlgs
14. gfca.05.03.4.33m_NoAlgs
15. gfca.05.01.4.51m_NoAlgs
16. gfca.05.02.4.20m_NoAlgs
17. gfca.05.02.4.43m_NoAlgs
18. gfca.05.03.4.05m_NoAlgs
19. gfca.05.02.4.11m_NoAlgs
20. gfca.05.01.4.45m_NoAlgs
21. gfca.05.03.4.16m_NoAlgs
22. gfca.05.04.4.28m_NoAlgs
23. gfca.05.03.4.37m_NoAlgs
24. gfca.05.04.4.17m_NoAlgs
25. gfca.05.01.4.27m_NoAlgs

1. A partial solution set (3) is given for the equation. Find the complete solution set.

   $$x^3 + 4x^2 - 9x - 36 = 0$$

   Select the correct answer.

   a. $1, -4, 2$
   b. $3, -1, -3$
   c. $3, 4, -3$
   d. $3, -4, -3$

2. A partial solution set $(6, 6)$ is given for the equation. Find the complete solution set.

   $$x^4 - 12x^3 + 37x^2 - 12x + 36 = 0$$

3. Find a polynomial with the given zeros $(5, 5, 2)$.

   Select the correct answer.

   a. $45x^3 - 12x^2 + 45x - 50$
   b. $x^3 + 12x^2 - 45x + 50$
   c. $12x^2 + 45x - 50$
   d. $x^3 - 12x^2 + 45x - 50$

4. Find a polynomial with the zeros $(i, \sqrt{3}, -i)$.

5. Use synthetic division to do the division.

   $$\frac{3x^3 - 14x^2 - 62x + 40}{x + 3}$$

6. The number (1 + *i*) is a root of the equation. Find the other roots of the equation.

   $x^4 - 2x^3 - 7x^2 + 18x - 18 = 0$

7. Let $P(x) = 7x^3 + 3x^2 - 7x + 10$. Use synthetic division to find the value $P(5)$.

   Select the correct answer.

   a. 924
   b. 1,850
   c. 925
   d. 926

8. Tell how many roots the equation has.

   $x^6 = 5$

   Select the correct answer.

   a. 6
   b. 5
   c. 9
   d. 7

9. One root of the equation How many other roots are there?

   $x(4x^4 - 3) = 5x$ is 0.

10. Write a third-degree polynomial equation with real coefficients and the roots 2, -*i*.

11. Write a third-degree polynomial equation with real coefficients and the roots 8 - *i*, -3.

    Select the correct answer.

    a. $x^3 + 195 = 0$
    b. $x^3 + 17x^2 + 15x - 13 = 0$
    c. $x^3 - 13x^2 + 17x + 195 = 0$
    d. $x^3 + 15x^2 - 195 = 0$

12. Write a fourth-degree polynomial equation with real coefficients and the roots 3, 1, 2 + i.

13. Find integer bounds for the roots of the equation.

    $4x^3 - 9x^2 - 58x = 0$

    Select the correct answer.

    a. $-3, 4$
    b. $-4, 3$
    c. $-3, 6$
    d. $-4, 5$

14. Find all rational roots of the equation.

    $x^3 - 5x^2 - x + 5 = 0$

    Select the correct answer(s).

    a. $x = 1$
    b. $x = 5$
    c. $x = 8$
    d. $x = 2$
    e. $x = -1$
    f. $x = 0$

15. Find all rational roots of the equation.

    $x^3 + 5x^2 - x - 5 = 0$

16. Find all rational roots of the equation.

    $x^4 - 15x^3 + 80x^2 - 180x + 144 = 0$

    Select the correct answer(s).

    a. $x = 2$
    b. $x = 5$
    c. $x = 3$
    d. $x = 4$
    e. $x = -6$
    f. $x = -3$
    g. $x = 6$

17. Find all rational roots of the equation.

    $x^5 - 5x^4 - 5x^3 + 25x^2 + 4x - 20 = 0$

18. The number $(1 + i)$ is a root of the equation. Find the other roots of the equation.

    $x^3 - 10x^2 + 18x - 16 = 0$

    Select the correct answer(s).

    a. $x = 1 - i$
    b. $x = 6$
    c. $x = 1 + i$
    d. $x = 8$

19. Solve the equation.

    $x^3 - \dfrac{53}{3}x^2 + \dfrac{124}{3}x - 20 = 0$

    Select the correct answer(s).

    a. $x = 2$
    b. $x = 15$
    c. $x = -1$
    d. $x = 1$
    e. $x = \dfrac{2}{3}$
    f. $x = -\dfrac{2}{3}$
    g. $x = -15$

20. Find the interval at which the equation $3x^3 - 11x^2 - 7x = 0$ has at least one real root.

    Select the correct answer

    a.   2 and 3
    b.   - 4 and - 5
    c.   - 3 and - 2
    d.   4 and 5
    e.   1 and 2

21. Find the negative root of the equation using the bisection method.

$$x^2 - 8 = 0$$

22. Find the negative root of the equation using the bisection method.

$$x^3 - x + 1 = 0.$$

Select the correct answer.

  a. -1.6
  b. -1.1
  c. -1.2
  d. -1.3
  e. -1.5

23. Use a graphing calculator to find the solution of the equation.

$$x^2 - 8x + 16 = 0$$

Select the correct answer.

  a. $x = 8$
  b. $x = -4$
  c. $x = -8$
  d. $x = -16$
  e. $x = 4$

24. Use a graphing calculator to find the real solutions of the equation.

$$x^3 - 3x^2 - 2x + 6 = 0$$

25. The length of the shipping crate is to be 2 feet greater than its height and 5 feet greater than its width, and its volume is to be 32 cubic feet. Use the bisection method or a graphing calculator to find the height of the crate to the nearest tenth of a foot.

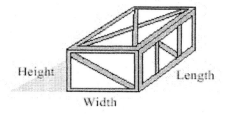

ANSWER KEY

Gustafson/Frisk - College Algebra 8E Chapter 5 Form E

1. d
2. $6, 6, -i, i$
3. d
4. $x^3 - \sqrt{3} \cdot x^2 + x - \sqrt{3}$
5. $3x^2 - 23x + 7 + \dfrac{19}{x+3}$
6. $3, -3, 1-i$
7. c
8. a
9. 4
10. $x^3 - (2x)^2 + x - 2 = 0$
11. c
12. $x^4 - 8x^3 + 24x^2 - 32x + 15 = 0$
13. c
14. a,b,e
15. $1, -1, -5$
16. a,c,d,g
17. $1, 2, -1, -2, 5$
18. a,d
19. a,b,e
20. d
21. $-2, 8$
22. d
23. e
24. $1, 4, -1, 4, 3$
25. $4, 2$

List of Problem Codes for BCA Testing

Gustafson/Frisk - College Algebra 8E Chapter 5 Form E

1. gfca.05.01.4.23m_NoAlgs
2. gfca.05.01.4.27_NoAlgs
3. gfca.05.01.4.33m_NoAlgs
4. gfca.05.01.4.37_NoAlgs
5. gfca.05.01.4.51_NoAlgs
6. gfca.05.03.4.36_NoAlgs
7. gfca.05.01.4.57m_NoAlgs
8. gfca.05.02.4.11m_NoAlgs
9. gfca.05.02.4.15_NoAlgs
10. gfca.05.02.4.17_NoAlgs
11. gfca.05.02.4.20m_NoAlgs
12. gfca.05.02.4.22_NoAlgs
13. gfca.05.02.4.43m_NoAlgs
14. gfca.05.03.4.05m_NoAlgs
15. gfca.05.03.4.09_NoAlgs
16. gfca.05.03.4.11m_NoAlgs
17. gfca.05.03.4.16_NoAlgs
18. gfca.05.03.4.33m_NoAlgs
19. gfca.05.03.4.37m_NoAlgs
20. gfca.05.04.4.07m_NoAlgs
21. gfca.05.04.4.16_NoAlgs
22. gfca.05.04.4.20m_NoAlgs
23. gfca.05.04.4.24m_NoAlgs
24. gfca.05.04.4.26_NoAlgs
25. gfca.05.04.4.28_NoAlgs

Gustafson/Frisk - College Algebra 8E Chapter 5 Form F

1. A partial solution set $(6, 6)$ is given for the equation. Find the complete solution set.

   $$x^4 - 12x^3 + 37x^2 - 12x + 36 = 0$$

2. Find all rational roots of the equation.

   $$x^3 + 5x^2 - x - 5 = 0$$

3. Tell how many roots the equation has.

   $$x^6 = 5$$

   Select the correct answer.

   a. 6
   b. 5
   c. 9
   d. 7

4. One root of the equation $x(4x^4 - 3) = 5x$ is 0. How many other roots are there?

5. Find all rational roots of the equation.

   $$x^4 - 15x^3 + 80x^2 - 180x + 144 = 0$$

   Select the correct answer(s).

   a. $x = 2$
   b. $x = 5$
   c. $x = 3$
   d. $x = 4$
   e. $x = -6$
   f. $x = -3$
   g. $x = 6$

6. Write a fourth-degree polynomial equation with real coefficients and the roots 3, 1, 2 + i.

7. Solve the equation.

$$x^3 - \frac{53}{3}x^2 + \frac{124}{3}x - 20 = 0$$

Select the correct answer(s).

a. $x = 2$
b. $x = 15$
c. $x = -1$
d. $x = 1$
e. $x = \frac{2}{3}$
f. $x = -\frac{2}{3}$
g. $x = -15$

8. A partial solution set (3) is given for the equation. Find the complete solution set.

$$x^3 + 4x^2 - 9x - 36 = 0$$

Select the correct answer.

a. $3, 4, -3$
b. $3, -4, -3$
c. $1, -4, 2$
d. $3, -1, -3$

9. Find a polynomial with the zeros $(i, \sqrt{3}, -i)$.

10. Use a graphing calculator to find the solution of the equation.

    $x^2 - 8x + 16 = 0$

    Select the correct answer.

    a. $x = 8$
    b. $x = -4$
    c. $x = -8$
    d. $x = -16$
    e. $x = 4$

11. Find the interval at which the equation $3x^3 - 11x^2 - 7x = 0$ has at least one real root.

    Select the correct answer.

    a. 2 and 3
    b. -4 and -5
    c. -3 and -2
    d. 4 and 5
    e. 1 and 2

12. Find a polynomial with the given zeros

    (5, 5, 2).

    Select the correct answer.

    a. $45x^3 - 12x^2 + 45x - 50$
    b. $x^3 + 12x^2 - 45x + 50$
    c. $12x^2 + 45x - 50$
    d. $x^3 - 12x^2 + 45x - 50$

13. Find the negative root of the equation using the bisection method.

    $x^2 - 8 = 0$

14. Find all rational roots of the equation.

    $x^3 - 5x^2 - x + 5 = 0$

    Select the correct answer(s).

    a. $x = 1$
    b. $x = 5$
    c. $x = 8$
    d. $x = 2$
    e. $x = -1$
    f. $x = 0$

15. Use synthetic division to do the division.

    $$\frac{3x^3 - 14x^2 - 62x + 40}{x + 3}$$

16. Write a third-degree polynomial equation with real coefficients and the roots 8 - *i*, -3.

    Select the correct answer.

    a. $x^3 + 195 = 0$
    b. $x^3 + 17x^2 + 15x - 13 = 0$
    c. $x^3 - 13x^2 + 17x + 195 = 0$
    d. $x^3 + 15x^2 - 195 = 0$

17. The number (1 + *i*) is a root of the equation.

    $x^4 - 2x^3 - 7x^2 + 18x - 18 = 0$

    Find the other roots of the equation.

18. The number (1 + *i*) is a root of the equation.

$$x^3 - 10x^2 + 18x - 16 = 0$$

Find the other roots of the equation.

Select the correct answer(s).

    a. $x = 1 - i$
    b. $x = 6$
    c. $x = 1 + i$
    d. $x = 8$

19. Find all rational roots of the equation.

$$x^5 - 5x^4 - 5x^3 + 25x^2 + 4x - 20 = 0$$

20. Let $P(x) = 7x^3 + 3x^2 - 7x + 10$. Use synthetic division to find the value $P(5)$.

Select the correct answer.

    a.   924
    b.   1,850
    c.   925
    d.   926

21. Find integer bounds for the roots of the equation.

$$4x^3 - 9x^2 - 58x = 0$$

Select the correct answer.

    a.   - 3, 4
    b.   - 4, 3
    c.   - 3, 6
    d.   - 4, 5

22. The length of the shipping crate is to be 2 feet greater than its height and 5 feet greater than its width, and its volume is to be 32 cubic feet. Use the bisection method or a graphing calculator to find the height of the crate to the nearest tenth of a foot.

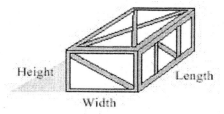

23. Write a third-degree polynomial equation with real coefficients and the roots 2, -*i*.

24. Find the negative root of the equation using the bisection method.

$$x^3 - x + 1 = 0.$$

Select the correct answer.

a. -1.6
b. -1.1
c. -1.2
d. -1.3
e. -1.5

25. Use a graphing calculator to find the real solutions of the equation.

$$x^3 - 3x^2 - 2x + 6 = 0$$

ANSWER KEY

Gustafson/Frisk - College Algebra 8E Chapter 5 Form F

1. 6, 6, −i, i
2. 1, −1, −5
3. a
4. 4
5. a,c,d,g
6. $x^4 - 8x^3 + 24x^2 - 32x + 15 = 0$
7. a,b,e
8. b
9. $x^3 - \sqrt{3} \cdot x^2 + x - \sqrt{3}$
10. e
11. d
12. d
13. −2, 8
14. a,b,e
15. $3x^2 - 23x + 7 + \dfrac{19}{x+3}$
16. c
17. 3, −3, 1−i
18. a,d
19. 1, 2, −1, −2, 5
20. c
21. c
22. 4, 2
23. $x^3 - (2x)^2 + x - 2 = 0$
24. d
25. 1, 4, −1, 4, 3

List of Problem Codes for BCA Testing

Gustafson/Frisk - College Algebra 8E Chapter 5 Form F

1. gfca.05.01.4.27_NoAlgs
2. gfca.05.03.4.09_NoAlgs
3. gfca.05.02.4.11m_NoAlgs
4. gfca.05.02.4.15_NoAlgs
5. gfca.05.03.4.11m_NoAlgs
6. gfca.05.02.4.22_NoAlgs
7. gfca.05.03.4.37m_NoAlgs
8. gfca.05.01.4.23m_NoAlgs
9. gfca.05.01.4.37_NoAlgs
10. gfca.05.04.4.24m_NoAlgs
11. gfca.05.04.4.07m_NoAlgs
12. gfca.05.01.4.33m_NoAlgs
13. gfca.05.04.4.16_NoAlgs
14. gfca.05.03.4.05m_NoAlgs
15. gfca.05.01.4.51_NoAlgs
16. gfca.05.02.4.20m_NoAlgs
17. gfca.05.03.4.36_NoAlgs
18. gfca.05.03.4.33m_NoAlgs
19. gfca.05.03.4.16_NoAlgs
20. gfca.05.01.4.57m_NoAlgs
21. gfca.05.02.4.43m_NoAlgs
22. gfca.05.04.4.28_NoAlgs
23. gfca.05.02.4.17_NoAlgs
24. gfca.05.04.4.20m_NoAlgs
25. gfca.05.04.4.26_NoAlgs

1. Find all rational roots of the equation.

   $x^4 - 15x^3 + 80x^2 - 180x + 144 = 0$

   Select the correct answer(s).

   a. $x = 4$
   b. $x = 3$
   c. $x = -6$
   d. $x = 5$
   e. $x = 2$
   f. $x = -3$
   g. $x = 6$

2. Find all rational roots of the equation.

   $x^5 - 4x^4 - 10x^3 + 40x^2 + 9x - 36 = 0$

   Select the correct answer(s).

   a. $x = 3$
   b. $x = -4$
   c. $x = -2$
   d. $x = 4$
   e. $x = -3$
   f. $x = -1$
   g. $x = 1$
   h. $x = 2$

3. Find integer bounds for the roots of the equation.

   $4x^3 - 9x^2 - 58x = 0$

   Select the correct answer.

   a. $-3, 6$
   b. $-4, 5$
   c. $-4, 3$
   d. $-3, 4$

4. The number $(1 + i)$ is a root of the equation. Find the other roots of the equation.

   $x^4 - 2x^3 - 7x^2 + 18x - 18 = 0$

5. One root of the equation $x(8x^5 - 7) = 9x$ is 0. How many other roots are there?

   Select the correct answer.

   a. 4
   b. 6
   c. 5
   d. 0

6. Find the positive root of the equation using the bisection method.

   $x^2 - 15 = 0$

   Select the correct answer.

   a. 3.8
   b. 3.9
   c. 4.3
   d. 3.7
   e. 3.6

7. Tell how many roots the equation has.

   $x^7 = 5$

8. Find the negative root of the equation using the bisection method.

   $x^3 - x + 1 = 0$

   Select the correct answer.

   a. −1.2
   b. −1.6
   c. −1.5
   d. −1.3
   e. −1.1

9. Find all rational roots of the equation.

   $x^4 + 8x^3 + 24x^2 + 32x + 16 = 0$

10. The number (1 + *i*) is a root of the equation. Find the other roots of the equation.

$x^3 - 10x^2 + 18x - 16 = 0$

Select the correct answer(s).

a. $x = 1 + i$
b. $x = 1 - i$
c. $x = 8$
d. $x = 6$

11. Let $P(x) = 3x^3 + 3x^2 - 5x + 7$. Use synthetic division to find the value $P(8)$.

12. Use synthetic division to do the division.

$$\frac{2x^3 - 14x^2 - 33x + 21}{x + 2}$$

Select the correct answer.

a. $x^2 - 18x + 2 - \dfrac{15}{x + 2}$

b. $2x^2 - 18x + 3$

c. $2x^2 - 18x + 3 + \dfrac{15}{x + 2}$

d. $2x^2 - 18x + 15 + \dfrac{21}{x + 2}$

13. Write a fourth-degree polynomial equation with real coefficients and the roots 2, 3, 4+*i*.

Select the correct answer.

a. $x^4 + 96x^3 - 102x^2 + 13 = 0$
b. $x^4 - 13x^3 + 61x^2 - 133x + 96 = 0$
c. $x^4 - 13x^3 + 63x^2 - 133x + 102 = 0$
d. $x^4 + 102 = 0$

14. Find all rational roots of the equation.

    $x^3 - 5x^2 - x + 5 = 0$

    Select the correct answer(s).

    a. $x = 5$
    b. $x = 1$
    c. $x = 2$
    d. $x = 8$
    e. $x = -1$
    f. $x = 0$

15. The length of the shipping crate is to be 2 feet greater than its height and 5 feet greater than its width, and its volume is to be 32 cubic feet. Use the bisection method or a graphing calculator to find the height of the crate to the nearest tenth of a foot.

    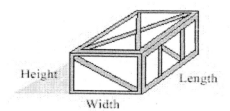

16. Write a third-degree polynomial equation with real coefficients and the roots 9 - i, -4.

17. Use a graphing calculator to find the solution of the equation.

    $x^2 - 8x + 16 = 0$

    Select the correct answer.

    a. $x = -8$
    b. $x = -16$
    c. $x = 4$
    d. $x = -4$
    e. $x = 8$

18. Find a polynomial with the zeros $\left(i, \sqrt{3}, -i\right)$.

19. Write a third-degree polynomial equation with real coefficients and the roots 7, -i.

    Select the correct answer.

    a. $x^3 - 49 = 0$
    b. $x^3 - 7 = 0$
    c. $x^3 - 8x^2 - x + 49 = 0$
    d. $x^3 - 7x^2 + x - 7 = 0$

20. Use synthetic division to express $P(x) = 2x^3 + 3x^2 - 8x + 29$ in the form $(divisor)(quotient) + remainder$ for the divisor $x + 3$

    Select the correct answer.

    a. $(x + 3)(26x^2 - x + 1) + 26$

    b. $(x - 3)(2x^2 - 3x + 1) - 26$

    c. $(x + 3)(2x^2 - 3x + 1) + 26$

    d. $(x + 3)(2x^2 - 3x + 26) + 1$

21. Find the negative root of the equation using the bisection method.

    $x^2 - 8 = 0$

22. Solve the equation.

$$x^3 - \frac{53}{3}x^2 + \frac{124}{3}x - 20 = 0$$

Select the correct answer(s).

- a. $x = -15$
- b. $x = \frac{2}{3}$
- c. $x = 15$
- d. $x = -1$
- e. $x = 1$
- f. $x = 2$
- g. $x = -\frac{2}{3}$

23. Find a polynomial with the given zeros

(5, 5, 2).

Select the correct answer.

- a. $45x^3 - 12x^2 + 45x - 50$
- b. $12x^2 + 45x - 50$
- c. $x^3 + 12x^2 - 45x + 50$
- d. $x^3 - 12x^2 + 45x - 50$

24. Use the bisection method or a graphing calculator to find the coordinates of the two points on the graph of $y = x^3$ that lie $r = 1$ unit from the origin. Give the result to the nearest hundredth.

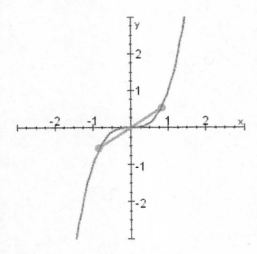

Select the correct answer.

a. $(0.76, 0.51), (-0.76, -0.51)$
b. $(0.62, 0.42), (-0.62, -0.42)$
c. $(0.83, 0.56), (-0.83, -0.56)$
d. $(1, 0.68), (-1, -0.68)$
e. $(1.04, 0.7), (-1.04, -0.7)$

25. Find the interval in which the equation has at least one real root.

$$x^4 - 17x^2 + 60 = 0$$

Select the correct answer.

a. 3 and 4
b. 1 and 2
c. 0 and 1
d. -2 and -1
e. -1 and 0

ANSWER KEY

Gustafson/Frisk - College Algebra 8E Chapter 5 Form G

1. a,b,e,g
2. a,d,e,f,g
3. a
4. $3, -3, 1-i$
5. c
6. b
7. 7
8. d
9. $-2, -2, -2, -2$
10. b,c
11. $1, 695$
12. c
13. c
14. a,b,e
15. $4, 2$
16. $x^3 - 14x^2 + 10x + 328 = 0$
17. c
18. $x^3 - \sqrt{3} \cdot x^2 + x - \sqrt{3}$
19. d
20. c
21. $-2, 8$
22. b,c,f
23. d
24. c
25. a

List of Problem Codes for BCA Testing

Gustafson/Frisk - College Algebra 8E Chapter 5 Form G

1. gfca.05.03.4.11m_NoAlgs
2. gfca.05.03.4.16m_NoAlgs
3. gfca.05.02.4.43m_NoAlgs
4. gfca.05.03.4.36_NoAlgs
5. gfca.05.02.4.15m_NoAlgs
6. gfca.05.04.4.17m_NoAlgs
7. gfca.05.02.4.11_NoAlgs
8. gfca.05.04.4.20m_NoAlgs
9. gfca.05.03.4.12_NoAlgs
10. gfca.05.03.4.33m_NoAlgs
11. gfca.05.01.4.57_NoAlgs
12. gfca.05.01.4.51m_NoAlgs
13. gfca.05.02.4.22m_NoAlgs
14. gfca.05.03.4.05m_NoAlgs
15. gfca.05.04.4.28_NoAlgs
16. gfca.05.02.4.20_NoAlgs
17. gfca.05.04.4.24m_NoAlgs
18. gfca.05.01.4.37_NoAlgs
19. gfca.05.02.4.17m_NoAlgs
20. gfca.05.01.4.45m_NoAlgs
21. gfca.05.04.4.16_NoAlgs
22. gfca.05.03.4.37m_NoAlgs
23. gfca.05.01.4.33m_NoAlgs
24. gfca.05.04.4.27m_NoAlgs
25. gfca.05.04.4.09m_NoAlgs

Gustafson/Frisk - College Algebra 8E Chapter 5 Form H

1. Let $P(x) = 3x^3 + 3x^2 - 5x + 7$. Use synthetic division to find the value $P(8)$.

2. Use synthetic division to express $P(x) = 2x^3 + 3x^2 - 8x + 29$ in the form (divisor)(quotient) + remainder for the divisor $x + 3$

   Select the correct answer.

   a. $(x + 3)(26x^2 - x + 1) + 26$
   b. $(x - 3)(2x^2 - 3x + 1) - 26$
   c. $(x + 3)(2x^2 - 3x + 1) + 26$
   d. $(x + 3)(2x^2 - 3x + 26) + 1$

3. Find the positive root of the equation using the bisection method.

   $x^2 - 15 = 0$

   Select the correct answer.

   a. 3.8
   b. 3.6
   c. 3.7
   d. 3.9
   e. 4.3

4. Write a third-degree polynomial equation with real coefficients and the roots 7, -i.

   Select the correct answer.

   a. $x^3 - 49 = 0$
   b. $x^3 - 7 = 0$
   c. $x^3 - 8x^2 - x + 49 = 0$
   d. $x^3 - 7x^2 + x - 7 = 0$

5. Find all rational roots of the equation.

   $x^4 + 8x^3 + 24x^2 + 32x + 16 = 0$

6. One root of the equation $x(8x^5 - 7) = 9x$ is 0. How many other roots are there?

   Select the correct answer.

   a. 4
   b. 6
   c. 5
   d. 0

7. Find all rational roots of the equation.

   $x^3 - 5x^2 - x + 5 = 0$

   Select the correct answer(s).

   a. $x = 5$
   b. $x = 1$
   c. $x = 2$
   d. $x = 8$
   e. $x = -1$
   f. $x = 0$

8. The length of the shipping crate is to be 2 feet greater than its height and 5 feet greater than its width, and its volume is to be 32 cubic feet. Use the bisection method or a graphing calculator to find the height of the crate to the nearest tenth of a foot.

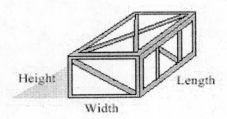

9. The number $(1 + i)$ is a root of the equation. Find the other roots of the equation.

   $x^4 - 2x^3 - 7x^2 + 18x - 18 = 0$

10. Use synthetic division to do the division.

$$\frac{2x^3 - 14x^2 - 33x + 21}{x + 2}$$

Select the correct answer.

a. $x^2 - 18x + 2 - \dfrac{15}{x + 2}$

b. $2x^2 - 18x + 3$

c. $2x^2 - 18x + 3 + \dfrac{15}{x + 2}$

d. $2x^2 - 18x + 15 + \dfrac{21}{x + 2}$

11. Write a third-degree polynomial equation with real coefficients and the roots $9 - i$, -4.

12. Find all rational roots of the equation.

$x^5 - 4x^4 - 10x^3 + 40x^2 + 9x - 36 = 0$

Select the correct answer(s).

a. $x = 3$
b. $x = -4$
c. $x = -2$
d. $x = 4$
e. $x = -3$
f. $x = -1$
g. $x = 1$
h. $x = 2$

13. Use a graphing calculator to find the solution of the equation.

$$x^2 - 8x + 16 = 0$$

Select the correct answer.

a. $x = 8$
b. $x = -4$
c. $x = 4$
d. $x = -8$
e. $x = -16$

14. Use the bisection method or a graphing calculator to find the coordinates of the two points on the graph of $y = x^3$ that lie $r = 1$ unit from the origin. Give the result to the nearest hundredth.

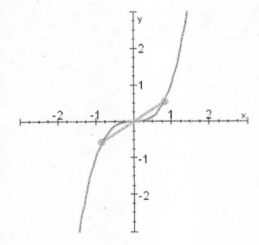

Select the correct answer.

a. $(0.76, 0.51), (-0.76, -0.51)$
b. $(0.62, 0.42), (-0.62, -0.42)$
c. $(0.83, 0.56), (-0.83, -0.56)$
d. $(1, 0.68), (-1, -0.68)$
e. $(1.04, 0.7), (-1.04, -0.7)$

15. Solve the equation.

$$x^3 - \frac{53}{3}x^2 + \frac{124}{3}x - 20 = 0$$

Select the correct answer(s).

a. $x = -15$
b. $x = \frac{2}{3}$
c. $x = 15$
d. $x = -1$
e. $x = 1$
f. $x = 2$
g. $x = -\frac{2}{3}$

16. Find the negative root of the equation using the bisection method.

$$x^2 - 8 = 0$$

17. Find all rational roots of the equation.

$$x^4 - 15x^3 + 80x^2 - 180x + 144 = 0$$

Select the correct answer(s).

a. $x = 4$
b. $x = 3$
c. $x = -6$
d. $x = 5$
e. $x = 2$
f. $x = -3$
g. $x = 6$

18. Find integer bounds for the roots of the equation.

$$4x^3 - 9x^2 - 58x = 0$$

Select the correct answer.

a. -3, 6
b. -4, 5
c. -4, 3
d. -3, 4

19. Tell how many roots the equation has.

    $x^7 = 5$

20. Find a polynomial with the given zeros

    (5, 5, 2).

    Select the correct answer.

    a. $45x^3 - 12x^2 + 45x - 50$
    b. $12x^2 + 45x - 50$
    c. $x^3 + 12x^2 - 45x + 50$
    d. $x^3 - 12x^2 + 45x - 50$

21. Write a fourth-degree polynomial equation with real coefficients and the roots 2, 3, 4+*i*.

    Select the correct answer.

    a. $x^4 + 96x^3 - 102x^2 + 13 = 0$
    b. $x^4 - 13x^3 + 61x^2 - 133x + 96 = 0$
    c. $x^4 - 13x^3 + 63x^2 - 133x + 102 = 0$
    d. $x^4 + 102 = 0$

22. Find a polynomial with the zeros $\left(i, \sqrt{3}, -i\right)$.

23. Find the negative root of the equation using the bisection method.

    $x^3 - x + 1 = 0$.

    Select the correct answer.

    a. -1.2
    b. -1.6
    c. -1.5
    d. -1.3
    e. -1.1

24. Find the interval in which the equation has at least one real root.

$$x^4 - 17x^2 + 60 = 0$$

Select the correct answer.

a.   3 and 4
b.   1 and 2
c.   0 and 1
d.   −2 and −1
e.   −1 and 0

25. The number (1 + *i*) is a root of the equation. Find the other roots of the equation.

$$x^3 - 10x^2 + 18x - 16 = 0$$

Select the correct answer(s).

a. $x = 1 + i$
b. $x = 1 - i$
c. $x = 8$
d. $x = 6$

ANSWER KEY

Gustafson/Frisk - College Algebra 8E Chapter 5 Form H

1. 1.695
2. c
3. d
4. d
5. $-2, -2, -2, -2$
6. c
7. a,b,e
8. 4.2
9. $3, -3, 1-i$
10. c
11. $x^3 - 14x^2 + 10x + 328 = 0$
12. a,d,e,f,g
13. c
14. c
15. b,c,f
16. -2.8
17. a,b,e,g
18. a
19. 7
20. d
21. c
22. $x^3 - \sqrt{3} \cdot x^2 + x - \sqrt{3}$
23. d
24. a
25. b,c

List of Problem Codes for BCA Testing

Gustafson/Frisk - College Algebra 8E Chapter 5 Form H

1. gfca.05.01.4.57_NoAlgs
2. gfca.05.01.4.45m_NoAlgs
3. gfca.05.04.4.17m_NoAlgs
4. gfca.05.02.4.17m_NoAlgs
5. gfca.05.03.4.12_NoAlgs
6. gfca.05.02.4.15m_NoAlgs
7. gfca.05.03.4.05m_NoAlgs
8. gfca.05.04.4.28_NoAlgs
9. gfca.05.03.4.36_NoAlgs
10. gfca.05.01.4.51m_NoAlgs
11. gfca.05.02.4.20_NoAlgs
12. gfca.05.03.4.16m_NoAlgs
13. gfca.05.04.4.24m_NoAlgs
14. gfca.05.04.4.27m_NoAlgs
15. gfca.05.03.4.37m_NoAlgs
16. gfca.05.04.4.16_NoAlgs
17. gfca.05.03.4.11m_NoAlgs
18. gfca.05.02.4.43m_NoAlgs
19. gfca.05.02.4.11_NoAlgs
20. gfca.05.01.4.33m_NoAlgs
21. gfca.05.02.4.22m_NoAlgs
22. gfca.05.01.4.37_NoAlgs
23. gfca.05.04.4.20m_NoAlgs
24. gfca.05.04.4.09m_NoAlgs
25. gfca.05.03.4.33m_NoAlgs

Gustafson/Frisk - College Algebra 8E Chapter 6 Form A

1. Solve the system of equations by graphing.

$$\begin{cases} 8x - 7y = 18 \\ 10x + 3y = 46 \end{cases}$$

2. Use a graphing calculator to approximate the solutions of the system. Give answers to the nearest tenth.

$$\begin{cases} 6.6x + 3.3y = 50.49 \\ -9x + 3y = -21.6 \end{cases}$$

3. Solve the system by the addition method, if possible.

$$\begin{cases} x + \dfrac{y}{7} = \dfrac{15}{7} \\ \dfrac{x+y}{3} = 3 - x \end{cases}$$

4. Solve the system, if possible.

$$\begin{cases} 4x + 7y + 2z = -36 \\ 7x - 3y + 10z = 135 \\ 5x - 5y - 8z = 19 \end{cases}$$

5. Solve the system by Gaussian elimination.

$$\begin{cases} 3x - 5y = -32 \\ 2x + y = -4 \end{cases}$$

6. Solve the system by Gauss-Jordan elimination.

$$\begin{cases} x - 2y = -2 \\ y = -1 \end{cases}$$

7. Minimize $P = 8x + y$ subject to the given constraints.

$$\begin{cases} x \geq 0 \\ y \geq 0 \\ 3y - x \leq 1 \\ y - 3x \geq -3 \end{cases}$$

8. Find the inverse of the matrix.

$$\begin{bmatrix} 8 & 1 \\ 1 & 8 \end{bmatrix}$$

9. Graph the linear inequality.

   $2x + 3y < 12$

   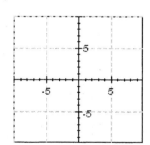

10. Solve the system using Gauss-Jordan elimination. If the system has infinitely many solutions, show a general solution (in terms of z).

$$\begin{cases} x + 2y + z = 15 \\ 3x - y - z = 13 \end{cases}$$

11. Find values of x and y, if any, that will make the matrices equal.

$$\begin{bmatrix} x+y & 5+x \\ -4 & 4y \end{bmatrix} = \begin{bmatrix} 5 & 8 \\ -4 & 8 \end{bmatrix}$$

12. Find the product.

$$\begin{bmatrix} 2 & 2 \\ 6 & 3 \end{bmatrix} \begin{bmatrix} 8 & 3 \\ 6 & 6 \end{bmatrix}$$

13. Find the product.

$$\begin{bmatrix} 6 \\ -8 \\ -8 \end{bmatrix} \begin{bmatrix} 6 & -4 & -8 \end{bmatrix}$$

14. Find the inverse of the matrix.

$$\begin{bmatrix} 1 & 0 & -3 \\ -1 & 1 & -3 \\ 2 & 1 & 1 \end{bmatrix}$$

15. Write the system:

$$\begin{cases} 8x + 13y = 20 \\ 6x + 15y = 36 \end{cases}$$

as a matrix equation $AX = B$, where A is the coefficient matrix of the system, X is a column matrix of variables, and B is the column matrix of constants. Multiply each side of the equation on the left by the matrix inverse of A to obtain an equivalent system of equations and solve it.

16. Graph the solution set of the system.

$$\begin{cases} y < 3 \\ x \geq 2 \end{cases}$$

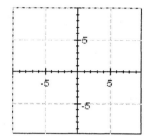

17. Evaluate the determinant.

$$\begin{vmatrix} 1 & -2 \\ -8 & 3 \end{vmatrix}$$

18. $$A = \begin{bmatrix} 9 & 2 & 10 \\ 8 & 4 & 6 \\ 1 & 10 & 1 \end{bmatrix}$$

Find the cofactor C_{12} of the matrix.

19. Use Cramer's rule to find the solution of the system, if possible.

$$\begin{cases} 9x - y = 30 \\ x + y = 0 \end{cases}$$

20. Decompose the fraction into partial fractions.

$$\frac{19x - 1}{x(x-1)}$$

21. Decompose the fraction into partial fractions.

$$\frac{-2x + 16}{x^2 - 4x - 5}$$

22. Two woodworkers, Tom and Carlos, earn a profit of $92 for making a table and $66 for making a chair. On average, Tom must work 3 hours and Carlos 2 hours to make a chair. Tom must work 2 hours and Carlos 6 hours to make a table. If neither wishes to work more than 42 hours per week, how many tables and how many chairs should they make each week to maximize their income? The information is summarized in the table below.

	Table	Chair	Time available
Profit (dollars)	92	66	
Tom's time (hours)	2	3	42
Carlos' time (hours)	6	2	42

23. Graph the solution set of the system.

$$\begin{cases} 4x + 2y \leq 6 \\ 2x - 4y \geq 10 \end{cases}$$

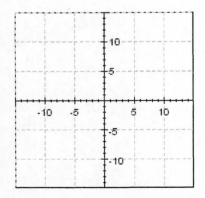

24. Maximize $P = x - 6y$ subject to the given constraints. Find the maximum value of P and the point at which that value occurs.

$$\begin{cases} x \leq 6 \\ y \leq 9 \\ x \geq 0 \\ y \geq 0 \\ x + y \leq 15 \end{cases}$$

25. Graph the solution set of the system.

$$\begin{cases} y < 3x + 2 \\ y < -2x + 3 \end{cases}$$

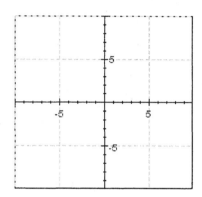

ANSWER KEY

Gustafson/Frisk - College Algebra 8E Chapter 6 Form A

1. $(4, 2)$
2. $(4, 5, 6, 3)$
3. $(2, 1)$
4. $x = 5, y = -10, z = 7$
5. $(-4, 4)$
6. $(-4, -1)$
7. $0, (0, 0)$
8. $\begin{bmatrix} \dfrac{8}{63} & -\dfrac{1}{63} \\ -\dfrac{1}{63} & \dfrac{8}{63} \end{bmatrix}$
9.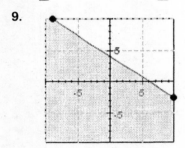
10. $x = \dfrac{41}{7} + \dfrac{z}{7}, y = \dfrac{32}{7} - \dfrac{4z}{7}, z$
11. $3, 2$
12. $\begin{pmatrix} 28 & 18 \\ 66 & 36 \end{pmatrix}$
13. $\begin{pmatrix} 36 & -24 & -48 \\ -48 & 32 & 64 \\ -48 & 32 & 64 \end{pmatrix}$
14. $\begin{pmatrix} 0.307692 & -0.230769 & 0.230769 \\ -0.384615 & 0.538462 & 0.461538 \\ -0.230769 & -0.076923 & 0.076923 \end{pmatrix}$
15. $(-4, 4)$

ANSWER KEY

Gustafson/Frisk - College Algebra 8E Chapter 6 Form A

16.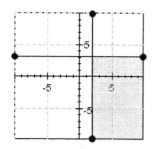

17. -13

18. -2

19. $x = 3, y = -3$

20. $\dfrac{1}{x} + \dfrac{18}{(x-1)}$

21. $\dfrac{1}{(x-5)} - \dfrac{3}{(x+1)}$

22. $3, 12, \$1,068$

23.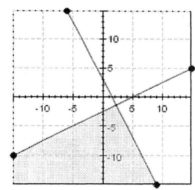

24. $6, (6, 0)$

25.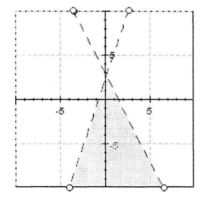

List of Problem Codes for BCA Testing

Gustafson/Frisk - College Algebra 8E Chapter 6 Form A

1. gfca.06.01.4.14_NoAlgs
2. gfca.06.01.4.20_NoAlgs
3. gfca.06.01.4.40_NoAlgs
4. gfca.06.01.4.57_NoAlgs
5. gfca.06.02.4.30_NoAlgs
6. gfca.06.02.4.41_NoAlgs
7. gfca.06.08.4.15_NoAlgs
8. gfca.06.04.4.06_NoAlgs
9. gfca.06.07.4.05_NoAlgs
10. gfca.06.02.4.63_NoAlgs
11. gfca.06.03.4.09_NoAlgs
12. gfca.06.03.4.23_NoAlgs
13. gfca.06.03.4.30_NoAlgs
14. gfca.06.04.4.09_NoAlgs
15. gfca.06.04.4.25_NoAlgs
16. gfca.06.07.4.17_NoAlgs
17. gfca.06.05.4.09_NoAlgs
18. gfca.06.05.4.15_NoAlgs
19. gfca.06.05.4.42_NoAlgs
20. gfca.06.06.4.05m_NoAlgs
21. gfca.06.06.4.09_NoAlgs
22. gfca.06.08.4.21_NoAlgs
23. gfca.06.07.4.28_NoAlgs
24. gfca.06.08.4.10_NoAlgs
25. gfca.06.07.4.22_NoAlgs

1. Graph the solution set of the system.

$$\begin{cases} 4x + 2y \le 6 \\ 2x - 4y \ge 10 \end{cases}$$

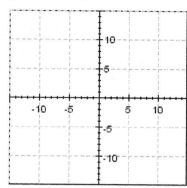

2. Solve the system by Gaussian elimination.

$$\begin{cases} 3x - 5y = -32 \\ 2x + y = -4 \end{cases}$$

3. Use Cramer's rule to find the solution of the system, if possible.

$$\begin{cases} 9x - y = 30 \\ x + y = 0 \end{cases}$$

4. Graph the solution set of the system.

$$\begin{cases} y < 3 \\ x \ge 2 \end{cases}$$

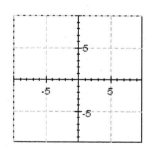

5. Find values of x and y, if any, that will make the matrices equal.

$$\begin{bmatrix} x+y & 5+x \\ -4 & 4y \end{bmatrix} = \begin{bmatrix} 5 & 8 \\ -4 & 8 \end{bmatrix}$$

6. Find the product.

$$\begin{bmatrix} 6 \\ -8 \\ -8 \end{bmatrix} \begin{bmatrix} 6 & -4 & -8 \end{bmatrix}$$

7. Find the product.

$$\begin{bmatrix} 2 & 2 \\ 6 & 3 \end{bmatrix} \begin{bmatrix} 8 & 3 \\ 6 & 6 \end{bmatrix}$$

8. Use a graphing calculator to approximate the solutions of the system. Give answers to the nearest tenth.

$$\begin{cases} 6.6x + 3.3y = 50.49 \\ -9x + 3y = -21.6 \end{cases}$$

9. Write the system

$$\begin{cases} 8x + 13y = 20 \\ 6x + 15y = 36 \end{cases}$$

as a matrix equation $AX = B$, where A is the coefficient matrix of the system, X is a column matrix of variables, and B is the column matrix of constants. Multiply each side of the equation on the left by the matrix inverse of A to obtain an equivalent system of equations and solve it.

10. Minimize $P = 8x + y$ subject to the given constraints.

$$\begin{cases} x \geq 0 \\ y \geq 0 \\ 3y - x \leq 1 \\ y - 3x \geq -3 \end{cases}$$

11. Solve the system, if possible.

$$\begin{cases} 4x + 7y + 2z = -36 \\ 7x - 3y + 10z = 135 \\ 5x - 5y - 8z = 19 \end{cases}$$

12. Solve the system by the addition method, if possible.

$$\begin{cases} x + \dfrac{y}{7} = \dfrac{15}{7} \\ \dfrac{x+y}{3} = 3 - x \end{cases}$$

13. Solve the system by Gauss-Jordan elimination.

$$\begin{cases} x - 2y = -2 \\ y = -1 \end{cases}$$

14. Evaluate the determinant.

$$\begin{vmatrix} 1 & -2 \\ -8 & 3 \end{vmatrix}$$

15. Find the inverse of the matrix.

$$\begin{bmatrix} 1 & 0 & -3 \\ -1 & 1 & -3 \\ 2 & 1 & 1 \end{bmatrix}$$

16. Decompose the fraction into partial fractions.

$$\frac{19x - 1}{x(x - 1)}$$

17. Graph the linear inequality.

    $2x + 3y < 12$

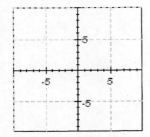

18. $A = \begin{bmatrix} 9 & 2 & 10 \\ 8 & 4 & 6 \\ 1 & 10 & 1 \end{bmatrix}$

    Find the cofactor C_{12} of the matrix.

19. Solve the system of equations by graphing.

$$\begin{cases} 8x - 7y = 18 \\ 10x + 3y = 46 \end{cases}$$

20. Solve the system using Gauss-Jordan elimination. If the system has infinitely many solutions, show a general solution (in terms of z).

$$\begin{cases} x + 2y + z = 15 \\ 3x - y - z = 13 \end{cases}$$

21. Decompose the fraction into partial fractions.

$$\frac{-2x + 16}{x^2 - 4x - 5}$$

22. Maximize $P = x - 6y$ subject to the given constraints. Find the maximum value of P and the point at which that value occurs.

$$\begin{cases} x \leq 6 \\ y \leq 9 \\ x \geq 0 \\ y \geq 0 \\ x + y \leq 15 \end{cases}$$

23. Two woodworkers, Tom and Carlos, earn a profit of $92 for making a table and $66 for making a chair. On average, Tom must work 3 hours and Carlos 2 hours to make a chair. Tom must work 2 hours and Carlos 6 hours to make a table. If neither wishes to work more than 42 hours per week, how many tables and how many chairs should they make each week to maximize their income? The information is summarized in the table below.

	Table	Chair	Time available
Profit (dollars)	92	66	
Tom's time (hours)	2	3	42
Carlos' time (hours)	6	2	42

24. Find the inverse of the matrix.

$$\begin{bmatrix} 8 & 1 \\ 1 & 8 \end{bmatrix}$$

25. Graph the solution set of the system.

$$\begin{cases} y < 3x + 2 \\ y < -2x + 3 \end{cases}$$

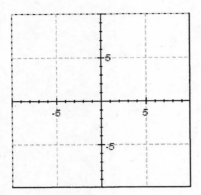

ANSWER KEY

Gustafson/Frisk - College Algebra 8E Chapter 6 Form B

1.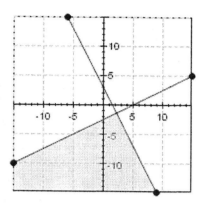

2. $(-4, 4)$

3. $x = 3, y = -3$

4.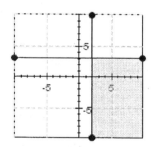

5. $3, 2$

6. $\begin{pmatrix} 36 & -24 & -48 \\ -48 & 32 & 64 \\ -48 & 32 & 64 \end{pmatrix}$

7. $\begin{pmatrix} 28 & 18 \\ 66 & 36 \end{pmatrix}$

8. $(4, 5, 6, 3)$

9. $(-4, 4)$

10. $0, (0, 0)$

11. $x = 5, y = -10, z = 7$

12. $(2, 1)$

13. $(-4, -1)$

14. -13

ANSWER KEY

Gustafson/Frisk - College Algebra 8E Chapter 6 Form B

15. $\begin{pmatrix} 0.307692 & -0.230769 & 0.230769 \\ -0.384615 & 0.538462 & 0.461538 \\ -0.230769 & -0.076923 & 0.076923 \end{pmatrix}$

16. $\dfrac{1}{x} + \dfrac{18}{(x-1)}$

17.

18. -2

19. $(4, 2)$

20. $x = \dfrac{41}{7} + \dfrac{z}{7},\ y = \dfrac{32}{7} - \dfrac{4z}{7},\ z$

21. $\dfrac{1}{(x-5)} - \dfrac{3}{(x+1)}$

22. $6, (6, 0)$

23. $3, 12, \$1.068$

24. $\begin{bmatrix} \dfrac{8}{63} & -\dfrac{1}{63} \\ -\dfrac{1}{63} & \dfrac{8}{63} \end{bmatrix}$

25.

List of Problem Codes for BCA Testing

Gustafson/Frisk - College Algebra 8E Chapter 6 Form B

1. gfca.06.07.4.28_NoAlgs
2. gfca.06.02.4.30_NoAlgs
3. gfca.06.05.4.42_NoAlgs
4. gfca.06.07.4.17_NoAlgs
5. gfca.06.03.4.09_NoAlgs
6. gfca.06.03.4.30_NoAlgs
7. gfca.06.03.4.23_NoAlgs
8. gfca.06.01.4.20_NoAlgs
9. gfca.06.04.4.25_NoAlgs
10. gfca.06.08.4.15_NoAlgs
11. gfca.06.01.4.57_NoAlgs
12. gfca.06.01.4.40_NoAlgs
13. gfca.06.02.4.41_NoAlgs
14. gfca.06.05.4.09_NoAlgs
15. gfca.06.04.4.09_NoAlgs
16. gfca.06.06.4.05m_NoAlgs
17. gfca.06.07.4.05_NoAlgs
18. gfca.06.05.4.15_NoAlgs
19. gfca.06.01.4.14_NoAlgs
20. gfca.06.02.4.63_NoAlgs
21. gfca.06.06.4.09_NoAlgs
22. gfca.06.08.4.10_NoAlgs
23. gfca.06.08.4.21_NoAlgs
24. gfca.06.04.4.06_NoAlgs
25. gfca.06.07.4.22_NoAlgs

1. Solve the system of equations by graphing.

$$\begin{cases} 2x - 9y = -62 \\ 4x + 5y = 60 \end{cases}$$

Select the correct answer.

a. (8, 5)
b. (-3, 14)
c. (-5, -8)
d. (13, -2)
e. (5, 8)

2. Use a graphing calculator to approximate the solutions of the system.

$$\begin{cases} 4.8x + 5.3y = 61.6 \\ -7x + 3y = -4 \end{cases}$$

Select the correct answer.

a. (8, 4)
b. (12, -3)
c. (-4, 13)
d. (4, 8)
e. (-4, -8)

3. Solve the system by the addition method, if possible.

$$\begin{cases} x + \dfrac{y}{5} = \dfrac{31}{5} \\ \dfrac{x+y}{7} = 7 - x \end{cases}$$

Select the correct answer.

a. (-6, -1)
b. (5, 8)
c. (6, 1)
d. (2, 6)
e. (7, 6)

4. Solve the system, if possible.

$$\begin{cases} 8x + 9y + 10z = -70 \\ 3x - 5y + 2z = 17 \\ 9x - 7y - 8z = -53 \end{cases}$$

Select the correct answer.

a. $x = -7, y = -6, z = 4$
b. $x = -5, y = -7, z = -4$
c. $x = -13, y = 0, z = -4$
d. $x = 7, y = 6, z = -4$

5. Solve the system using Gaussian elimination.

$$\begin{cases} 3x - 5y = -7 \\ 2x + y = -22 \end{cases}$$

Select the correct answer.

a. $(9, 4)$
b. $(-9, -4)$
c. $(-9, 4)$
d. $(9, -4)$

6. Solve the system by Gauss-Jordan elimination.

$$\begin{cases} x - 2y = 21 \\ y = -6 \end{cases}$$

Select the correct answer.

a. $(9, -6)$
b. $(-10, -6)$
c. $(10, -6)$
d. $(-9, -6)$

7. Solve the system using Gauss-Jordan elimination.

$$\begin{cases} x + 2y + z = 12 \\ 3x - y - z = 9 \end{cases}$$

Select the correct answer.

a. $\left(\dfrac{30}{7} + \dfrac{2z}{7}, \dfrac{27}{7} + \dfrac{4z}{7}, z \right)$

b. $\left(\dfrac{30}{7} - \dfrac{z}{7}, \dfrac{27}{7} - \dfrac{2z}{7}, z \right)$

c. $\left(\dfrac{30}{7} - \dfrac{z}{7}, \dfrac{27}{7} + \dfrac{4z}{7}, z \right)$

d. $\left(\dfrac{30}{7} + \dfrac{z}{7}, \dfrac{27}{7} - \dfrac{4z}{7}, z \right)$

e. $\left(\dfrac{30}{7} + \dfrac{3z}{7}, \dfrac{27}{7} - \dfrac{z}{7}, z \right)$

8. Find values of x and y, if any, that will make the matrices equal.

$$\begin{bmatrix} x + y & 7 + x \\ -3 & 3y \end{bmatrix} = \begin{bmatrix} 7 & 12 \\ -3 & 6 \end{bmatrix}$$

Select the correct answer.

a.   x = 2, y = 5
b.   x = 5, y = 2
c.   x = -5, y = -2
d.   no solution

9. Decompose the fraction into partial fractions.

$$\frac{19x + 17}{(x + 1)(x - 1)}$$

Select the correct answer.

a. $\dfrac{1}{x - 18} + \dfrac{19}{x + 1}$

b. $\dfrac{1}{x + 1} + \dfrac{18}{x - 1}$

c. $\dfrac{1}{x - 1} + \dfrac{18}{x + 1}$

d. $\dfrac{1}{x + 1} + \dfrac{19}{x - 1}$

10. Decompose the fraction into partial fractions.

$$\frac{3x^2 + x + 27}{x(x^2 + 9)}$$

Select the correct answer.

a. $\dfrac{3}{x} - \dfrac{2}{x^2 + 9}$

b. $\dfrac{9}{x} - \dfrac{1}{x^2 + 3}$

c. $\dfrac{3}{x} + \dfrac{1}{x^2 + 9}$

d. $\dfrac{3}{x} + \dfrac{2}{x - 9}$

11. Two woodworkers, Tom and Carlos, earn a profit of $81 for making a table and $77 for making a chair. On average, Tom must work 3 hours and Carlos 2 hours to make a chair. Tom must work 2 hours and Carlos 6 hours to make a table. If neither wishes to work more than 42 hours per week, how many tables and how many chairs should they make each week to maximize their income? The information is summarized in Illustration.

	Table	Chair	Time available
Profit (dollars)	81	77	
Tom's time (hours)	2	3	42
Carlos' time (hours)	6	2	42

Select the correct answer.

a. 3 tables and 12 chairs: $1,167
b. no tables and 21 chairs: $1,701
c. 12 tables and 3 chairs: $1,203

12. Find the product

$$\begin{bmatrix} 8 & 4 \\ 6 & 6 \end{bmatrix} \begin{bmatrix} 7 & 7 \\ 7 & 5 \end{bmatrix}$$

Select the correct answer.

a. $\begin{bmatrix} 84 & 76 \\ 84 & 72 \end{bmatrix}$

b. $\begin{bmatrix} 56 & 76 \\ 84 & 72 \end{bmatrix}$

c. $\begin{bmatrix} 84 & 76 \\ 42 & 30 \end{bmatrix}$

d. $\begin{bmatrix} 56 & 28 \\ 42 & 30 \end{bmatrix}$

13. Find the product

$$\begin{bmatrix} 1 \\ -9 \\ -9 \end{bmatrix} \begin{bmatrix} 4 & -1 & -9 \end{bmatrix}$$

Select the correct answer.

a. $\begin{bmatrix} 1 & 1 & -9 \\ 9 & 9 & 81 \\ 9 & 1 & 81 \end{bmatrix}$

b. $\begin{bmatrix} 4 & -1 & -9 \\ -36 & 9 & -36 \\ -9 & -1 & 4 \end{bmatrix}$

c. 4

d. $\begin{bmatrix} 4 & -1 & -9 \\ -36 & 9 & 81 \\ -36 & 9 & 81 \end{bmatrix}$

14. Minimize $P = 9y + x$ subject to the given constraints.

$$\begin{cases} x \leq 10 \\ y \geq 0 \\ x + y \geq 5 \\ 10y - x \leq 5 \end{cases}$$

Select the correct answer.

a. $P = 5$ at $(5, 0)$
b. $P = -5$ at $(-5, 0)$
c. $P = -4$ at $(10, -5)$

15. Find the inverse of the matrix.

$$\begin{bmatrix} 8 & 1 \\ 1 & 8 \end{bmatrix}$$

Select the correct answer.

a. $\begin{bmatrix} \dfrac{8}{63} & 0 \\ 0 & \dfrac{8}{63} \end{bmatrix}$

b. $\begin{bmatrix} -\dfrac{8}{63} & -\dfrac{1}{63} \\ -\dfrac{1}{63} & -\dfrac{8}{63} \end{bmatrix}$

c. $\begin{bmatrix} -\dfrac{8}{63} & \dfrac{1}{63} \\ \dfrac{1}{63} & -\dfrac{8}{63} \end{bmatrix}$

d. $\begin{bmatrix} \dfrac{8}{63} & \dfrac{1}{63} \\ \dfrac{1}{63} & \dfrac{8}{63} \end{bmatrix}$

e. $\begin{bmatrix} \dfrac{8}{63} & -\dfrac{1}{63} \\ -\dfrac{1}{63} & \dfrac{8}{63} \end{bmatrix}$

16. Find the inverse of the matrix.

$$\begin{bmatrix} 1 & 0 & 5 \\ 1 & 1 & 5 \\ -5 & 1 & 1 \end{bmatrix}$$

Select the correct answer.

a. $\begin{bmatrix} -\dfrac{4}{26} & \dfrac{5}{26} & -\dfrac{5}{26} \\ -1 & 1 & 0 \\ \dfrac{6}{26} & -\dfrac{1}{26} & \dfrac{1}{26} \end{bmatrix}$

b. $\begin{bmatrix} -29 & 5 & -5 \\ -1 & 1 & 0 \\ 6 & 1 & 1 \end{bmatrix}$

c. $\begin{bmatrix} 29 & 5 & -5 \\ -1 & 1 & 1 \\ 6 & -1 & 1 \end{bmatrix}$

d. $\begin{bmatrix} 4 & \dfrac{5}{26} & -\dfrac{5}{26} \\ 1 & 1 & 0 \\ \dfrac{6}{26} & -\dfrac{1}{26} & \dfrac{1}{26} \end{bmatrix}$

e. $\begin{bmatrix} 29 & 5 & -5 \\ -1 & 1 & -1 \\ 6 & -1 & 1 \end{bmatrix}$

17. $A = \begin{bmatrix} 1 & 2 & 10 \\ 10 & 3 & 3 \\ 3 & 6 & 2 \end{bmatrix}$

Find the cofactor C_{12} of the matrix.

Select the correct answer.

a. $C_{12} = 11$
b. $C_{12} = -11$
c. $C_{12} = -12$
d. $C_{12} = 12$

18. Find the graph of the linear inequality.

    $3x + 2y < 6$

    a.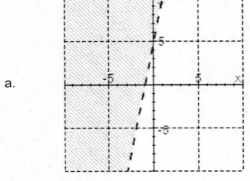

    b.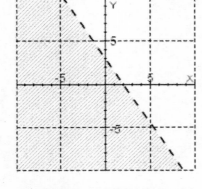

    c.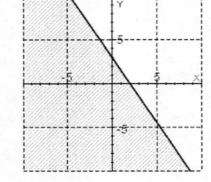

19. Find the solution set of the system.

$$\begin{cases} y < 2 \\ x \geq 4 \end{cases}$$

a.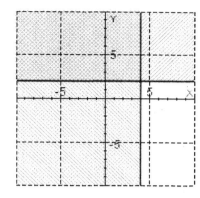

b.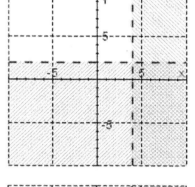

c.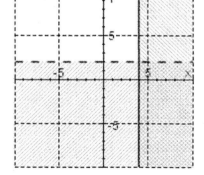

20. Find the solution set of the system.

$$\begin{cases} y < 6x + 3 \\ y < -3x + 6 \end{cases}$$

a.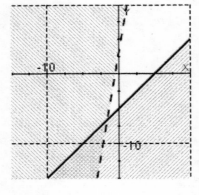

b.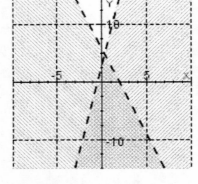

c.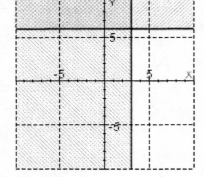

21. Find the solution set of the system.

$$\begin{cases} 4x + 4y \le 4 \\ 4x - 4y \ge 5 \end{cases}$$

a.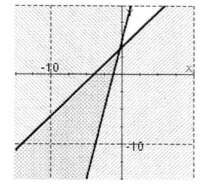

b.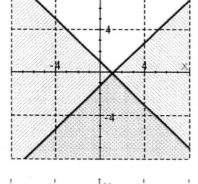

c.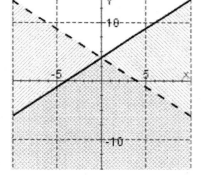

22. Write the system

$$\begin{cases} 6x + 15y = 57 \\ 14x + 8y = 52 \end{cases}$$

as a matrix equation $AX = B$, where A is the coefficient matrix of the system, X is a column matrix of variables, and B is the column matrix of constants. Multiply each side of the equation on the left by the matrix inverse of A to obtain an equivalent system of equations and solve it.

Select the correct answer.

a. (2, 3)
b. (3, 5)
c. (0, 3)
d. (2, 5)
e. (2, 0)

23. Evaluate the determinant.

$$\begin{vmatrix} 2 & -10 \\ -1 & 2 \end{vmatrix}$$

Select the correct answer.

a. $D = 14$
b. $D = -14$
c. $D = -6$
d. $D = 6$

24. Use Cramer's rule to find the solution of the system, if possible

$$\begin{cases} 6x - y = 7 \\ x + y = 0 \end{cases}$$

Select the correct answer.

a. $x = 6, y = -6$
b. $x = 1, y = -1$
c. $x = 6, y = -7$
d. $x = 7, y = -6$
e. The system is inconsistent.
f. The equations are dependent.

25. Maximize $P = 7x - 6y$ subject to the given constraints.

$$\begin{cases} x \leq 3 \\ x \geq -3 \\ y - x \leq 3 \\ x - y \leq 3 \end{cases}$$

Select the correct answer.

a. $P = 28$ at $(4, 0)$
b. $P = 21$ at $(3, 0)$
c. $P = 39$ at $(3, 6)$

ANSWER KEY

Gustafson/Frisk - College Algebra 8E Chapter 6 Form C

1. e
2. d
3. c
4. a
5. b
6. a
7. d
8. b
9. b
10. c
11. a
12. a
13. d
14. a
15. e
16. a
17. b
18. b
19. c
20. b
21. b
22. a
23. c
24. b
25. b

List of Problem Codes for BCA Testing

Gustafson/Frisk - College Algebra 8E Chapter 6 Form C

1. gfca.06.01.4.14m_NoAlgs
2. gfca.06.01.4.20m_NoAlgs
3. gfca.06.01.4.40m_NoAlgs
4. gfca.06.01.4.57m_NoAlgs
5. gfca.06.02.4.30m_NoAlgs
6. gfca.06.02.4.41m_NoAlgs
7. gfca.06.02.4.63m_NoAlgs
8. gfca.06.03.4.09m_NoAlgs
9. gfca.06.06.4.07_NoAlgs
10. gfca.06.06.4.17_NoAlgs
11. gfca.06.08.4.21m_NoAlgs
12. gfca.06.03.4.23m_NoAlgs
13. gfca.06.03.4.30m_NoAlgs
14. gfca.06.08.4.16m_NoAlgs
15. gfca.06.04.4.05m_NoAlgs
16. gfca.06.04.4.09m_NoAlgs
17. gfca.06.05.4.15m_NoAlgs
18. gfca.06.07.4.05m_NoAlgs
19. gfca.06.07.4.17m_NoAlgs
20. gfca.06.07.4.22m_NoAlgs
21. gfca.06.07.4.28m_NoAlgs
22. gfca.06.04.4.25m_NoAlgs
23. gfca.06.05.4.09m_NoAlgs
24. gfca.06.05.4.42m_NoAlgs
25. gfca.06.08.4.11m_NoAlgs

1. Find the graph of the linear inequality.

   $3x + 2y < 6$

   a.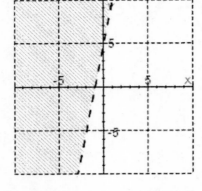

   b.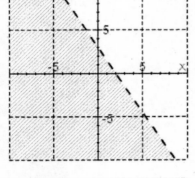

   c.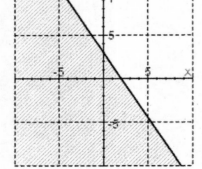

2. Find the solution set of the system.

$$\begin{cases} y < 2 \\ x \geq 4 \end{cases}$$

a.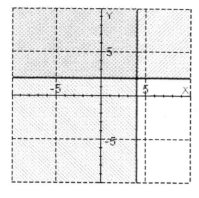

b.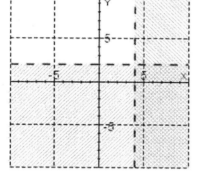

c.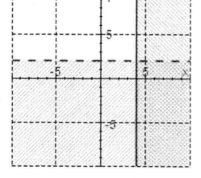

3. Decompose the fraction into partial fractions.

$$\frac{3x^2 + x + 27}{x(x^2 + 9)}$$

Select the correct answer.

a. $\quad \dfrac{3}{x} - \dfrac{2}{x^2 + 9}$

b. $\quad \dfrac{9}{x} - \dfrac{1}{x^2 + 3}$

c. $\quad \dfrac{3}{x} + \dfrac{1}{x^2 + 9}$

d. $\quad \dfrac{3}{x} + \dfrac{2}{x - 9}$

4. Find values of x and y, if any, that will make the matrices equal.

$$\begin{bmatrix} x+y & 7+x \\ -3 & 3y \end{bmatrix} = \begin{bmatrix} 7 & 12 \\ -3 & 6 \end{bmatrix}$$

Select the correct answer.

a. $x = 5, y = 2$
b. $x = 2, y = 5$
c. $x = -5, y = -2$
d. no solution

5. Minimize $P = 9y + x$ subject to the given constraints.

$$\begin{cases} x \leq 10 \\ y \geq 0 \\ x + y \geq 5 \\ 10y - x \leq 5 \end{cases}$$

Select the correct answer.

a. $P = 5$ at $(5, 0)$
b. $P = -5$ at $(-5, 0)$
c. $P = -4$ at $(10, -5)$

6. Solve the system using Gauss-Jordan elimination.

$$\begin{cases} x + 2y + z = 12 \\ 3x - y - z = 9 \end{cases}$$

Select the correct answer.

a. $\left(\dfrac{30}{7} + \dfrac{2z}{7}, \dfrac{27}{7} + \dfrac{4z}{7}, z \right)$

b. $\left(\dfrac{30}{7} - \dfrac{z}{7}, \dfrac{27}{7} - \dfrac{2z}{7}, z \right)$

c. $\left(\dfrac{30}{7} - \dfrac{z}{7}, \dfrac{27}{7} + \dfrac{4z}{7}, z \right)$

d. $\left(\dfrac{30}{7} + \dfrac{z}{7}, \dfrac{27}{7} - \dfrac{4z}{7}, z \right)$

e. $\left(\dfrac{30}{7} + \dfrac{3z}{7}, \dfrac{27}{7} - \dfrac{z}{7}, z \right)$

7. Find the solution set of the system.

$$\begin{cases} y < 6x + 3 \\ y < -3x + 6 \end{cases}$$

a.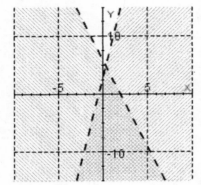

b.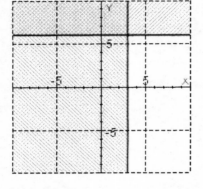

c.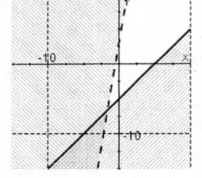

8. Find the inverse of the matrix.

$$\begin{bmatrix} 1 & 0 & 5 \\ 1 & 1 & 5 \\ -5 & 1 & 1 \end{bmatrix}$$

Select the correct answer.

a. $\begin{bmatrix} 29 & 5 & -5 \\ -1 & 1 & 1 \\ 6 & -1 & 1 \end{bmatrix}$

b. $\begin{bmatrix} 4 & \frac{5}{26} & -\frac{5}{26} \\ 1 & 1 & 0 \\ \frac{6}{26} & -\frac{1}{26} & \frac{1}{26} \end{bmatrix}$

c. $\begin{bmatrix} 29 & 5 & -5 \\ -1 & 1 & -1 \\ 6 & -1 & 1 \end{bmatrix}$

d. $\begin{bmatrix} -29 & 5 & -5 \\ -1 & 1 & 0 \\ 6 & 1 & 1 \end{bmatrix}$

e. $\begin{bmatrix} -\frac{4}{26} & \frac{5}{26} & -\frac{5}{26} \\ -1 & 1 & 0 \\ \frac{6}{26} & -\frac{1}{26} & \frac{1}{26} \end{bmatrix}$

9. Maximize $P = 7x - 6y$ subject to the given constraints.

$$\begin{cases} x \le 3 \\ x \ge -3 \\ y - x \le 3 \\ x - y \le 3 \end{cases}$$

Select the correct answer.

a. $P = 39$ at $(3, 6)$
b. $P = 21$ at $(3, 0)$
c. $P = 28$ at $(4, 0)$

10. $A = \begin{bmatrix} 1 & 2 & 10 \\ 10 & 3 & 3 \\ 3 & 6 & 2 \end{bmatrix}$

Find the cofactor C_{12} of the matrix.

Select the correct answer.

a. $C_{12} = 11$
b. $C_{12} = -11$
c. $C_{12} = -12$
d. $C_{12} = 12$

11. Solve the system, if possible.

$$\begin{cases} 8x + 9y + 10z = -70 \\ 3x - 5y + 2z = 17 \\ 9x - 7y - 8z = -53 \end{cases}$$

Select the correct answer.

a. $x = 7, y = 6, z = -4$
b. $x = -7, y = -6, z = 4$
c. $x = -5, y = -7, z = -4$
d. $x = -13, y = 0, z = -4$

12. Evaluate the determinant.

$$\begin{vmatrix} 2 & -10 \\ -1 & 2 \end{vmatrix}$$

Select the correct answer.

a. $D = -14$
b. $D = 14$
c. $D = 6$
d. $D = -6$

13. Two woodworkers, Tom and Carlos, earn a profit of $81 for making a table and $77 for making a chair. On average, Tom must work 3 hours and Carlos 2 hours to make a chair. Tom must work 2 hours and Carlos 6 hours to make a table. If neither wishes to work more than 42 hours per week, how many tables and how many chairs should they make each week to maximize their income? The information is summarized in Illustration.

	Table	Chair	Time available
Profit (dollars)	81	77	
Tom's time (hours)	2	3	42
Carlos' time (hours)	6	2	42

Select the correct answer.
a. 3 tables and 12 chairs: $1,167
b. 12 tables and 3 chairs: $1,203
c. no tables and 21 chairs: $1,701

14. Find the product.

$$\begin{bmatrix} 1 \\ -9 \\ -9 \end{bmatrix} \begin{bmatrix} 4 & -1 & -9 \end{bmatrix}$$

Select the correct answer.

a. 4

b. $\begin{bmatrix} 4 & -1 & -9 \\ -36 & 9 & -36 \\ -9 & -1 & 4 \end{bmatrix}$

c. $\begin{bmatrix} 1 & 1 & -9 \\ 9 & 9 & 81 \\ 9 & 1 & 81 \end{bmatrix}$

d. $\begin{bmatrix} 4 & -1 & -9 \\ -36 & 9 & 81 \\ -36 & 9 & 81 \end{bmatrix}$

15. Find the inverse of the matrix.

$$\begin{bmatrix} 8 & 1 \\ 1 & 8 \end{bmatrix}$$

Select the correct answer.

a. $\begin{bmatrix} \dfrac{8}{63} & -\dfrac{1}{63} \\ -\dfrac{1}{63} & \dfrac{8}{63} \end{bmatrix}$

b. $\begin{bmatrix} \dfrac{8}{63} & \dfrac{1}{63} \\ \dfrac{1}{63} & \dfrac{8}{63} \end{bmatrix}$

c. $\begin{bmatrix} -\dfrac{8}{63} & \dfrac{1}{63} \\ \dfrac{1}{63} & -\dfrac{8}{63} \end{bmatrix}$

d. $\begin{bmatrix} \dfrac{8}{63} & 0 \\ 0 & \dfrac{8}{63} \end{bmatrix}$

16. Find the product

$$\begin{bmatrix} 8 & 4 \\ 6 & 6 \end{bmatrix} \begin{bmatrix} 7 & 7 \\ 7 & 5 \end{bmatrix}$$

Select the correct answer.

a. $\begin{bmatrix} 84 & 76 \\ 84 & 72 \end{bmatrix}$

b. $\begin{bmatrix} 56 & 76 \\ 84 & 72 \end{bmatrix}$

c. $\begin{bmatrix} 84 & 76 \\ 42 & 30 \end{bmatrix}$

d. $\begin{bmatrix} 56 & 28 \\ 42 & 30 \end{bmatrix}$

17. Use Cramer's rule to find the solution of the system, if possible.

$$\begin{cases} 6x - y = 7 \\ x + y = 0 \end{cases}$$

Select the correct answer.

a. $x = 6, y = -7$
b. $x = 7, y = -6$
c. $x = 1, y = -1$
d. $x = 6, y = -6$
e. The system is inconsistent.
f. The equations are dependent.

18. Decompose the fraction into partial fractions.

$$\frac{19x + 17}{(x + 1)(x - 1)}$$

Select the correct answer.

a. $\dfrac{1}{x - 18} + \dfrac{19}{x + 1}$

b. $\dfrac{1}{x + 1} + \dfrac{18}{x - 1}$

c. $\dfrac{1}{x - 1} + \dfrac{18}{x + 1}$

d. $\dfrac{1}{x + 1} + \dfrac{19}{x - 1}$

19. Solve the system by Gauss-Jordan elimination.

$$\begin{cases} x - 2y = 21 \\ y = -6 \end{cases}$$

Select the correct answer.

a. $(-10, -6)$
b. $(9, -6)$
c. $(-9, -6)$
d. $(10, -6)$

20. Solve the system by the addition method, if possible.

$$\begin{cases} x + \dfrac{y}{5} = \dfrac{31}{5} \\ \dfrac{x+y}{7} = 7 - x \end{cases}$$

Select the correct answer.

a. (2, 6)
b. (6, 1)
c. (-6, -1)
d. (5, 8)
e. (7, 6)

21. Solve the system of equations by graphing.

$$\begin{cases} 2x - 9y = -62 \\ 4x + 5y = 60 \end{cases}$$

Select the correct answer.

a. (8, 5)
b. (-3, 14)
c. (-5, -8)
d. (13, -2)
e. (5, 8)

22. Write the system

$$\begin{cases} 6x + 15y = 57 \\ 14x + 8y = 52 \end{cases}$$

as a matrix equation $AX = B$, where A is the coefficient matrix of the system, X is a column matrix of variables, and B is the column matrix of constants. Multiply each side of the equation on the left by the matrix inverse of A to obtain an equivalent system of equations and solve it.

Select the correct answer.

a. (0, 3)
b. (2, 3)
c. (2, 0)
d. (3, 5)
e. (2, 5)

23. Use a graphing calculator to approximate the solutions of the system.

$$\begin{cases} 4.8x + 5.3y = 61.6 \\ -7x + 3y = -4 \end{cases}$$

Select the correct answer.

a. (4, 8)
b. (-4, -8)
c. (8, 4)
d. (-4, 13)
e. (12, -3)

24. Solve the system using Gaussian elimination.

$$\begin{cases} 3x - 5y = -7 \\ 2x + y = -22 \end{cases}$$

Select the correct answer.

a. (9, 4)
b. (-9, -4)
c. (-9, 4)
d. (9, -4)

25. Find the solution set of the system.

$$\begin{cases} 4x + 4y \le 4 \\ 4x - 4y \ge 5 \end{cases}$$

a.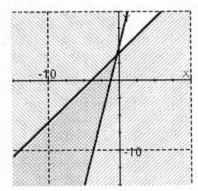

b.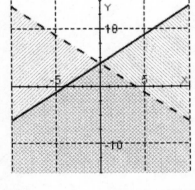

c.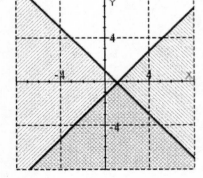

ANSWER KEY

Gustafson/Frisk - College Algebra 8E Chapter 6 Form D

1. b
2. c
3. c
4. a
5. a
6. d
7. a
8. e
9. b
10. b
11. b
12. d
13. a
14. d
15. a
16. a
17. c
18. b
19. b
20. b
21. e
22. b
23. a
24. b
25. c

List of Problem Codes for BCA Testing

Gustafson/Frisk - College Algebra 8E Chapter 6 Form D

1. gfca.06.07.4.05m_NoAlgs
2. gfca.06.07.4.17m_NoAlgs
3. gfca.06.06.4.17_NoAlgs
4. gfca.06.03.4.09m_NoAlgs
5. gfca.06.08.4.16m_NoAlgs
6. gfca.06.02.4.63m_NoAlgs
7. gfca.06.07.4.22m_NoAlgs
8. gfca.06.04.4.09m_NoAlgs
9. gfca.06.08.4.11m_NoAlgs
10. gfca.06.05.4.15m_NoAlgs
11. gfca.06.01.4.57m_NoAlgs
12. gfca.06.05.4.09m_NoAlgs
13. gfca.06.03.4.30m_NoAlgs
14. gfca.06.08.4.21m_NoAlgs
15. gfca.06.04.4.05m_NoAlgs
16. gfca.06.03.4.23m_NoAlgs
17. gfca.06.05.4.42m_NoAlgs
18. gfca.06.06.4.07_NoAlgs
19. gfca.06.02.4.41m_NoAlgs
20. gfca.06.01.4.40m_NoAlgs
21. gfca.06.01.4.14m_NoAlgs
22. gfca.06.04.4.25m_NoAlgs
23. gfca.06.01.4.20m_NoAlgs
24. gfca.06.02.4.30m_NoAlgs.
25. gfca.06.07.4.28m_NoAlg

1. Solve the system of equations by graphing.

$$\begin{cases} 8x - 7y = 18 \\ 10x + 3y = 46 \end{cases}$$

2. Use a graphing calculator to approximate the solutions of the system.

$$\begin{cases} 4.8x + 5.3y = 61.6 \\ -7x + 3y = -4 \end{cases}$$

Select the correct answer.

a. (-4, -8)
b. (12, -3)
c. (4, 8)
d. (8, 4)
e. (-4, 13)

3. Solve the system by the addition method, if possible.

$$\begin{cases} x + \dfrac{y}{5} = \dfrac{31}{5} \\ \dfrac{x+y}{7} = 7 - x \end{cases}$$

Select the correct answer.

a. (-6, -1)
b. (5, 8)
c. (6, 1)
d. (2, 6)
e. (7, 6)

4. Solve the system, if possible.

$$\begin{cases} 4x + 7y + 2z = -36 \\ 7x - 3y + 10z = 135 \\ 5x - 5y - 8z = 19 \end{cases}$$

5. Solve the system using Gaussian elimination.

$$\begin{cases} 3x - 5y = -7 \\ 2x + y = -22 \end{cases}$$

Select the correct answer.

a. $(9, 4)$
b. $(-9, 4)$
c. $(-9, -4)$
d. $(9, -4)$

6. Solve the system by Gauss-Jordan elimination.

$$\begin{cases} x - 2y = -2 \\ y = -1 \end{cases}$$

7. Solve the system using Gauss-Jordan elimination. If the system has infinitely many solutions, show a general solution (in terms of z).

$$\begin{cases} x + 2y + z = 15 \\ 3x - y - z = 13 \end{cases}$$

8. Find values of x and y, if any, that will make the matrices equal.

$$\begin{bmatrix} x+y & 7+x \\ -3 & 3y \end{bmatrix} = \begin{bmatrix} 7 & 12 \\ -3 & 6 \end{bmatrix}$$

Select the correct answer.

a. $x = 5, y = 2$
b. $x = -5, y = -2$
c. $x = 2, y = 5$
d. no solution

9. Find the product.

$$\begin{bmatrix} 2 & 2 \\ 6 & 3 \end{bmatrix} \begin{bmatrix} 8 & 3 \\ 6 & 6 \end{bmatrix}$$

10. Find the inverse of the matrix.

$$\begin{bmatrix} 1 & 0 & 5 \\ 1 & 1 & 5 \\ -5 & 1 & 1 \end{bmatrix}$$

Select the correct answer.

a. $\begin{bmatrix} -\dfrac{4}{26} & \dfrac{5}{26} & -\dfrac{5}{26} \\ -1 & 1 & 0 \\ \dfrac{6}{26} & -\dfrac{1}{26} & \dfrac{1}{26} \end{bmatrix}$

b. $\begin{bmatrix} 29 & 5 & -5 \\ -1 & 1 & 1 \\ 6 & -1 & 1 \end{bmatrix}$

c. $\begin{bmatrix} 29 & 5 & -5 \\ -1 & 1 & -1 \\ 6 & -1 & 1 \end{bmatrix}$

d. $\begin{bmatrix} -29 & 5 & -5 \\ -1 & 1 & 0 \\ 6 & 1 & 1 \end{bmatrix}$

e. $\begin{bmatrix} 4 & \dfrac{5}{26} & -\dfrac{5}{26} \\ 1 & 1 & 0 \\ \dfrac{6}{26} & -\dfrac{1}{26} & \dfrac{1}{26} \end{bmatrix}$

11. Write the system

$$\begin{cases} 8x + 13y = 20 \\ 6x + 15y = 36 \end{cases}$$

as a matrix equation $AX = B$, where A is the coefficient matrix of the system, X is a column matrix of variables, and B is the column matrix of constants. Multiply each side of the equation on the left by the matrix inverse of A to obtain an equivalent system of equations and solve it.

12. Evaluate the determinant.

$$\begin{vmatrix} 2 & -10 \\ -1 & 2 \end{vmatrix}$$

Select the correct answer.

a. $D = -14$
b. $D = -6$
c. $D = 14$
d. $D = 6$

13.
$$A = \begin{bmatrix} 9 & 2 & 10 \\ 8 & 4 & 6 \\ 1 & 10 & 1 \end{bmatrix}$$

Find the cofactor C_{12} of the matrix.

14. Use Cramer's rule to find the solution of the system, if possible.

$$\begin{cases} 6x - y = 7 \\ x + y = 0 \end{cases}$$

Select the correct answer.

a. $x = 7, y = -6$
b. $x = 1, y = -1$
c. $x = 6, y = -6$
d. $x = 6, y = -7$
e. The system is inconsistent.
f. The equations are dependent.

15. Decompose the fraction into partial fractions.

$$\frac{19x - 1}{x(x - 1)}$$

16. Decompose the fraction into partial fractions.

$$\frac{3x^2 + x + 27}{x(x^2 + 9)}$$

Select the correct answer.

a. $\dfrac{3}{x} - \dfrac{2}{x^2 + 9}$

b. $\dfrac{9}{x} - \dfrac{1}{x^2 + 3}$

c. $\dfrac{3}{x} + \dfrac{1}{x^2 + 9}$

d. $\dfrac{3}{x} + \dfrac{2}{x - 9}$

17. Graph the linear inequality.

$2x + 3y < 12$

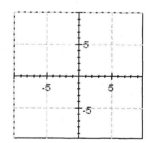

18. Find the solution set of the system.

$$\begin{cases} y < 2 \\ x \geq 4 \end{cases}$$

a.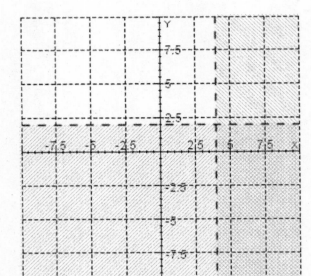

b.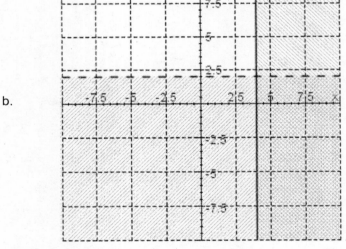

19. Graph the solution set of the system.

$$\begin{cases} y < 3x + 2 \\ y < -2x + 3 \end{cases}$$

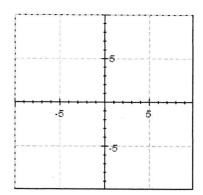

20. Find the solution set of the system.

$$\begin{cases} 4x + 4y \leq 4 \\ 4x - 4y \geq 5 \end{cases}$$

a.

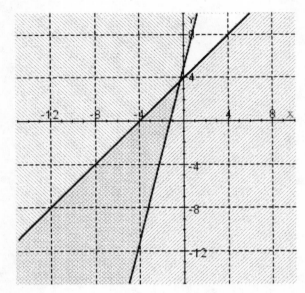

b.

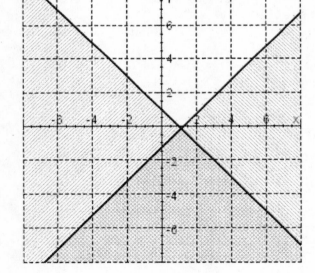

21. Maximize $P = x - 6y$ subject to the given constraints. Find the maximum value of P and the point at which that value occurs.

$$\begin{cases} x \leq 6 \\ y \leq 9 \\ x \geq 0 \\ y \geq 0 \\ x + y \leq 15 \end{cases}$$

22. Minimize $P = 9y + x$ subject to the given constraints.

$$\begin{cases} x \leq 10 \\ y \geq 0 \\ x + y \geq 5 \\ 10y - x \leq 5 \end{cases}$$

Select the correct answer.

a. $P = -5$ at $(-5, 0)$
b. $P = -4$ at $(10, -5)$
c. $P = 5$ at $(5, 0)$

23. Two woodworkers, Tom and Carlos, earn a profit of $81 for making a table and $77 for making a chair. On average, Tom must work 3 hours and Carlos 2 hours to make a chair. Tom must work 2 hours and Carlos 6 hours to make a table. If neither wishes to work more than 42 hours per week, how many tables and how many chairs should they make each week to maximize their income? The information is summarized in Illustration.

	Table	Chair	Time available
Profit (dollars)	81	77	
Tom's time (hours)	2	3	42
Carlos' time (hours)	6	2	42

Select the correct answer.

a.   3 tables and 12 chairs: $1,167
b.   no tables and 21 chairs: $1,701
c.   12 tables and 3 chairs: $1,203

24. Find the product

$$\begin{bmatrix} 1 \\ -9 \\ -9 \end{bmatrix} \begin{bmatrix} 4 & -1 & -9 \end{bmatrix}$$

Select the correct answer.

a. 4

b. $\begin{bmatrix} 4 & -1 & -9 \\ -36 & 9 & 81 \\ -36 & 9 & 81 \end{bmatrix}$

c. $\begin{bmatrix} 1 & 1 & -9 \\ 9 & 9 & 81 \\ 9 & 1 & 81 \end{bmatrix}$

d. $\begin{bmatrix} 4 & -1 & -9 \\ -36 & 9 & -36 \\ -9 & -1 & 4 \end{bmatrix}$

25. Find the inverse of the matrix.

$$\begin{bmatrix} -2 & -2 \\ -5 & -2 \end{bmatrix}$$

ANSWER KEY

Gustafson/Frisk - College Algebra 8E Chapter 6 Form E

1. $(4, 2)$

2. c

3. c

4. $x = 5, y = -10, z = 7$

5. c

6. $(-4, -1)$

7. $x = \frac{41}{7} + \frac{z}{7}, y = \frac{32}{7} - \frac{4z}{7}, z$

8. a

9. $\begin{pmatrix} 28 & 18 \\ 66 & 36 \end{pmatrix}$

10. a

11. $(-4, 4)$

12. b

13. -2

14. b

15. $\frac{1}{x} + \frac{13}{(x-1)}$

16. c

17.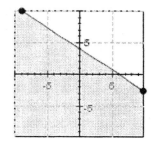

18. b

ANSWER KEY

Gustafson/Frisk - College Algebra 8E Chapter 6 Form E

19.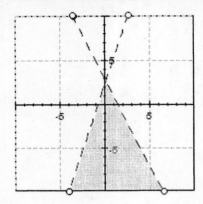

20. b

21. 6, (6, 0)

22. c

23. a

24. d

25. $\begin{pmatrix} 0.333333 & -0.333333 \\ -0.833333 & 0.333333 \end{pmatrix}$

List of Problem Codes for BCA Testing

Gustafson/Frisk - College Algebra 8E Chapter 6 Form E

1. gfca.06.01.4.14_NoAlgs
2. gfca.06.01.4.20m_NoAlgs
3. gfca.06.01.4.40m_NoAlgs
4. gfca.06.01.4.57_NoAlgs
5. gfca.06.02.4.30m_NoAlgs
6. gfca.06.02.4.41_NoAlgs
7. gfca.06.02.4.63_NoAlgs
8. gfca.06.03.4.09m_NoAlgs
9. gfca.06.03.4.23_NoAlgs
10. gfca.06.04.4.09m_NoAlgs
11. gfca.06.04.4.25_NoAlgs
12. gfca.06.05.4.09m_NoAlgs
13. gfca.06.05.4.15_NoAlgs
14. gfca.06.05.4.42m_NoAlgs
15. gfca.06.06.4.05_NoAlgs
16. gfca.06.06.4.17m_NoAlgs
17. gfca.06.07.4.05_NoAlgs
18. gfca.06.07.4.17m_NoAlgs
19. gfca.06.07.4.22_NoAlgs
20. gfca.06.07.4.28m_NoAlgs
21. gfca.06.08.4.10_NoAlgs
22. gfca.06.08.4.16m_NoAlgs
23. gfca.06.08.4.21m_NoAlgs
24. gfca.06.03.4.30m_NoAlgs
25. gfca.06.04.4.07_NoAlgs

1. Graph the linear inequality.

   $2x + 3y < 12$

   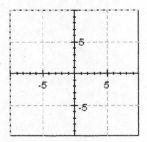

2. Find the product

   $$\begin{bmatrix} 2 & 2 \\ 6 & 3 \end{bmatrix} \begin{bmatrix} 8 & 3 \\ 6 & 6 \end{bmatrix}$$

3. Evaluate the determinant.

   $$\begin{vmatrix} 2 & -10 \\ -1 & 2 \end{vmatrix}$$

   Select the correct answer.

   a. $D = -14$
   b. $D = -6$
   c. $D = 14$
   d. $D = 6$

4. Solve the system by Gauss-Jordan elimination.

   $$\begin{cases} x - 2y = -2 \\ y = -1 \end{cases}$$

5. Find the inverse of the matrix.

$$\begin{bmatrix} 1 & 0 & 5 \\ 1 & 1 & 5 \\ -5 & 1 & 1 \end{bmatrix}$$

Select the correct answer.

a. $\begin{bmatrix} -\dfrac{4}{26} & \dfrac{5}{26} & -\dfrac{5}{26} \\ -1 & 1 & 0 \\ \dfrac{6}{26} & -\dfrac{1}{26} & \dfrac{1}{26} \end{bmatrix}$

b. $\begin{bmatrix} 29 & 5 & -5 \\ -1 & 1 & 1 \\ 6 & -1 & 1 \end{bmatrix}$

c. $\begin{bmatrix} 29 & 5 & -5 \\ -1 & 1 & -1 \\ 6 & -1 & 1 \end{bmatrix}$

d. $\begin{bmatrix} -29 & 5 & -5 \\ -1 & 1 & 0 \\ 6 & 1 & 1 \end{bmatrix}$

e. $\begin{bmatrix} 4 & \dfrac{5}{26} & -\dfrac{5}{26} \\ 1 & 1 & 0 \\ \dfrac{6}{26} & -\dfrac{1}{26} & \dfrac{1}{26} \end{bmatrix}$

6. Solve the system using Gauss-Jordan elimination. If the system has infinitely many solutions, show a general solution (in terms of z).

$$\begin{cases} x + 2y + z = 15 \\ 3x - y - z = 13 \end{cases}$$

7. Maximize $P = x - 6y$ subject to the given constraints. Find the maximum value of P and the point at which that value occurs.

$$\begin{cases} x \leq 6 \\ y \leq 9 \\ x \geq 0 \\ y \geq 0 \\ x + y \leq 15 \end{cases}$$

8. $A = \begin{bmatrix} 9 & 2 & 10 \\ 8 & 4 & 6 \\ 1 & 10 & 1 \end{bmatrix}$

Find the cofactor C_{12} of the matrix.

9. Decompose the fraction into partial fractions.

$$\frac{3x^2 + x + 27}{x(x^2 + 9)}$$

Select the correct answer.

a. $\dfrac{3}{x} - \dfrac{2}{x^2 + 9}$

b. $\dfrac{9}{x} - \dfrac{1}{x^2 + 3}$

c. $\dfrac{3}{x} + \dfrac{1}{x^2 + 9}$

d. $\dfrac{3}{x} + \dfrac{2}{x - 9}$

10. Find values of x and y, if any, that will make the matrices equal.

$$\begin{bmatrix} x+y & 7+x \\ -3 & 3y \end{bmatrix} = \begin{bmatrix} 7 & 12 \\ -3 & 6 \end{bmatrix}$$

Select the correct answer.

a.   x = 5, y = 2
b.   x = -5, y = -2
c.   x = 2, y = 5
d.   no solution

11. Solve the system, if possible.

$$\begin{cases} 4x + 7y + 2z = -36 \\ 7x - 3y + 10z = 135 \\ 5x - 5y - 8z = 19 \end{cases}$$

12. Solve the system of equations by graphing.

$$\begin{cases} 8x - 7y = 18 \\ 10x + 3y = 46 \end{cases}$$

13. Two woodworkers, Tom and Carlos, earn a profit of $81 for making a table and $77 for making a chair. On average, Tom must work 3 hours and Carlos 2 hours to make a chair. Tom must work 2 hours and Carlos 6 hours to make a table. If neither wishes to work more than 42 hours per week, how many tables and how many chairs should they make each week to maximize their income? The information is summarized in Illustration.

	Table	Chair	Time available
Profit (dollars)	81	77	
Tom's time (hours)	2	3	42
Carlos' time (hours)	6	2	42

Select the correct answer.

a.   3 tables and 12 chairs: $1,167
b.   no tables and 21 chairs: $1,701
c.   12 tables and 3 chairs: $1,203

14. Solve the system using Gaussian elimination.

$$\begin{cases} 3x - 5y = -7 \\ 2x + y = -22 \end{cases}$$

Select the correct answer.

a. (9, 4)
b. (-9, 4)
c. (-9, -4)
d. (9, -4)

15. Graph the solution set of the system.

$$\begin{cases} y < 3x + 2 \\ y < -2x + 3 \end{cases}$$

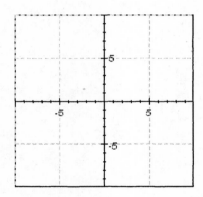

16. Minimize $P = 9y + x$ subject to the given constraints.

$$\begin{cases} x \leq 10 \\ y \geq 0 \\ x + y \geq 5 \\ 10y - x \leq 5 \end{cases}$$

Select the correct answer.

a. $P = -5$ at $(-5, 0)$
b. $P = -4$ at $(10, -5)$
c. $P = 5$ at $(5, 0)$

17. Find the solution set of the system.

$$\begin{cases} y < 2 \\ x \geq 4 \end{cases}$$

a.

b.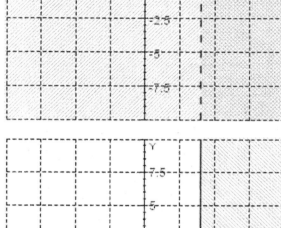

18. Find the solution set of the system.

$$\begin{cases} 4x + 4y \leq 4 \\ 4x - 4y \geq 5 \end{cases}$$

a.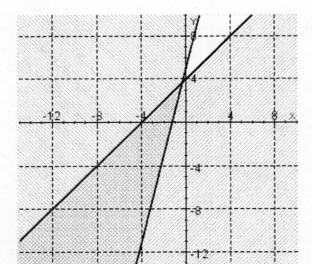

b.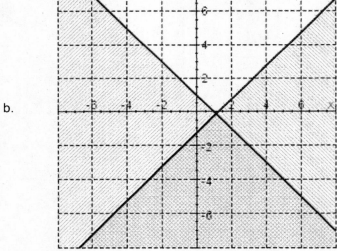

19. Solve the system by the addition method, if possible.

$$\begin{cases} x + \dfrac{y}{5} = \dfrac{31}{5} \\ \dfrac{x+y}{7} = 7 - x \end{cases}$$

Select the correct answer.

a. (-6, -1)
b. (5, 8)
c. (6, 1)
d. (2, 6)
e. (7, 6)

20. Write the system:

$$\begin{cases} 8x + 13y = 20 \\ 6x + 15y = 36 \end{cases}$$

as a matrix equation $AX = B$, where A is the coefficient matrix of the system, X is a column matrix of variables, and B is the column matrix of constants. Multiply each side of the equation on the left by the matrix inverse of A to obtain an equivalent system of equations and solve it.

21. Use a graphing calculator to approximate the solutions of the system.

$$\begin{cases} 4.8x + 5.3y = 61.6 \\ -7x + 3y = -4 \end{cases}$$

Select the correct answer.

a. (-4, -8)
b. (8, 4)
c. (12, -3)
d. (-4, 13)
e. (4, 8)

22. Decompose the fraction into partial fractions.

$$\frac{19x - 1}{x(x - 1)}$$

23. Use Cramer's rule to find the solution of the system, if possible.

$$\begin{cases} 6x - y = 7 \\ x + y = 0 \end{cases}$$

Select the correct answer.

a. $x = 7, y = -6$
b. $x = 1, y = -1$
c. $x = 6, y = -6$
d. $x = 6, y = -7$
e. The system is inconsistent.
f. The equations are dependent.

24. Find the product.

$$\begin{bmatrix} 1 \\ -9 \\ -9 \end{bmatrix} \begin{bmatrix} 4 & -1 & -9 \end{bmatrix}$$

Select the correct answer.

a. 4

b. $\begin{bmatrix} 4 & -1 & -9 \\ -36 & 9 & 81 \\ -36 & 9 & 81 \end{bmatrix}$

c. $\begin{bmatrix} 1 & 1 & -9 \\ 9 & 9 & 81 \\ 9 & 1 & 81 \end{bmatrix}$

d. $\begin{bmatrix} 4 & -1 & -9 \\ -36 & 9 & -36 \\ -9 & -1 & 4 \end{bmatrix}$

25. Find the inverse of the matrix.

$$\begin{bmatrix} -2 & -2 \\ -5 & -2 \end{bmatrix}$$

ANSWER KEY

Gustafson/Frisk - College Algebra 8E Chapter 6 Form F

1.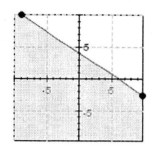

2. $\begin{pmatrix} 28 & 18 \\ 66 & 36 \end{pmatrix}$

3. b

4. $(-4, -1)$

5. a

6. $x = \frac{41}{7} + \frac{z}{7}, y = \frac{32}{7} - \frac{4z}{7}, z$

7. $6, (6, 0)$

8. -2

9. c

10. a

11. $x = 5, y = -10, z = 7$

12. $(4, 2)$

13. a

14. c

ANSWER KEY

Gustafson/Frisk - College Algebra 8E Chapter 6 Form F

15.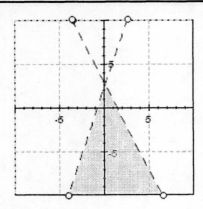

16. c

17. b

18. b

19. c

20. $(-4, 4)$

21. e

22. $\dfrac{1}{x} + \dfrac{18}{(x-1)}$

23. b

24. b

25. $\begin{pmatrix} 0.333333 & -0.333333 \\ -0.833333 & 0.333333 \end{pmatrix}$

List of Problem Codes for BCA Testing

Gustafson/Frisk - College Algebra 8E Chapter 6 Form F

1. gfca.06.07.4.05_NoAlgs
2. gfca.06.03.4.23_NoAlgs
3. gfca.06.05.4.09m_NoAlgs
4. gfca.06.02.4.41_NoAlgs
5. gfca.06.04.4.09m_NoAlgs
6. gfca.06.02.4.63_NoAlgs
7. gfca.06.08.4.10_NoAlgs
8. gfca.06.05.4.15_NoAlgs
9. gfca.06.06.4.17m_NoAlgs
10. gfca.06.03.4.09m_NoAlgs
11. gfca.06.01.4.57_NoAlgs
12. gfca.06.01.4.14_NoAlgs
13. gfca.06.08.4.21m_NoAlgs
14. gfca.06.02.4.30m_NoAlgs
15. gfca.06.07.4.22_NoAlgs
16. gfca.06.08.4.16m_NoAlgs
17. gfca.06.07.4.17m_NoAlgs
18. gfca.06.07.4.28m_NoAlgs
19. gfca.06.01.4.40m_NoAlgs
20. gfca.06.04.4.25_NoAlgs
21. gfca.06.01.4.20m_NoAlgs
22. gfca.06.06.4.05_NoAlgs
23. gfca.06.05.4.42m_NoAlgs
24. gfca.06.03.4.30m_NoAlgs
25. gfca.06.04.4.07_NoAlgs

Gustafson/Frisk - College Algebra 8E Chapter 6 Form G

1. Solve the system of equations by graphing.

$$\begin{cases} 2x - 9y = -62 \\ 4x + 5y = 60 \end{cases}$$

Select the correct answer.

a. (-3, 14)
b. (-5, -8)
c. (8, 5)
d. (5, 8)
e. (13, -2)

2. Solve the system by the addition method, if possible.

$$\begin{cases} x + \dfrac{y}{5} = \dfrac{31}{5} \\ \dfrac{x+y}{7} = 7 - x \end{cases}$$

Select the correct answer.

a. (6, 1)
b. (7, 6)
c. (5, 8)
d. (2, 6)
e. (-6, -1)

3. Solve the system, if possible.

$$\begin{cases} 4x + 7y + 2z = -36 \\ 7x - 3y + 10z = 135 \\ 5x - 5y - 8z = 19 \end{cases}$$

4. Solve the system by Gauss-Jordan elimination.

$$\begin{cases} x - 2y = -2 \\ y = -1 \end{cases}$$

5. Solve the system using Gaussian elimination.

$$\begin{cases} 3x - 5y = -7 \\ 2x + y = -22 \end{cases}$$

Select the correct answer.

a. $(-9, -4)$
b. $(9, -4)$
c. $(-9, 4)$
d. $(9, 4)$

6. Solve the system using Gauss-Jordan elimination.

$$\begin{cases} x + 2y + z = 12 \\ 3x - y - z = 9 \end{cases}$$

Select the correct answer.

a. $\left(\dfrac{30}{7} - \dfrac{z}{7}, \dfrac{27}{7} - \dfrac{2z}{7}, z\right)$

b. $\left(\dfrac{30}{7} + \dfrac{2z}{7}, \dfrac{27}{7} + \dfrac{4z}{7}, z\right)$

c. $\left(\dfrac{30}{7} + \dfrac{z}{7}, \dfrac{27}{7} - \dfrac{4z}{7}, z\right)$

d. $\left(\dfrac{30}{7} - \dfrac{z}{7}, \dfrac{27}{7} + \dfrac{4z}{7}, z\right)$

e. $\left(\dfrac{30}{7} + \dfrac{3z}{7}, \dfrac{27}{7} - \dfrac{z}{7}, z\right)$

7. Find values of x and y, if any, that will make the matrices equal.

$$\begin{bmatrix} x+y & 7+x \\ -3 & 3y \end{bmatrix} = \begin{bmatrix} 7 & 12 \\ -3 & 6 \end{bmatrix}$$

Select the correct answer.

a. $x = 5, y = 2$
b. $x = -5, y = -2$
c. $x = 2, y = 5$
d. no solution

8. Find the product.

$$\begin{bmatrix} 8 & 4 \\ 6 & 6 \end{bmatrix} \begin{bmatrix} 7 & 7 \\ 7 & 5 \end{bmatrix}$$

Select the correct answer.

a. $\begin{bmatrix} 56 & 76 \\ 84 & 72 \end{bmatrix}$

b. $\begin{bmatrix} 84 & 76 \\ 84 & 72 \end{bmatrix}$

c. $\begin{bmatrix} 84 & 76 \\ 42 & 30 \end{bmatrix}$

d. $\begin{bmatrix} 56 & 28 \\ 42 & 30 \end{bmatrix}$

9. Find the product.

$$\begin{bmatrix} 6 \\ -8 \\ -8 \end{bmatrix} \begin{bmatrix} 6 & -4 & -8 \end{bmatrix}$$

10. Minimize $P = 9y + x$ subject to the given constraints.

$$\begin{cases} x \leq 10 \\ y \geq 0 \\ x + y \geq 5 \\ 10y - x \leq 5 \end{cases}$$

Select the correct answer.

a. $P = 5$ at $(5, 0)$
b. $P = -4$ at $(10, -5)$
c. $P = -5$ at $(-5, 0)$

11. Find the inverse of the matrix.

$$\begin{bmatrix} 8 & 1 \\ 1 & 8 \end{bmatrix}$$

Select the correct answer.

a. $\begin{bmatrix} -\dfrac{8}{63} & -\dfrac{1}{63} \\ -\dfrac{1}{63} & -\dfrac{8}{63} \end{bmatrix}$

b. $\begin{bmatrix} \dfrac{8}{63} & 0 \\ 0 & \dfrac{8}{63} \end{bmatrix}$

c. $\begin{bmatrix} -\dfrac{8}{63} & \dfrac{1}{63} \\ \dfrac{1}{63} & -\dfrac{8}{63} \end{bmatrix}$

d. $\begin{bmatrix} \dfrac{8}{63} & \dfrac{1}{63} \\ \dfrac{1}{63} & \dfrac{8}{63} \end{bmatrix}$

e. $\begin{bmatrix} \dfrac{8}{63} & -\dfrac{1}{63} \\ -\dfrac{1}{63} & \dfrac{8}{63} \end{bmatrix}$

12. Two woodworkers, Tom and Carlos, earn a profit of $92 for making a table and $66 for making a chair. On average, Tom must work 3 hours and Carlos 2 hours to make a chair. Tom must work 2 hours and Carlos 6 hours to make a table. If neither wishes to work more than 42 hours per week, how many tables and how many chairs should they make each week to maximize their income? The information is summarized in the table below.

	Table	Chair	Time available
Profit (dollars)	92	66	
Tom's time (hours)	2	3	42
Carlos' time (hours)	6	2	42

13. Find the inverse of the matrix.

$$\begin{bmatrix} 1 & 0 & 5 \\ 1 & 1 & 5 \\ -5 & 1 & 1 \end{bmatrix}$$

Select the correct answer.

a. $\begin{bmatrix} -29 & 5 & -5 \\ -1 & 1 & 0 \\ 6 & 1 & 1 \end{bmatrix}$

b. $\begin{bmatrix} 4 & \frac{5}{26} & -\frac{5}{26} \\ 1 & 1 & 0 \\ \frac{6}{26} & -\frac{1}{26} & \frac{1}{26} \end{bmatrix}$

c. $\begin{bmatrix} 29 & 5 & -5 \\ -1 & 1 & -1 \\ 6 & -1 & 1 \end{bmatrix}$

d. $\begin{bmatrix} -\frac{4}{26} & \frac{5}{26} & -\frac{5}{26} \\ -1 & 1 & 0 \\ \frac{6}{26} & -\frac{1}{26} & \frac{1}{26} \end{bmatrix}$

e. $\begin{bmatrix} 29 & 5 & -5 \\ -1 & 1 & 1 \\ 6 & -1 & 1 \end{bmatrix}$

14. Write the system:

$$\begin{cases} 8x + 13y = 20 \\ 6x + 15y = 36 \end{cases}$$

as a matrix equation $AX = B$, where A is the coefficient matrix of the system, X is a column matrix of variables, and B is the column matrix of constants. Multiply each side of the equation on the left by the matrix inverse of A to obtain an equivalent system of equations and solve it.

15. Evaluate the determinant.

$$\begin{vmatrix} 2 & -10 \\ -1 & 2 \end{vmatrix}$$

Select the correct answer.

a. $D = 6$
b. $D = 14$
c. $D = -14$
d. $D = -6$

16. $A = \begin{bmatrix} 1 & 2 & 10 \\ 10 & 3 & 3 \\ 3 & 6 & 2 \end{bmatrix}$

Find the cofactor C_{12} of the matrix.

Select the correct answer.

a. $C_{12} = -11$
b. $C_{12} = 12$
c. $C_{12} = -12$
d. $C_{12} = 11$

17. Use Cramer's rule to find the solution of the system, if possible.

$$\begin{cases} 9x - y = 30 \\ x + y = 0 \end{cases}$$

18. Decompose the fraction into partial fractions.

$$\frac{2x^2 + 10x + 28}{(x + 1)(x^2 + 4x + 13)}$$

19. Minimize $P = 8x + y$ subject to the given constraints.

$$\begin{cases} x \geq 0 \\ y \geq 0 \\ 3y - x \leq 1 \\ y - 3x \geq -3 \end{cases}$$

20. Decompose the fraction into partial fractions.

$$\frac{19x + 17}{(x + 1)(x - 1)}$$

Select the correct answer.

a. $\dfrac{1}{x - 18} + \dfrac{19}{x + 1}$

b. $\dfrac{1}{x + 1} + \dfrac{18}{x - 1}$

c. $\dfrac{1}{x - 1} + \dfrac{18}{x + 1}$

d. $\dfrac{1}{x + 1} + \dfrac{19}{x - 1}$

21. Maximize $P = x - 6y$ subject to the given constraints. Find the maximum value of P and the point at which that value occurs.

$$\begin{cases} x \leq 6 \\ y \leq 9 \\ x \geq 0 \\ y \geq 0 \\ x + y \leq 15 \end{cases}$$

22. Find the graph of the linear inequality.

    $3x + 2y < 6$

a.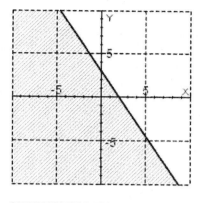

b.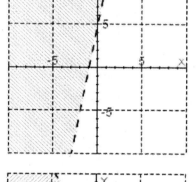

c.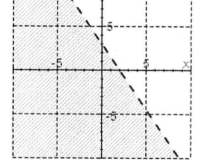

23. Find the solution set of the system.

$$\begin{cases} y < 2 \\ x \geq 4 \end{cases}$$

a.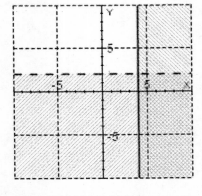

b.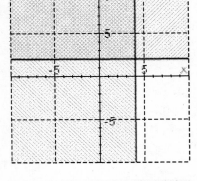

c.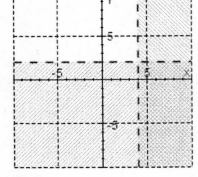

24. Find the solution set of the system.

$$\begin{cases} 4x + 4y \le 4 \\ 4x - 4y \ge 5 \end{cases}$$

a.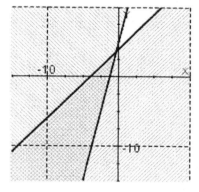

b.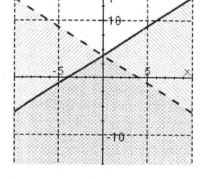

c.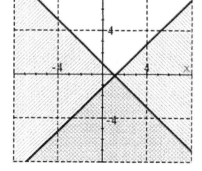

25. Maximize $P = 7x - 6y$ subject to the given constraints.

$$\begin{cases} x \leq 3 \\ x \geq -3 \\ y - x \leq 3 \\ x - y \leq 3 \end{cases}$$

Select the correct answer.

a. $P = 28$ at $(4, 0)$
b. $P = 39$ at $(3, 6)$
c. $P = 21$ at $(3, 0)$

ANSWER KEY

Gustafson/Frisk - College Algebra 8E Chapter 6 Form G

1. d
2. a
3. $x = 5, y = -10, z = 7$
4. $(-4, -1)$
5. a
6. c
7. a
8. b
9. $\begin{pmatrix} 36 & -24 & -48 \\ -48 & 32 & 64 \\ -48 & 32 & 64 \end{pmatrix}$
10. a
11. e
12. 3, 12, $1,068
13. d
14. $(-4, 4)$
15. d
16. a
17. $x = 3, y = -3$
18. $\dfrac{2}{(x+1)} + \dfrac{2}{(x^2+4x+13)}$
19. 0, (0, 0)
20. b
21. 6, (6, 0)
22. c
23. a
24. c
25. c

List of Problem Codes for BCA Testing

Gustafson/Frisk - College Algebra 8E Chapter 6 Form G

1. gfca.06.01.4.14m_NoAlgs
2. gfca.06.01.4.40m_NoAlgs
3. gfca.06.01.4.57_NoAlgs
4. gfca.06.02.4.41_NoAlgs
5. gfca.06.02.4.30m_NoAlgs
6. gfca.06.02.4.63m_NoAlgs
7. gfca.06.03.4.09m_NoAlgs
8. gfca.06.03.4.23m_NoAlgs
9. gfca.06.03.4.30_NoAlgs
10. gfca.06.08.4.16m_NoAlgs
11. gfca.06.04.4.05m_NoAlgs
12. gfca.06.08.4.21_NoAlgs
13. gfca.06.04.4.09m_NoAlgs
14. gfca.06.04.4.25_NoAlgs
15. gfca.06.05.4.09m_NoAlgs
16. gfca.06.05.4.15m_NoAlgs
17. gfca.06.05.4.42_NoAlgs
18. gfca.06.06.4.19_NoAlgs
19. gfca.06.08.4.15_NoAlgs
20. gfca.06.06.4.07m_NoAlgs
21. gfca.06.08.4.10_NoAlgs
22. gfca.06.07.4.05m_NoAlgs
23. gfca.06.07.4.17m_NoAlgs
24. gfca.06.07.4.28m_NoAlgs
25. gfca.06.08.4.11m_NoAlgs

1. Maximize $P = 7x - 6y$ subject to the given constraints.

$$\begin{cases} x \leq 3 \\ x \geq -3 \\ y - x \leq 3 \\ x - y \leq 3 \end{cases}$$

   Select the correct answer.

   a. $P = 39$ at $(3, 6)$
   b. $P = 21$ at $(3, 0)$
   c. $P = 28$ at $(4, 0)$

2. Find the product

$$\begin{bmatrix} 8 & 4 \\ 6 & 6 \end{bmatrix} \begin{bmatrix} 7 & 7 \\ 7 & 5 \end{bmatrix}$$

   Select the correct answer.

   a. $\begin{bmatrix} 56 & 76 \\ 84 & 72 \end{bmatrix}$

   b. $\begin{bmatrix} 84 & 76 \\ 84 & 72 \end{bmatrix}$

   c. $\begin{bmatrix} 84 & 76 \\ 42 & 30 \end{bmatrix}$

   d. $\begin{bmatrix} 56 & 28 \\ 42 & 30 \end{bmatrix}$

3. Use Cramer's rule to find the solution of the system, if possible.

$$\begin{cases} 9x - y = 30 \\ x + y = 0 \end{cases}$$

4. Find the graph of the linear inequality.

$3x + 2y < 6$

a.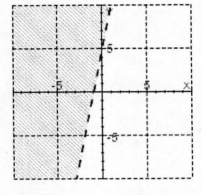

b.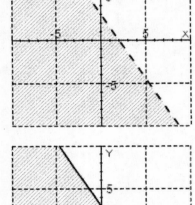

c.

5. Find the inverse of the matrix.

$$\begin{bmatrix} 1 & 0 & 5 \\ 1 & 1 & 5 \\ -5 & 1 & 1 \end{bmatrix}$$

Select the correct answer.

a. $\begin{bmatrix} 4 & \dfrac{5}{26} & -\dfrac{5}{26} \\ 1 & 1 & 0 \\ \dfrac{6}{26} & -\dfrac{1}{26} & \dfrac{1}{26} \end{bmatrix}$

b. $\begin{bmatrix} -29 & 5 & -5 \\ -1 & 1 & 0 \\ 6 & 1 & 1 \end{bmatrix}$

c. $\begin{bmatrix} 29 & 5 & -5 \\ -1 & 1 & -1 \\ 6 & -1 & 1 \end{bmatrix}$

d. $\begin{bmatrix} 29 & 5 & -5 \\ -1 & 1 & 1 \\ 6 & -1 & 1 \end{bmatrix}$

e. $\begin{bmatrix} -\dfrac{4}{26} & \dfrac{5}{26} & -\dfrac{5}{26} \\ -1 & 1 & 0 \\ \dfrac{6}{26} & -\dfrac{1}{26} & \dfrac{1}{26} \end{bmatrix}$

6. Evaluate the determinant.

$$\begin{vmatrix} 2 & -10 \\ -1 & 2 \end{vmatrix}$$

Select the correct answer.

a. $D = -6$
b. $D = -14$
c. $D = 6$
d. $D = 14$

7. Solve the system using Gauss-Jordan elimination.

$$\begin{cases} x + 2y + z = 12 \\ 3x - y - z = 9 \end{cases}$$

Select the correct answer.

a. $\left(\dfrac{30}{7} - \dfrac{z}{7}, \dfrac{27}{7} - \dfrac{2z}{7}, z\right)$

b. $\left(\dfrac{30}{7} + \dfrac{2z}{7}, \dfrac{27}{7} + \dfrac{4z}{7}, z\right)$

c. $\left(\dfrac{30}{7} + \dfrac{z}{7}, \dfrac{27}{7} - \dfrac{4z}{7}, z\right)$

d. $\left(\dfrac{30}{7} - \dfrac{z}{7}, \dfrac{27}{7} + \dfrac{4z}{7}, z\right)$

e. $\left(\dfrac{30}{7} + \dfrac{3z}{7}, \dfrac{27}{7} - \dfrac{z}{7}, z\right)$

8. Find values of x and y, if any, that will make the matrices equal.

$$\begin{bmatrix} x+y & 7+x \\ -3 & 3y \end{bmatrix} = \begin{bmatrix} 7 & 12 \\ -3 & 6 \end{bmatrix}$$

Select the correct answer.

a. $x = 5, y = 2$
b. $x = -5, y = -2$
c. $x = 2, y = 5$
d. no solution

9. Write the system

$$\begin{cases} 8x + 13y = 20 \\ 6x + 15y = 36 \end{cases}$$

as a matrix equation $AX = B$, where A is the coefficient matrix of the system, X is a column matrix of variables, and B is the column matrix of constants. Multiply each side of the equation on the left by the matrix inverse of A to obtain an equivalent system of equations and solve it.

10. Maximize P = x - 6y subject to the given constraints. Find the maximum value of P and the point at which that value occurs.

$$\begin{cases} x \leq 6 \\ y \leq 9 \\ x \geq 0 \\ y \geq 0 \\ x + y \leq 15 \end{cases}$$

11. Two woodworkers, Tom and Carlos, earn a profit of $92 for making a table and $66 for making a chair. On average, Tom must work 3 hours and Carlos 2 hours to make a chair. Tom must work 2 hours and Carlos 6 hours to make a table. If neither wishes to work more than 42 hours per week, how many tables and how many chairs should they make each week to maximize their income? The information is summarized in the table below.

	Table	Chair	Time available
Profit (dollars)	92	66	
Tom's time (hours)	2	3	42
Carlos' time (hours)	6	2	42

12. Minimize P = 9y + x subject to the given constraints.

$$\begin{cases} x \leq 10 \\ y \geq 0 \\ x + y \geq 5 \\ 10y - x \leq 5 \end{cases}$$

Select the correct answer.

a. P = -4 at (10, -5)
b. P = 5 at (5, 0)
c. P = -5 at (-5, 0)

13. Solve the system by Gauss-Jordan elimination.

$$\begin{cases} x - 2y = -2 \\ y = -1 \end{cases}$$

14. Decompose the fraction into partial fractions.

$$\frac{19x + 17}{(x + 1)(x - 1)}$$

Select the correct answer.

a. $\dfrac{1}{x - 18} + \dfrac{19}{x + 1}$

b. $\dfrac{1}{x + 1} + \dfrac{18}{x - 1}$

c. $\dfrac{1}{x - 1} + \dfrac{18}{x + 1}$

d. $\dfrac{1}{x + 1} + \dfrac{19}{x - 1}$

15. Solve the system using Gaussian elimination.

$$\begin{cases} 3x - 5y = -7 \\ 2x + y = -22 \end{cases}$$

Select the correct answer.

a. $(-9, 4)$
b. $(9, 4)$
c. $(-9, -4)$
d. $(9, -4)$

16. Solve the system by the addition method, if possible.

    $$\begin{cases} x + \dfrac{y}{5} = \dfrac{31}{5} \\ \dfrac{x+y}{7} = 7 - x \end{cases}$$

    Select the correct answer.

    a. (2, 6)
    b. (-6, -1)
    c. (7, 6)
    d. (6, 1)
    e. (5, 8)

17. Solve the system of equations by graphing.

    $$\begin{cases} 2x - 9y = -62 \\ 4x + 5y = 60 \end{cases}$$

    Select the correct answer.

    a. (13, -2)
    b. (-3, 14)
    c. (-5, -8)
    d. (5, 8)
    e. (8, 5)

18. $$A = \begin{bmatrix} 1 & 2 & 10 \\ 10 & 3 & 3 \\ 3 & 6 & 2 \end{bmatrix}$$

    Find the cofactor C_{12} of the matrix.

    Select the correct answer.

    a. $C_{12} = 12$
    b. $C_{12} = -11$
    c. $C_{12} = -12$
    d. $C_{12} = 11$

19. Minimize $P = 8x + y$ subject to the given constraints.

$$\begin{cases} x \geq 0 \\ y \geq 0 \\ 3y - x \leq 1 \\ y - 3x \geq -3 \end{cases}$$

20. Solve the system, if possible.

$$\begin{cases} 4x + 7y + 2z = -36 \\ 7x - 3y + 10z = 135 \\ 5x - 5y - 8z = 19 \end{cases}$$

21. Find the product.

$$\begin{bmatrix} 6 \\ -8 \\ -8 \end{bmatrix} \begin{bmatrix} 6 & -4 & -8 \end{bmatrix}$$

22. Decompose the fraction into partial fractions.

$$\frac{2x^2 + 10x + 28}{(x+1)(x^2 + 4x + 13)}$$

23. Find the solution set of the system.

$$\begin{cases} 4x + 4y \le 4 \\ 4x - 4y \ge 5 \end{cases}$$

a.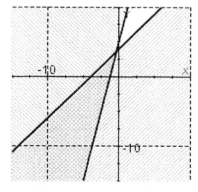

b.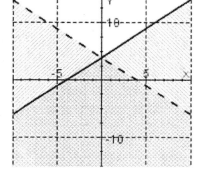

c.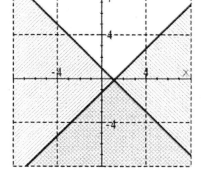

24. Find the inverse of the matrix.

$$\begin{bmatrix} 8 & 1 \\ 1 & 8 \end{bmatrix}$$

Select the correct answer.

a. $\begin{bmatrix} \dfrac{8}{63} & -\dfrac{1}{63} \\ -\dfrac{1}{63} & \dfrac{8}{63} \end{bmatrix}$

b. $\begin{bmatrix} -\dfrac{8}{63} & -\dfrac{1}{63} \\ -\dfrac{1}{63} & -\dfrac{8}{63} \end{bmatrix}$

c. $\begin{bmatrix} -\dfrac{8}{63} & \dfrac{1}{63} \\ \dfrac{1}{63} & -\dfrac{8}{63} \end{bmatrix}$

d. $\begin{bmatrix} \dfrac{8}{63} & \dfrac{1}{63} \\ \dfrac{1}{63} & \dfrac{8}{63} \end{bmatrix}$

e. $\begin{bmatrix} \dfrac{8}{63} & 0 \\ 0 & \dfrac{8}{63} \end{bmatrix}$

25. Find the solution set of the system.

$$\begin{cases} y < 2 \\ x \geq 4 \end{cases}$$

a.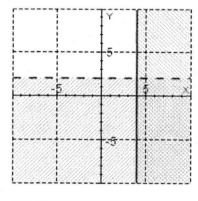

b.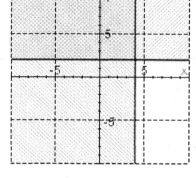

c.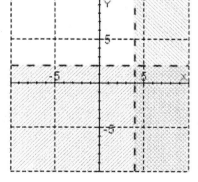

ANSWER KEY

Gustafson/Frisk - College Algebra 8E Chapter 6 Form H

1. b
2. b
3. $x = 3, y = -3$
4. b
5. e
6. a
7. c
8. a
9. $(-4, 4)$
10. $6, (6, 0)$
11. $3, 12, \$1,068$
12. b
13. $(-4, -1)$
14. b
15. c
16. d
17. d
18. b
19. $0, (0, 0)$
20. $x = 5, y = -10, z = 7$
21. $\begin{pmatrix} 36 & -24 & -48 \\ -48 & 32 & 64 \\ -48 & 32 & 64 \end{pmatrix}$
22. $\dfrac{2}{(x+1)} + \dfrac{2}{(x^2+4x+13)}$
23. c
24. a
25. a

List of Problem Codes on BCA

Gustafson/Frisk - College Algebra 8E Chapter 6 Form H

1. gfca.06.08.4.11m_NoAlgs
2. gfca.06.03.4.23m_NoAlgs
3. gfca.06.05.4.42_NoAlgs
4. gfca.06.07.4.05m_NoAlgs
5. gfca.06.04.4.09m_NoAlgs
6. gfca.06.05.4.09m_NoAlgs
7. gfca.06.02.4.63m_NoAlgs
8. gfca.06.03.4.09m_NoAlgs
9. gfca.06.04.4.25_NoAlgs
10. gfca.06.08.4.10_NoAlgs
11. gfca.06.08.4.21_NoAlgs
12. gfca.06.08.4.16m_NoAlgs
13. gfca.06.02.4.41_NoAlgs
14. gfca.06.06.4.07m_NoAlgs
15. gfca.06.02.4.30m_NoAlgs
16. gfca.06.01.4.40m_NoAlgs
17. gfca.06.01.4.14m_NoAlgs
18. gfca.06.05.4.15m_NoAlgs
19. gfca.06.08.4.15_NoAlgs
20. gfca.06.01.4.57_NoAlgs
21. gfca.06.03.4.30_NoAlgs
22. gfca.06.06.4.19_NoAlgs
23. gfca.06.07.4.28m_NoAlgs
24. gfca.06.04.4.05m_NoAlgs
25. gfca.06.07.4.17m_NoAlgs

Gustafson/Frisk - College Algebra 8E Chapter 7 Form A

1. Give the circle's radius.

   $x^2 + y^2 - 9 = 0$

2. Write the equation of the circle graphed below. Express the equation in standard form.

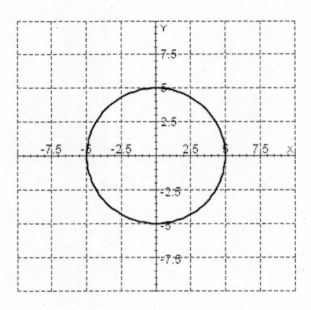

3. Write the equation of the circle with a radius of 7 and its center at the intersection of $3x + y = 3$ and $-2x - 4y = 8$. Express the equation in standard form.

4. Solve the system of equations algebraically for real values of x and y.

   $$\begin{cases} x^2 + y^2 = 85 \\ y = x^2 - 79 \end{cases}$$

5. Solve the system of equations algebraically for real values of x and y.

   $$\begin{cases} 2x^2 + y^2 = 187 \\ x^2 - 4y^2 = -19 \end{cases}$$

6. Solve the system of equations algebraically for real values of x and y.

$$\begin{cases} x^2 - y^2 = -32 \\ 4x^2 + 3x^2 = 124 \end{cases}$$

7. Graph the following equation.

$$x^2 - 2x + y^2 = 16$$

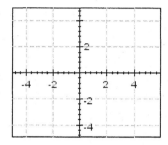

8. Find the equation of the parabola with vertex at (5, 3) and focus at (-2, 3). Express the equation in standard form.

9. Change the equation to standard form.

$$x^2 - 7y + 11 = -4x$$

10. Find the equation, in the form $(x - h)^2 + (y - k)^2 = r^2$, of the circle passing through the given points.

    (0, 9), (5, 4), and (4, 7)

11. Write the equation of the ellipse that has its center at the origin, focus at (4, 0) and vertex at (5, 0). Express the equation in standard form.

12. Write the equation of the ellipse that has its center at the origin with focus at (0, 5) and major axis equal to 20. Express the equation in standard form.

13. Write the equation of the ellipse with center at (1, 5), a = 4, b = 3 and major axis parallel to the y-axis. Express the equation in standard form.

14. Write the equation of the ellipse with center at (5, 5), a = 4, b = 5 and major axis parallel to the x-axis. Express the equation in standard form.

15. Solve the system of equations algebraically for real values of x and y.

$$\begin{cases} x + y = 12 \\ x^2 + y^2 = 72 \end{cases}$$

16. Graph the ellipse.

$$\frac{x^2}{16} + \frac{(y+2)^2}{36} = 1$$

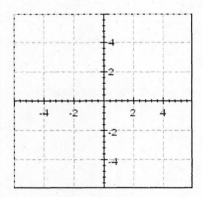

17. Graph the ellipse.

    $x^2 + 4y^2 - 4x + 8y + 4 = 0$

    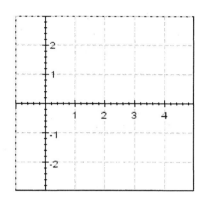

18. Write the equation of the hyperbola with center (1, 2), $a = 3$ and $b = 5$. Transverse axis is horizontal.

19. Write the equation of the hyperbola with center (2, 10), vertex (2, 12) and passing through $(26, 10 + 2\sqrt{10})$.

20. Write the equation of the hyperbola with center (2, -1), $a^2 = 9$ and $b^2 = 16$.

21. Find the area of the fundamental rectangle of the hyperbola.

    $4(x - 5)^2 - 36(y - 1)^2 = 144$

22. Graph the hyperbola.

    $\dfrac{y^2}{4} - \dfrac{x^2}{9} = 1$

    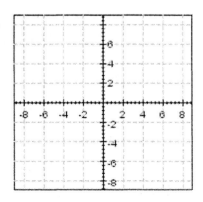

23. Graph the hyperbola.

$$9(y+2)^2 - 4(x-1)^2 = 36$$

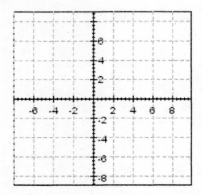

24. Solve the system of equations algebraically for real values of *x* and *y*.

$$\begin{cases} \dfrac{1}{x} + \dfrac{6}{y} = 1 \\ \dfrac{2}{x} - \dfrac{1}{y} = \dfrac{1}{7} \end{cases}$$

25. Solve the system of equations by graphing.

$$\begin{cases} x^2 - 12x - y = -32 \\ x^2 - 12x + y = -32 \end{cases}$$

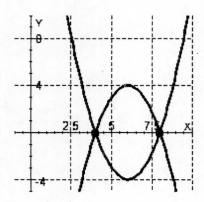

ANSWER KEY

Gustafson/Frisk - College Algebra 8E Chapter 7 Form A

1. 3

2. $y^2 + x^2 = 25$

3. $(x-2)^2 + (y+3)^2 = 49$

4. $(9, 2), (-9, 2)$

5. $(9, 5), (-9, 5), (9, -5), (-9, -5)$

6. $(2, 6), (-2, 6), (2, -6), (-2, -6)$

7.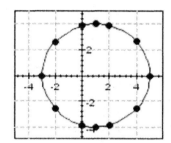

8. $(y-3)^2 = -28(x-5)$

9. $(x+2)^2 = 7(y-1)$

10. $x^2 + (y-4)^2 = 25$

11. $\dfrac{x^2}{25} + \dfrac{y^2}{9} = 1$

12. $\dfrac{x^2}{75} + \dfrac{y^2}{100} = 1$

13. $\dfrac{(x-1)^2}{9} + \dfrac{(y-5)^2}{16} = 1$

ANSWER KEY

Gustafson/Frisk - College Algebra 8E Chapter 7 Form A

14. $\dfrac{(x-5)^2}{16} + \dfrac{(y-5)^2}{25} = 1$

15. $(6, 6)$

16.
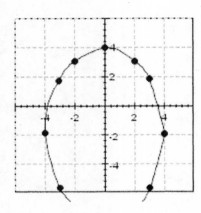

17.
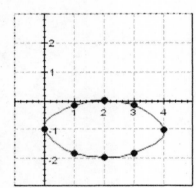

18. $\dfrac{(x-1)^2}{9} - \dfrac{(y-2)^2}{25} = 1$

19. $\dfrac{(y-10)^2}{4} - \dfrac{(x-2)^2}{64} = 1$

20. $\dfrac{(x-2)^2}{9} - \dfrac{(y+1)^2}{16} = 1$

21. 48

ANSWER KEY

Gustafson/Frisk - College Algebra 8E Chapter 7 Form A

22.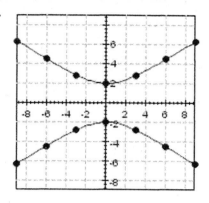

23.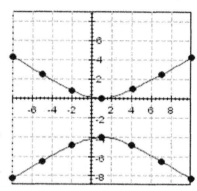

24. $(7, 7)$

25. $(4, 0), (8, 0)$

List of Problem Codes for BCA Testing

Gustafson/Frisk - College Algebra 8E Chapter 7 Form A

1. gfca.07.01.4.02_NoAlgs
2. gfca.07.01.4.11_NoAlgs
3. gfca.07.01.4.15_NoAlgs
4. gfca.07.04.4.23_NoAlgs
5. gfca.07.04.4.27_NoAlgs
6. gfca.07.04.4.35_NoAlgs
7. gfca.07.01.4.18_NoAlgs
8. gfca.07.01.4.26_NoAlgs
9. gfca.07.01.4.43_NoAlgs
10. gfca.07.01.4.65_NoAlgs
11. gfca.07.02.4.09_NoAlgs
12. gfca.07.02.4.13_NoAlgs
13. gfca.07.02.4.15_NoAlgs
14. gfca.07.02.4.17_NoAlgs
15. gfca.07.04.4.19_NoAlgs
16. gfca.07.02.4.27_NoAlgs
17. gfca.07.02.4.29_NoAlgs
18. gfca.07.03.4.09_NoAlgs
19. gfca.07.03.4.11_NoAlgs
20. gfca.07.03.4.15_NoAlgs
21. gfca.07.03.4.19_NoAlgs
22. gfca.07.03.4.28_NoAlgs
23. gfca.07.03.4.32_NoAlgs
24. gfca.07.04.4.37_NoAlgs
25. gfca.07.04.4.11_NoAlgs

1. Give the circle's radius.

   $x^2 + y^2 - 9 = 0$

2. Write the equation of the ellipse with center at (5, 5), a = 4, b = 5 and major axis parallel to the x-axis. Express the equation in standard form.

3. Find the equation of the parabola with vertex at (5, 3) and focus at (- 2, 3). Express the equation in standard form.

4. Write the equation of the ellipse that has its center at the origin, focus at (4, 0) and vertex at (5, 0). Express the equation in standard form.

5. Write the equation of the ellipse that has its center at the origin with focus at (0, 5) and major axis equal to 20. Express the equation in standard form.

6. Write the equation of the circle with a radius of 7 and its center at the intersection of 3x + y = 3 and - 2x - 4y = 8. Express the equation in standard form.

7. Write the equation of the circle graphed below. Express the equation in standard form.

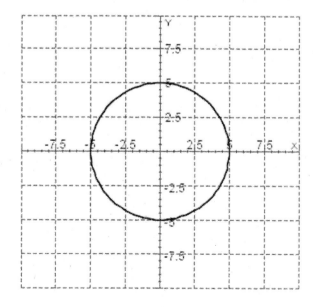

8. Graph the hyperbola.

$$\frac{y^2}{4} - \frac{x^2}{9} = 1$$

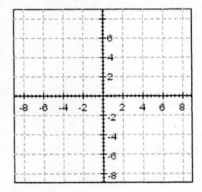

9. Write the equation of the hyperbola with center (1, 2), *a* = 3 and *b* = 5. Transverse axis is horizontal.

10. Write the equation of the ellipse with center at (1, 5), a = 4, b = 3 and major axis parallel to the y-axis. Express the equation in standard form.

11. Solve the system of equations algebraically for real values of *x* and *y*.

$$\begin{cases} 2x^2 + y^2 = 187 \\ x^2 - 4y^2 = -19 \end{cases}$$

12. Solve the system of equations by graphing.

$$\begin{cases} x^2 - 12x - y = -32 \\ x^2 - 12x + y = -32 \end{cases}$$

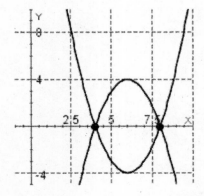

13. Solve the system of equations algebraically for real values of x and y.

$$\begin{cases} x^2 - y^2 = -32 \\ 4x^2 + 3x^2 = 124 \end{cases}$$

14. Write the equation of the hyperbola with center (2, 10), vertex (2, 12) and passing through $(26, 10 + 2\sqrt{10})$.

15. Find the equation, in the form $(x - h)^2 + (y - k)^2 = r^2$, of the circle passing through the given points.

    (0, 9), (5, 4), and (4, 7)

16. Solve the system of equations algebraically for real values of x and y.

$$\begin{cases} x^2 + y^2 = 85 \\ y = x^2 - 79 \end{cases}$$

17. Change the equation to standard form.

    $x^2 - 7y + 11 = -4x$

18. Graph the hyperbola.
    $9(y + 2)^2 - 4(x - 1)^2 = 36$

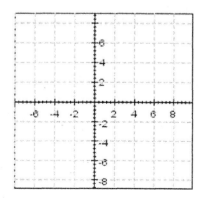

19. Solve the system of equations algebraically for real values of x and y.

$$\begin{cases} x + y = 12 \\ x^2 + y^2 = 72 \end{cases}$$

20. Find the area of the fundamental rectangle of the hyperbola.

$4(x - 5)^2 - 36(y - 1)^2 = 144$

21. Write the equation of the hyperbola with center (2, -1), $a^2 = 9$ and $b^2 = 16$.

22. Graph the ellipse.

$x^2 + 4y^2 - 4x + 8y + 4 = 0$

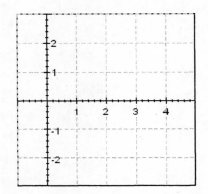

23. Graph the ellipse.

$$\frac{x^2}{16} + \frac{(y + 2)^2}{36} = 1$$

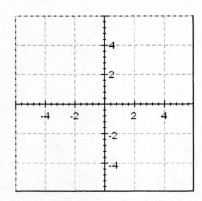

24. Graph the following equation.

$$x^2 - 2x + y^2 = 15$$

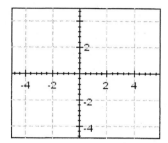

25. Solve the system of equations algebraically for real values of x and y.

$$\begin{cases} \dfrac{1}{x} + \dfrac{6}{y} = 1 \\ \dfrac{2}{x} - \dfrac{1}{y} = \dfrac{1}{7} \end{cases}$$

ANSWER KEY

Gustafson/Frisk - College Algebra 8E Chapter 7 Form B

1. 3

2. $\dfrac{(x-5)^2}{16} + \dfrac{(y-5)^2}{25} = 1$

3. $(y-3)^2 = -28(x-5)$

4. $\dfrac{x^2}{25} + \dfrac{y^2}{9} = 1$

5. $\dfrac{x^2}{75} + \dfrac{y^2}{100} = 1$

6. $(x-2)^2 + (y+3)^2 = 49$

7. $y^2 + x^2 = 25$

8.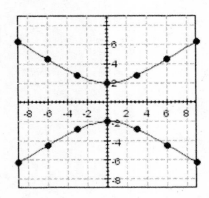

9. $\dfrac{(x-1)^2}{9} - \dfrac{(y-2)^2}{25} = 1$

10. $\dfrac{(x-1)^2}{9} + \dfrac{(y-5)^2}{16} = 1$

11. $(9, 5), (-9, 5), (9, -5), (-9, -5)$

12. $(4, 0), (8, 0)$

13. $(2, 6), (-2, 6), (2, -6), (-2, -6)$

ANSWER KEY

Gustafson/Frisk - College Algebra 8E Chapter 7 Form B

14. $\dfrac{(y-10)^2}{4} - \dfrac{(x-2)^2}{64} = 1$

15. $x^2 + (y-4)^2 = 25$

16. $(9, 2), (-9, 2)$

17. $(x+2)^2 = 7(y-1)$

18.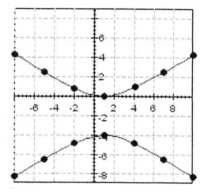

19. $(6, 6)$

20. 48

21. $\dfrac{(x-2)^2}{9} - \dfrac{(y+1)^2}{16} = 1$

22.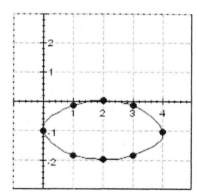

ANSWER KEY

Gustafson/Frisk - College Algebra 8E Chapter 7 Form B

23.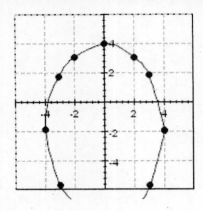

24.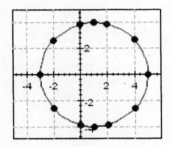

25. $(7, 7)$

List of Problem Codes for BCA Testing

Gustafson/Frisk - College Algebra 8E Chapter 7 Form B

1. gfca.07.01.4.02_NoAlgs
2. gfca.07.02.4.17_NoAlgs
3. gfca.07.01.4.26_NoAlgs
4. gfca.07.02.4.09_NoAlgs
5. gfca.07.02.4.13_NoAlgs
6. gfca.07.01.4.15_NoAlgs
7. gfca.07.01.4.11_NoAlgs
8. gfca.07.03.4.28_NoAlgs
9. gfca.07.03.4.09_NoAlgs
10. gfca.07.02.4.15_NoAlgs
11. gfca.07.04.4.27_NoAlgs
12. gfca.07.04.4.11_NoAlgs
13. gfca.07.04.4.35_NoAlgs
14. gfca.07.03.4.11_NoAlgs
15. gfca.07.01.4.65_NoAlgs
16. gfca.07.04.4.23_NoAlgs
17. gfca.07.01.4.43_NoAlgs
18. gfca.07.03.4.32_NoAlgs
19. gfca.07.04.4.19_NoAlgs
20. gfca.07.03.4.19_NoAlgs
21. gfca.07.03.4.15_NoAlgs
22. gfca.07.02.4.29_NoAlgs
23. gfca.07.02.4.27_NoAlgs
24. gfca.07.01.4.18_NoAlgs
25. gfca.07.04.4.37_NoAlgs

1. Give the coordinates of the circle's center and its radius.

   $x^2 + y^2 - 4 = 0$

   Select the correct answer.

   a. $(1, 1); r = 4$
   b. $(0, 0); r = 2$
   c. $(0, 0); r = 4$
   d. $(1, 1); r = 2$

2. Find the equation of the circle graphed below.

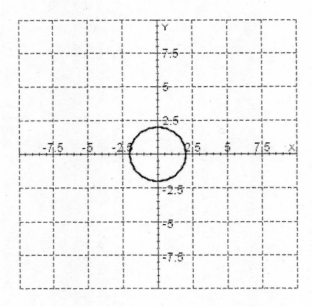

   Select the correct answer.

   a. $x^2 + y^2 = 2$

   b. $y^2 = x^2 + 4$

   c. $x^2 + y^2 = 4$

   d. $x^2 + y^2 = 1$

3. Graph the following equation.

$$x^2 - 2x + y^2 = 15$$

Select the correct answer.

a.

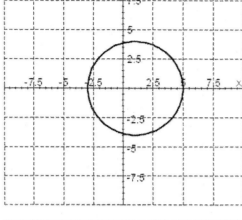

b.

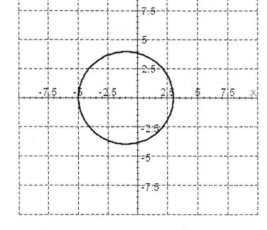

4. Find the equation of the parabola with vertex at (3, 1) and focus at (- 3, 1).

Select the correct answer.

a. $(y - 1)^2 = -24(x - 3)$

b. $(y - 1)^2 = 6(x - 3)$

c. $(y - 1)^2 = 24(x - 3)$

d. $(y + 1)^2 = 24(x + 3)$

5. Change the equation to standard form.

$$x^2 - 9y + 61 = -8x$$

Select the correct answer.

a. $(x + 4)^2 = 9(y + 5)$

b. $(x + 4)^2 = y - 5$

c. $(x + 4)^2 = 9(y - 5)$

d. $(x - 4)^2 = 9(y + 5)$

6. Find the equation, in the form $(x - h)^2 + (y - k)^2 = r^2$, of the circle passing through the given points.

(0, 12), (5, 7), and (4, 10)

Select the correct answer.

a. $x^2 - (y - 7)^2 = 25$

b. $x^2 + (y - 7)^2 = 25$

c. $x^2 + (y - 7)^2 = 5$

d. $x^2 + (y + 7)^2 = 5$

7. To draw an ellipse that is 40 centimeters wide and 10 centimeters tall, how far apart should the two thumbtacks be?

Select the correct answer.

a. 38.7 cm
b. 38.2 cm
c. 40 cm
d. 10 cm

8. Write the equation of the ellipse that has its center at the origin, focus at (3, 0), and major axis equal to 18.

   Select the correct answer.

   a. $\dfrac{x^2}{72} + \dfrac{y^2}{81} = 1$

   b. $\dfrac{x^2}{81} + \dfrac{y^2}{72} = 1$

   c. $\dfrac{x^2}{81} + \dfrac{y^2}{72} = -1$

   d. $\dfrac{x^2}{81} - \dfrac{y^2}{72} = 1$

9. Write the equation of the ellipse with b=2, center at (2, 4) and that passes through (2, 10) and (2, -2).

   Select the correct answer.

   a. $\dfrac{(x-2)^2}{36} + \dfrac{(y-4)^2}{4} = -1$

   b. $\dfrac{(x-2)^2}{4} + \dfrac{(y-4)^2}{36} = 1$

   c. $\dfrac{(x-2)^2}{36} + \dfrac{(y-4)^2}{4} = 1$

   d. $\dfrac{(x-2)^2}{36} - \dfrac{(y-4)^2}{4} = 1$

10. Solve the system of equations algebraically for real values of x and y.

    $\begin{cases} x^2 - 7x - y = -6 \\ x^2 - 7x + y = -6 \end{cases}$

    Select the correct answer.

    a. (0, 4), (0, 1)
    b. (6, 0), (1, 0)
    c. (4, 0), (1, 0)
    d. (0, 4), (6, 0)

11. Write the equation of the ellipse with foci at (4, 6), (8, 6), and b = 4.

    Select the correct answer.

    a. $\dfrac{(x-6)^2}{20} + \dfrac{(y-6)^2}{16} = -1$

    b. $\dfrac{(x-6)^2}{16} + \dfrac{(y-6)^2}{20} = 1$

    c. $\dfrac{(x-6)^2}{20} + \dfrac{(y-6)^2}{16} = 1$

    d. $\dfrac{(x-6)^2}{20} - \dfrac{(y-6)^2}{16} = 1$

12. Find the equation of the circle with a radius of 7 and its center at the intersection of $2x + y = 2$ and $-2x - 5y = 6$.

    Select the correct answer.

    a. $(x+2)^2 + (y+2)^2 = 49$

    b. $(x-2)^2 + (y-2)^2 = 7$

    c. $(x-2)^2 + (y+2)^2 = 49$

    d. $(x+2)^2 + (y+2)^2 = 7$

13. Find the area of the fundamental rectangle of the hyperbola
    $64(x-5)^2 - 4(y-3)^2 = 256$

    Select the correct answer.

    a. 64 square units
    b. 16 square units
    c. 128 square units

14. Find the graph of the following ellipse.

$$\frac{x^2}{36} + \frac{(y+1)^2}{9} = 1$$

Select the correct answer.

a.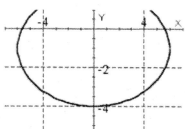

b.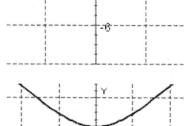

c.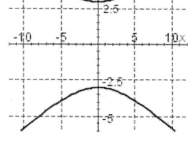

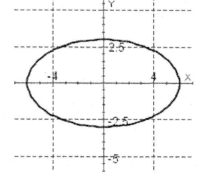

15. Find the equation of the hyperbola with vertices (2, 0), (-2, 0) and focus (7, 0).

Select the correct answer.

a. $\dfrac{x^2}{4} - \dfrac{y^2}{45} = 1$

b. $\dfrac{x^2}{4} + \dfrac{y^2}{45} = 1$

c. $\dfrac{y^2}{4} - \dfrac{x^2}{45} = 1$

16. Find the equation of the hyperbola with center (1, 1), vertex (3, 1) and focus (7, 1).

Select the correct answer.

a. $\dfrac{x^2}{4} - \dfrac{(y-1)^2}{32} = 1$

b. $\dfrac{(x-1)^2}{4} - \dfrac{(y-1)^2}{36} = 1$

c. $\dfrac{(x-1)^2}{4} - \dfrac{(y-1)^2}{32} = 1$

17. Find the equation of the hyperbola with foci (0, 13), (0, -13) and $\dfrac{c}{a} = \dfrac{13}{2}$.

Select the correct answer.

a. $\dfrac{y^2}{4} - \dfrac{x^2}{165} = 1$

b. $\dfrac{y^3}{4} - \dfrac{x}{165} = 1$

c. $\dfrac{(y-2)^2}{4} - \dfrac{x^2}{165} = 1$

18. Graph the hyperbola

    $4x^2 - 25y^2 = 100$

    Select the correct answer.

a.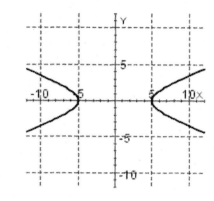

b.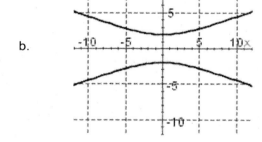

c.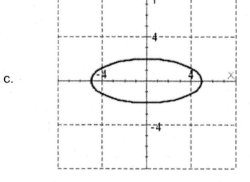

19. Find the equation of the curve on which point P lies. The difference of the distances between P (x, y) and the points (-5, 5) and (13, 5) is 6.

    Select the correct answer.

    a. $\dfrac{(x-4)^2}{9} - \dfrac{(y-5)^2}{72} = 1$

    b. $\dfrac{(x-12)^2}{81} - \dfrac{(y-10)^2}{74} = 1$

    c. $\dfrac{(x+4)^2}{9} - \dfrac{(y+5)^2}{72} = 1$

20. Solve the system of equations by graphing.

    $$\begin{cases} x^2 + y^2 = 2 \\ x + y = 2 \end{cases}$$

    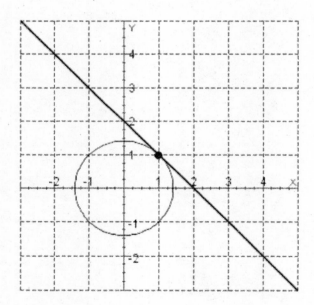

    Select the correct answer.

    a. (1, 2)
    b. (2, 1)
    c. (2, 2)
    d. (1, 1)

21. Solve the system of equations by graphing.

$$\begin{cases} y^2 + x^2 = 34 \\ y + x = 8 \end{cases}$$

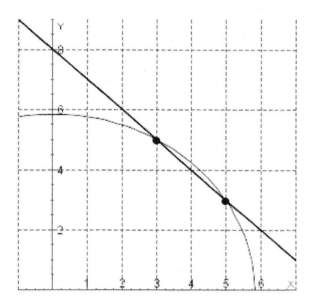

Select the correct answer.

a. (2, 3), (3, 2)
b. (5, 3), (3, 5)
c. (5, 5), (3, 3)
d. (2, 5), (3, 2)

22. Solve the system of equations algebraically for real values of x and y.

$$\begin{cases} y = x + 16 \\ y = x^2 + x \end{cases}$$

Select the correct answer.

a. (4, 20), (-4, 12)
b. (4, 72), (-4, 56)
c. (8, 20), (-8, 12)
d. (8, 72), (-8, 56)

23. Solve the system of equations algebraically for real values of x and y.

$$\begin{cases} x^2 + y^2 = 53 \\ x + y = 9 \end{cases}$$

Select the correct answer.

a. (2, 7), (7, 2)
b. (2, 7), (2, 4)
c. (7, 2), (7, 2)
d. (4, 7), (7, 4)

24. Solve the system of equations algebraically for real values of x and y.

$$\begin{cases} x^2 + 3y^2 = 200 \\ y = x^2 \end{cases}$$

Select the correct answer.

a. $(\sqrt{8}, 8), (-\sqrt{8}, 8)$
b. $(-\sqrt{8}, 11), (-\sqrt{8}, 11)$
c. $(\sqrt{8}, 11), (-\sqrt{11}, 8)$
d. $(\sqrt{8}, 8), (\sqrt{11}, -11)$

25. Write the equation of the ellipse that has its center at the origin, focus at (4, 0) and vertex at (9, 0).

Select the correct answer.

a. $\dfrac{x^2}{81} + \dfrac{y^2}{65} = -1$

b. $\dfrac{x^2}{65} + \dfrac{y^2}{81} = 1$

c. $\dfrac{x^2}{81} - \dfrac{y^2}{65} = 1$

d. $\dfrac{x^2}{81} + \dfrac{y^2}{65} = 1$

ANSWER KEY

Gustafson/Frisk - College Algebra 8E Chapter 7 Form C

1. b
2. c
3. a
4. a
5. c
6. b
7. a
8. b
9. b
10. b
11. c
12. c
13. a
14. a
15. a
16. c
17. a
18. a
19. a
20. d
21. b
22. a
23. a
24. a
25. d

List of Problem Codes for BCA Testing

Gustafson/Frisk - College Algebra 8E Chapter 7 Form C

1. gfca.07.01.4.02m_NoAlgs
2. gfca.07.01.4.11m_NoAlgs
3. gfca.07.01.4.18m_NoAlgs
4. gfca.07.01.4.26m_NoAlgs
5. gfca.07.01.4.43m_NoAlgs
6. gfca.07.01.4.65m_NoAlgs
7. gfca.07.02.4.08m_NoAlgs
8. gfca.07.02.4.14m_NoAlgs
9. gfca.07.02.4.16m_NoAlgs
10. gfca.07.04.4.32m_NoAlgs
11. gfca.07.02.4.19m_NoAlgs
12. gfca.07.01.4.15m_NoAlgs
13. gfca.07.03.4.19m_NoAlgs
14. gfca.07.02.4.27m_NoAlgs
15. gfca.07.03.4.07m_NoAlgs
16. gfca.07.03.4.10m_NoAlgs
17. gfca.07.03.4.12m_NoAlgs
18. gfca.07.03.4.29m_NoAlgs
19. gfca.07.03.4.39m_NoAlgs
20. gfca.07.04.4.04m_NoAlgs
21. gfca.07.04.4.06m_NoAlgs
22. gfca.07.04.4.13m_NoAlgs
23. gfca.07.04.4.21m_NoAlgs
24. gfca.07.04.4.25m_NoAlgs
25. gfca.07.02.4.09m_NoAlgs

Gustafson/Frisk - College Algebra 8E Chapter 7 Form D

1. Find the equation of the circle graphed below.

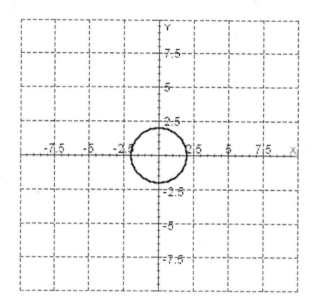

Select the correct answer.

a. $x^2 + y^2 = 4$

b. $x^2 + y^2 = 1$

c. $y^2 = x^2 + 4$

d. $x^2 + y^2 = 2$

2. Change the equation to standard form.

$x^2 - 9y + 61 = -8x$

Select the correct answer.

a. $(x + 4)^2 = 9(y + 5)$

b. $(x + 4)^2 = y - 5$

c. $(x - 4)^2 = 9(y + 5)$

d. $(x + 4)^2 = 9(y - 5)$

3. Graph the hyperbola

$$4x^2 - 25y^2 = 100$$

Select the correct answer.

a.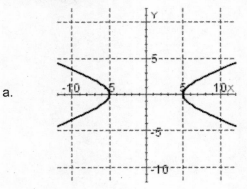

b.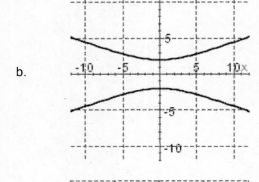

c.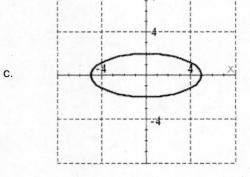

4. Solve the system of equations by graphing.

$$\begin{cases} y^2 + x^2 = 34 \\ y + x = 8 \end{cases}$$

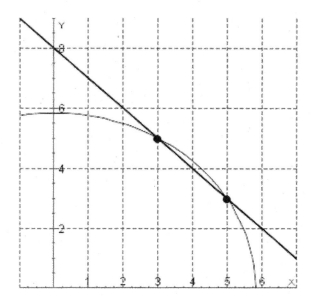

Select the correct answer.

a. (2, 3), (3, 2)
b. (5, 3), (3, 5)
c. (5, 5), (3, 3)
d. (2, 5), (3, 2)

5. Solve the system of equations algebraically for real values of x and y.

$$\begin{cases} x^2 + y^2 = 53 \\ x + y = 9 \end{cases}$$

Select the correct answer.

a. (7, 2), (7, 2)
b. (4, 7), (7, 4)
c. (2, 7), (7, 2)
d. (2, 7), (2, 4)

6. Solve the system of equations algebraically for real values of x and y.

$$\begin{cases} x^2 + 3y^2 = 200 \\ y = x^2 \end{cases}$$

Select the correct answer.

a. $(\sqrt{8}, 8), (-\sqrt{8}, 8)$
b. $(-\sqrt{8}, 11), (-\sqrt{8}, 11)$
c. $(\sqrt{8}, 11), (-\sqrt{11}, 8)$
d. $(\sqrt{8}, 8), (\sqrt{11}, -11)$

7. Solve the system of equations by graphing.

$$\begin{cases} x^2 + y^2 = 2 \\ x + y = 2 \end{cases}$$

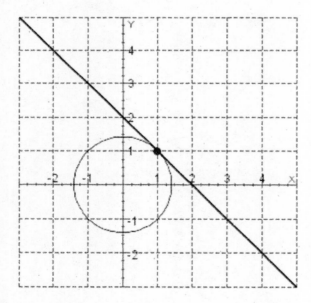

Select the correct answer.

a. (2, 2)
b. (2, 1)
c. (1, 2)
d. (1, 1)

8. Write the equation of the ellipse with b=2, center at (2, 4) and that passes through (2, 10) and (2, -2).

   Select the correct answer.

   a. $\dfrac{(x-2)^2}{36} + \dfrac{(y-4)^2}{4} = 1$

   b. $\dfrac{(x-2)^2}{36} + \dfrac{(y-4)^2}{4} = -1$

   c. $\dfrac{(x-2)^2}{4} + \dfrac{(y-4)^2}{36} = 1$

   d. $\dfrac{(x-2)^2}{36} - \dfrac{(y-4)^2}{4} = 1$

9. Write the equation of the ellipse with foci at (4, 6), (8, 6), and b = 4.

   Select the correct answer.

   a. $\dfrac{(x-6)^2}{20} + \dfrac{(y-6)^2}{16} = -1$

   b. $\dfrac{(x-6)^2}{16} + \dfrac{(y-6)^2}{20} = 1$

   c. $\dfrac{(x-6)^2}{20} + \dfrac{(y-6)^2}{16} = 1$

   d. $\dfrac{(x-6)^2}{20} - \dfrac{(y-6)^2}{16} = 1$

10. Solve the system of equations algebraically for real values of x and y.

    $$\begin{cases} x^2 - 7x - y = -6 \\ x^2 - 7x + y = -6 \end{cases}$$

    Select the correct answer.

    a. (0, 4), (0, 1)
    b. (6, 0), (1, 0)
    c. (4, 0), (1, 0)
    d. (0, 4), (6, 0)

11. Find the equation, in the form $(x-h)^2 + (y-k)^2 = r^2$, of the circle passing through the given points.

    (0, 12), (5, 7), and (4, 10)

    Select the correct answer.

    a. $x^2 - (y-7)^2 = 25$

    b. $x^2 + (y-7)^2 = 25$

    c. $x^2 + (y-7)^2 = 5$

    d. $x^2 + (y+7)^2 = 5$

12. To draw an ellipse that is 40 centimeters wide and 10 centimeters tall, how far apart should the two thumbtacks be?

    Select the correct answer.

    a. 10 cm
    b. 38.7 cm
    c. 40 cm
    d. 38.2 cm

13. Find the equation of the parabola with vertex at (3, 1) and focus at (-3, 1).

    Select the correct answer.

    a. $(y-1)^2 = 24(x-3)$

    b. $(y-1)^2 = 6(x-3)$

    c. $(y+1)^2 = 24(x+3)$

    d. $(y-1)^2 = -24(x-3)$

14. Solve the system of equations algebraically for real values of x and y.

    $$\begin{cases} y = x + 16 \\ y = x^2 + x \end{cases}$$

    Select the correct answer.

    a. (4, 20), (-4, 12)
    b. (4, 72), (-4, 56)
    c. (8, 20), (-8, 12)
    d. (8, 72), (-8, 56)

15. Find the area of the fundamental rectangle of the hyperbola

    $64(x - 5)^2 - 4(y - 3)^2 = 256$

    Select the correct answer.

    a.   16 square units
    b.   64 square units
    c.   128 square units

16. Write the equation of the ellipse that has its center at the origin, focus at (3, 0), and major axis equal to 18.

    Select the correct answer.

    a. $\dfrac{x^2}{81} - \dfrac{y^2}{72} = 1$

    b. $\dfrac{x^2}{81} + \dfrac{y^2}{72} = -1$

    c. $\dfrac{x^2}{72} + \dfrac{y^2}{81} = 1$

    d. $\dfrac{x^2}{81} + \dfrac{y^2}{72} = 1$

17. Write the equation of the ellipse that has its center at the origin, focus at (4, 0) and vertex at (9, 0).

    Select the correct answer.

    a. $\dfrac{x^2}{81} + \dfrac{y^2}{65} = -1$

    b. $\dfrac{x^2}{81} + \dfrac{y^2}{65} = 1$

    c. $\dfrac{x^2}{65} + \dfrac{y^2}{81} = 1$

    d. $\dfrac{x^2}{81} - \dfrac{y^2}{65} = 1$

18. Find the graph of the following ellipse.

$$\frac{x^2}{36} + \frac{(y+1)^2}{9} = 1$$

Select the correct answer.

a.
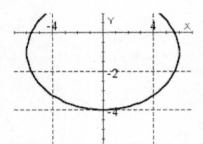

b.
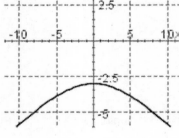

c.
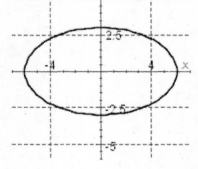

19. Find the equation of the hyperbola with foci (0, 13), (0, -13) and $\frac{c}{a} = \frac{13}{2}$.

    Select the correct answer.

    a. $\dfrac{(y-2)^2}{4} - \dfrac{x^2}{165} = 1$

    b. $\dfrac{y^3}{4} - \dfrac{x}{165} = 1$

    c. $\dfrac{y^2}{4} - \dfrac{x^2}{165} = 1$

20. Find the equation of the circle with a radius of 7 and its center at the intersection of $2x + y = 2$ and $-2x - 5y = 6$.

    Select the correct answer.

    a. $(x+2)^2 + (y+2)^2 = 49$

    b. $(x-2)^2 + (y+2)^2 = 49$

    c. $(x-2)^2 + (y-2)^2 = 7$

    d. $(x+2)^2 + (y+2)^2 = 7$

21. Find the equation of the curve on which point P lies. The difference of the distances between $P(x, y)$ and the points (-5, 5) and (13, 5) is 6.

    Select the correct answer.

    a. $\dfrac{(x-4)^2}{9} - \dfrac{(y-5)^2}{72} = 1$

    b. $\dfrac{(x-12)^2}{81} - \dfrac{(y-10)^2}{74} = 1$

    c. $\dfrac{(x+4)^2}{9} - \dfrac{(y+5)^2}{72} = 1$

22. Graph the following equation.

$$x^2 - 2x + y^2 = 15$$

Select the correct answer.

a.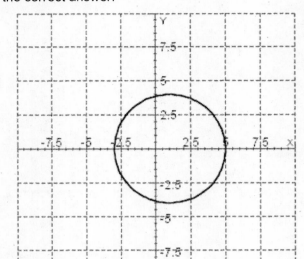

b.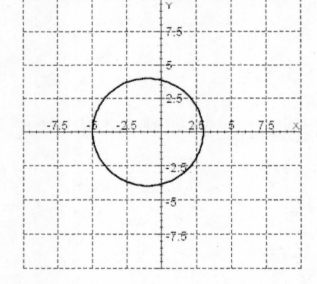

23. Give the coordinates of the circle's center and its radius.

$$x^2 + y^2 - 4 = 0$$

Select the correct answer.

a. $(1, 1); r = 4$
b. $(0, 0); r = 2$
c. $(0, 0); r = 4$
d. $(1, 1); r = 2$

24. Find the equation of the hyperbola with center (1, 1), vertex (3, 1) and focus (7, 1).

    Select the correct answer.

    a. $\dfrac{x^2}{4} - \dfrac{(y-1)^2}{32} = 1$

    b. $\dfrac{(x-1)^2}{4} - \dfrac{(y-1)^2}{36} = 1$

    c. $\dfrac{(x-1)^2}{4} - \dfrac{(y-1)^2}{32} = 1$

25. Find the equation of the hyperbola with vertices (2, 0), (-2, 0) and focus (7, 0).

    Select the correct answer.

    a. $\dfrac{x^2}{4} - \dfrac{y^2}{45} = 1$

    b. $\dfrac{x^2}{4} + \dfrac{y^2}{45} = 1$

    c. $\dfrac{y^2}{4} - \dfrac{x^2}{45} = 1$

ANSWER KEY

Gustafson/Frisk - College Algebra 8E Chapter 7 Form D

1. a
2. d
3. a
4. b
5. c
6. a
7. d
8. c
9. c
10. b
11. b
12. b
13. d
14. a
15. b
16. d
17. b
18. a
19. c
20. b
21. a
22. a
23. b
24. c
25. a

List of Problem Codes for BCA Testing

Gustafson/Frisk - College Algebra 8E Chapter 7 Form D

1. gfca.07.01.4.11m_NoAlgs
2. gfca.07.01.4.43m_NoAlgs
3. gfca.07.03.4.29m_NoAlgs
4. gfca.07.04.4.06m_NoAlgs
5. gfca.07.04.4.21m_NoAlgs
6. gfca.07.04.4.25m_NoAlgs
7. gfca.07.04.4.04m_NoAlgs
8. gfca.07.02.4.16m_NoAlgs
9. gfca.07.02.4.19m_NoAlgs
10. gfca.07.04.4.32m_NoAlgs
11. gfca.07.01.4.65m_NoAlgs
12. gfca.07.02.4.08m_NoAlgs
13. gfca.07.01.4.26m_NoAlgs
14. gfca.07.04.4.13m_NoAlgs
15. gfca.07.03.4.19m_NoAlgs
16. gfca.07.02.4.14m_NoAlgs
17. gfca.07.02.4.09m_NoAlgs
18. gfca.07.02.4.27m_NoAlgs
19. gfca.07.03.4.12m_NoAlgs
20. gfca.07.01.4.15m_NoAlgs
21. gfca.07.03.4.39m_NoAlgs
22. gfca.07.01.4.18m_NoAlgs
23. gfca.07.01.4.02m_NoAlgs
24. gfca.07.03.4.10m_NoAlgs
25. gfca.07.03.4.07m_NoAlgs

1. Give the circle's radius.

   $x^2 + y^2 - 9 = 0$

2. Find the equation of the circle graphed below.

   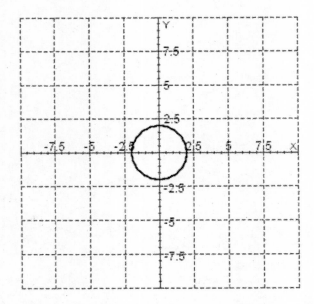

   Select the correct answer.

   a. $x^2 + y^2 = 2$

   b. $x^2 + y^2 = 1$

   c. $y^2 = x^2 + 4$

   d. $x^2 + y^2 = 4$

3. Change the equation to standard form.

   $x^2 - 7y + 11 = -4x$

4. Write the equation of the ellipse with center at (5, 5), a = 4, b = 5 and major axis parallel to the x-axis. Express the equation in standard form.

5. Find the area of the fundamental rectangle of the hyperbola

    $4(x - 5)^2 - 36(y - 1)^2 = 144$

6. Change the equation to standard form.

    $x^2 - 9y + 61 = -8x$

    Select the correct answer.

    a. $(x - 4)^2 = 9(y + 5)$

    b. $(x + 4)^2 = y - 5$

    c. $(x + 4)^2 = 9(y - 5)$

    d. $(x + 4)^2 = 9(y + 5)$

7. Find the equation, in the form $(x - h)^2 + (y - k)^2 = r^2$, of the circle passing through the given points.

    (0, 9), (5, 4), and (4, 7)

8. To draw an ellipse that is 40 centimeters wide and 10 centimeters tall, how far apart should the two thumbtacks be?

    Select the correct answer.

    a. 38.7 cm
    b. 10 cm
    c. 38.2 cm
    d. 40 cm

9. Write the equation of the ellipse that has its center at the origin with focus at (0, 5) and major axis equal to 20. Express the equation in standard form.

10. Write the equation of the ellipse with b=2, center at (2, 4) and that passes through (2, 10) and (2, -2).

    Select the correct answer.

    a. $\dfrac{(x-2)^2}{4} + \dfrac{(y-4)^2}{36} = 1$

    b. $\dfrac{(x-2)^2}{36} + \dfrac{(y-4)^2}{4} = 1$

    c. $\dfrac{(x-2)^2}{36} - \dfrac{(y-4)^2}{4} = 1$

    d. $\dfrac{(x-2)^2}{36} + \dfrac{(y-4)^2}{4} = -1$

11. Graph the ellipse.

    $x^2 + 4y^2 - 4x + 8y + 4 = 0$

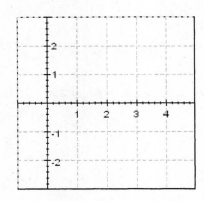

12. Find the equation of the hyperbola with vertices (2, 0), (-2, 0) and focus (7, 0).

    Select the correct answer.

    a. $\dfrac{x^2}{4} - \dfrac{y^2}{45} = 1$

    b. $\dfrac{y^2}{4} - \dfrac{x^2}{45} = 1$

    c. $\dfrac{x^2}{4} + \dfrac{y^2}{45} = 1$

13. Write the equation of the hyperbola with center (1, 2), $a = 3$ and $b = 5$. Transverse axis is horizontal.

14. Find the equation of the hyperbola with foci (0, 13), (0, -13) and $\dfrac{c}{a} = \dfrac{13}{2}$.

    Select the correct answer.

    a. $\dfrac{(y-2)^2}{4} - \dfrac{x^2}{165} = 1$

    b. $\dfrac{y^2}{4} - \dfrac{x^2}{165} = 1$

    c. $\dfrac{y^3}{4} - \dfrac{x}{165} = 1$

15. Write the equation of the ellipse that has its center at the origin, focus at (4, 0) and vertex at (9, 0).

    Select the correct answer.

    a. $\dfrac{x^2}{81} - \dfrac{y^2}{65} = 1$

    b. $\dfrac{x^2}{65} + \dfrac{y^2}{81} = 1$

    c. $\dfrac{x^2}{81} + \dfrac{y^2}{65} = 1$

    d. $\dfrac{x^2}{81} + \dfrac{y^2}{65} = -1$

16. Graph the hyperbola.

$$4x^2 - 25y^2 = 100$$

Select the correct answer.

a.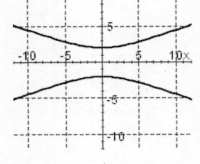

b.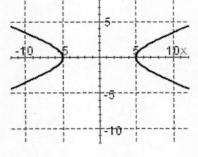

c.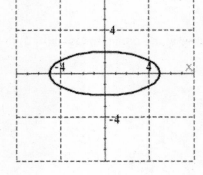

17. Find the equation of the curve on which point P lies. The difference of the distances between P(x, y) and the points (-5, 5) and (13, 5) is 6.

    Select the correct answer.

    a. $\dfrac{(x-4)^2}{9} - \dfrac{(y-5)^2}{72} = 1$

    b. $\dfrac{(x-12)^2}{81} - \dfrac{(y-10)^2}{74} = 1$

    c. $\dfrac{(x+4)^2}{9} - \dfrac{(y+5)^2}{72} = 1$

18. Solve the system of equations by graphing.

    $\begin{cases} x^2 + y^2 = 2 \\ x + y = 2 \end{cases}$

    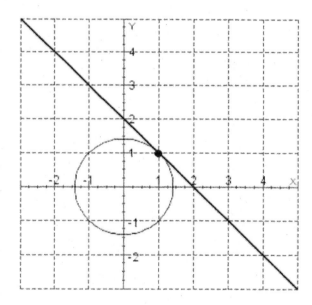

    Select the correct answer.

    a. (2, 2)
    b. (2, 1)
    c. (1, 2)
    d. (1, 1)

19. Solve the system of equations by graphing.

$$\begin{cases} x^2 - 12x - y = -32 \\ x^2 - 12x + y = -32 \end{cases}$$

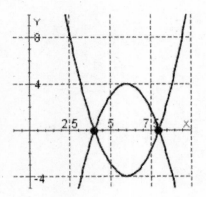

20. Solve the system of equations algebraically for real values of x and y.

$$\begin{cases} y = x + 16 \\ y = x^2 + x \end{cases}$$

Select the correct answer.

a. (8, 72), (-8, 56)
b. (4, 72), (-4, 56)
c. (4, 20), (-4, 12)
d. (8, 20), (-8, 12)

21. Solve the system of equations algebraically for real values of x and y.

$$\begin{cases} x^2 + y^2 = 85 \\ y = x^2 - 79 \end{cases}$$

22. Solve the system of equations algebraically for real values of x and y.

$$\begin{cases} x^2 + 3y^2 = 200 \\ y = x^2 \end{cases}$$

Select the correct answer.

a. $(\sqrt{8}, 11), (-\sqrt{11}, 8)$
b. $(\sqrt{8}, 8), (\sqrt{11}, -11)$
c. $(-\sqrt{8}, 11), (-\sqrt{8}, 11)$
d. $(\sqrt{8}, 8), (-\sqrt{8}, 8)$

23. Solve the system of equations algebraically for real values of x and y.

$$\begin{cases} x^2 - y^2 = -32 \\ 4x^2 + 3x^2 = 124 \end{cases}$$

24. Graph the following equation.

$$x^2 - 2x + y^2 = 15$$

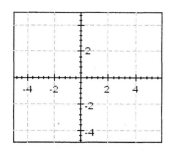

25. Graph the following equation.

$$x^2 - 2x + y^2 = 15$$

Select the correct answer.

a.
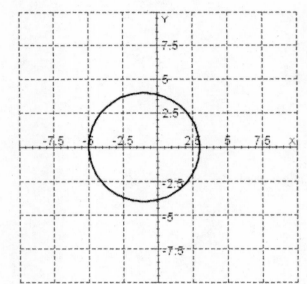

b.
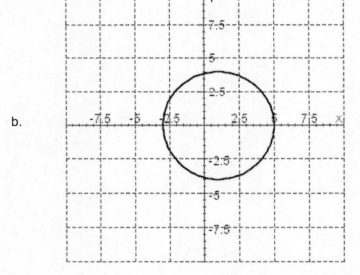

ANSWER KEY

Gustafson/Frisk - College Algebra 8E Chapter 7 Form E

1. 3

2. d

3. $(x+2)^2 = 7(y-1)$

4. $\dfrac{(x-5)^2}{16} + \dfrac{(y-5)^2}{25} = 1$

5. 48

6. c

7. $x^2 + (y-4)^2 = 25$

8. a

9. $\dfrac{x^2}{75} + \dfrac{y^2}{100} = 1$

10. a

11.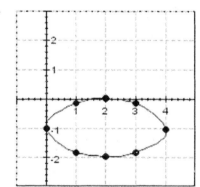

12. a

13. $\dfrac{(x-1)^2}{9} - \dfrac{(y-2)^2}{25} = 1$

14. b

15. c

ANSWER KEY

Gustafson/Frisk - College Algebra 8E Chapter 7 Form E

16. b

17. a

18. d

19. $(4, 0), (8, 0)$

20. c

21. $(9, 2), (-9, 2)$

22. d

23. $(2, 6), (-2, 6), (2, -6), (-2, -6)$

24.

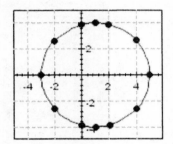

25. b

List of Problem Codes for BCA Testing

Gustafson/Frisk - College Algebra 8E Chapter 7 Form E

1. gfca.07.01.4.02_NoAlgs
2. gfca.07.01.4.11m_NoAlgs
3. gfca.07.01.4.43_NoAlgs
4. gfca.07.02.4.17_NoAlgs
5. gfca.07.03.4.19_NoAlgs
6. gfca.07.01.4.43m_NoAlgs
7. gfca.07.01.4.65_NoAlgs
8. gfca.07.02.4.08m_NoAlgs
9. gfca.07.02.4.13_NoAlgs
10. gfca.07.02.4.16m_NoAlgs
11. gfca.07.02.4.29_NoAlgs
12. gfca.07.03.4.07m_NoAlgs
13. gfca.07.03.4.09_NoAlgs
14. gfca.07.03.4.12m_NoAlgs
15. gfca.07.02.4.09m_NoAlgs
16. gfca.07.03.4.29m_NoAlgs
17. gfca.07.03.4.39m_NoAlgs
18. gfca.07.04.4.04m_NoAlgs
19. gfca.07.04.4.11_NoAlgs
20. gfca.07.04.4.13m_NoAlgs
21. gfca.07.04.4.23_NoAlgs
22. gfca.07.04.4.25m_NoAlgs
23. gfca.07.04.4.35_NoAlgs
24. gfca.07.01.4.18_NoAlgs
25. gfca.07.01.4.18m_NoAlgs

1. Change the equation to standard form.

   $x^2 - 7y + 11 = -4x$

2. Solve the system of equations by graphing.

   $\begin{cases} x^2 + y^2 = 2 \\ x + y = 2 \end{cases}$

   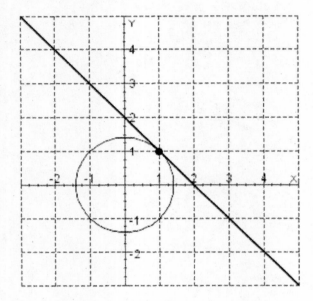

   Select the correct answer.

   a. (2, 2)
   b. (2, 1)
   c. (1, 2)
   d. (1, 1)

3. To draw an ellipse that is 40 centimeters wide and 10 centimeters tall, how far apart should the two thumbtacks be?

   Select the correct answer.

   a. 38.7 cm
   b. 10 cm
   c. 38.2 cm
   d. 40 cm

4. Find the equation of the hyperbola with vertices (2, 0), (- 2, 0) and focus (7, 0).

   Select the correct answer.

   a. $\dfrac{x^2}{4} - \dfrac{y^2}{45} = 1$

   b. $\dfrac{y^2}{4} - \dfrac{x^2}{45} = 1$

   c. $\dfrac{x^2}{4} + \dfrac{y^2}{45} = 1$

5. Change the equation to standard form.

   $x^2 - 9y + 61 = -8x$

   Select the correct answer.

   a. $(x + 4)^2 = 9(y + 5)$

   b. $(x - 4)^2 = 9(y + 5)$

   c. $(x + 4)^2 = 9(y - 5)$

   d. $(x + 4)^2 = y - 5$

6. Write the equation of the hyperbola with center (1, 2), a = 3 and b = 5. Transverse axis is horizontal.

7. Find the equation, in the form $(x - h)^2 + (y - k)^2 = r^2$, of the circle passing through the given points.

   (0, 9), (5, 4), and (4, 7)

8. Write the equation of the ellipse with center at (5, 5), a = 4, b = 5 and major axis parallel to the x-axis. Express the equation in standard form.

9. Give the circle's radius.

$$x^2 + y^2 - 9 = 0$$

10. Solve the system of equations algebraically for real values of *x* and *y*.

$$\begin{cases} x^2 + y^2 = 85 \\ y = x^2 - 79 \end{cases}$$

11. Write the equation of the ellipse that has its center at the origin with focus at (0, 5) and major axis equal to 20. Express the equation in standard form.

12. Graph the following equation.

$$x^2 - 2x + y^2 = 15$$

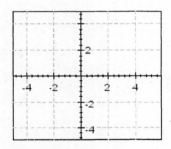

13. Solve the system of equations algebraically for real values of x and y.

$$\begin{cases} x^2 + 3y^2 = 200 \\ y = x^2 \end{cases}$$

Select the correct answer.

a. $(\sqrt{8}, 11), (-\sqrt{11}, 8)$
b. $(\sqrt{8}, 8), (\sqrt{11}, -11)$
c. $(-\sqrt{8}, 11), (-\sqrt{8}, 11)$
d. $(\sqrt{8}, 8), (-\sqrt{8}, 8)$

14. Solve the system of equations algebraically for real values of x and y.

$$\begin{cases} x^2 - y^2 = -32 \\ 4x^2 + 3x^2 = 124 \end{cases}$$

15. Find the equation of the curve on which point P lies. The difference of the distances between P (x, y) and the points (-5, 5) and (13, 5) is 6.

Select the correct answer.

a. $\dfrac{(x-4)^2}{9} - \dfrac{(y-5)^2}{72} = 1$

b. $\dfrac{(x-12)^2}{81} - \dfrac{(y-10)^2}{74} = 1$

c. $\dfrac{(x+4)^2}{9} - \dfrac{(y+5)^2}{72} = 1$

16. Graph the ellipse.

$$x^2 + 4y^2 - 4x + 8y + 4 = 0$$

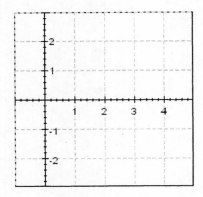

17. Find the area of the fundamental rectangle of the hyperbola.

$$4(x - 5)^2 - 36(y - 1)^2 = 144$$

18. Write the equation of the ellipse that has its center at the origin, focus at (4, 0) and vertex at (9, 0).

    Select the correct answer.

    a. $\dfrac{x^2}{81} + \dfrac{y^2}{65} = 1$

    b. $\dfrac{x^2}{65} + \dfrac{y^2}{81} = 1$

    c. $\dfrac{x^2}{81} - \dfrac{y^2}{65} = 1$

    d. $\dfrac{x^2}{81} + \dfrac{y^2}{65} = -1$

19. Graph the following equation.

$$x^2 - 2x + y^2 = 15$$

Select the correct answer.

a.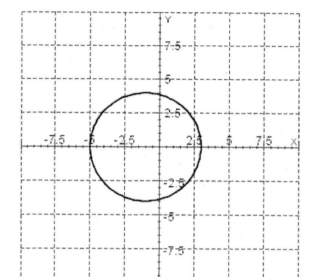

b.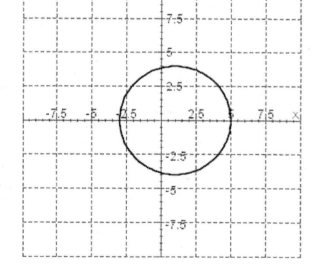

20. Graph the hyperbola.

$$4x^2 - 25y^2 = 100$$

Select the correct answer.

a.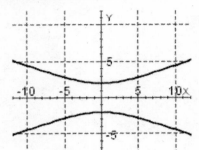

b.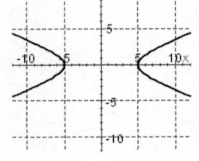

c.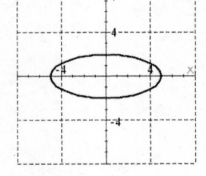

21. Solve the system of equations by graphing.

$$\begin{cases} x^2 - 12x - y = -32 \\ x^2 - 12x + y = -32 \end{cases}$$

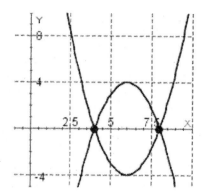

22. Write the equation of the ellipse with b=2, center at (2, 4) and that passes through (2, 10) and (2, -2).

    Select the correct answer.

    a. $\dfrac{(x-2)^2}{4} + \dfrac{(y-4)^2}{36} = 1$

    b. $\dfrac{(x-2)^2}{36} + \dfrac{(y-4)^2}{4} = 1$

    c. $\dfrac{(x-2)^2}{36} - \dfrac{(y-4)^2}{4} = 1$

    d. $\dfrac{(x-2)^2}{36} + \dfrac{(y-4)^2}{4} = -1$

23. Find the equation of the hyperbola with foci (0, 13), (0, -13) and $\frac{c}{a} = \frac{13}{2}$

Select the correct answer.

a. $\dfrac{(y-2)^2}{4} - \dfrac{x^2}{165} = 1$

b. $\dfrac{y^2}{4} - \dfrac{x^2}{165} = 1$

c. $\dfrac{y^3}{4} - \dfrac{x}{165} = 1$

24. Find the equation of the circle graphed below.

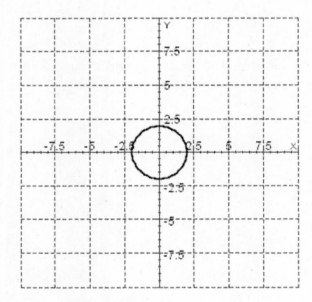

Select the correct answer.

a. $x^2 + y^2 = 2$

b. $x^2 + y^2 = 1$

c. $y^2 = x^2 + 4$

d. $x^2 + y^2 = 4$

25. Solve the system of equations algebraically for real values of x and y.

$$\begin{cases} y = x + 16 \\ y = x^2 + x \end{cases}$$

Select the correct answer.

a.   (8, 72), (-8, 56)
b.   (4, 72), (-4, 56)
c.   (4, 20), (-4, 12)
d.   (8, 20), (-8, 12)

ANSWER KEY

Gustafson/Frisk - College Algebra 8E Chapter 7 Form F

1. $(x+2)^2 = 7(y-1)$

2. d

3. a

4. a

5. c

6. $\dfrac{(x-1)^2}{9} - \dfrac{(y-2)^2}{25} = 1$

7. $x^2 + (y-4)^2 = 25$

8. $\dfrac{(x-5)^2}{16} + \dfrac{(y-5)^2}{25} = 1$

9. 3

10. $(9, 2), (-9, 2)$

11. $\dfrac{x^2}{75} + \dfrac{y^2}{100} = 1$

12.
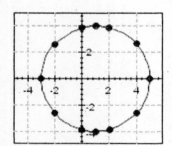

13. d

14. $(2, 6), (-2, 6), (2, -6), (-2, -6)$

15. a

Page 1

ANSWER KEY

Gustafson/Frisk - College Algebra 8E Chapter 7 Form F

16.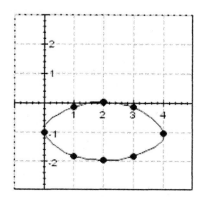

17. 4 8

18. a

19. b

20. b

21. $(4, 0), (8, 0)$

22. a

23. b

24. d

25. c

List of Problem Codes for BCA Testing

Gustafson/Frisk - College Algebra 8E Chapter 7 Form F

1. gfca.07.01.4.43_NoAlgs
2. gfca.07.04.4.04m_NoAlgs
3. gfca.07.02.4.08m_NoAlgs
4. gfca.07.03.4.07m_NoAlgs
5. gfca.07.01.4.43m_NoAlgs
6. gfca.07.03.4.09_NoAlgs
7. gfca.07.01.4.65_NoAlgs
8. gfca.07.02.4.17_NoAlgs
9. gfca.07.01.4.02_NoAlgs
10. gfca.07.04.4.23_NoAlgs
11. gfca.07.02.4.13_NoAlgs
12. gfca.07.01.4.18_NoAlgs
13. gfca.07.04.4.25m_NoAlgs
14. gfca.07.04.4.35_NoAlgs
15. gfca.07.03.4.39m_NoAlgs
16. gfca.07.02.4.29_NoAlgs
17. gfca.07.03.4.19_NoAlgs
18. gfca.07.02.4.09_NoAlgs
19. gfca.07.01.4.18m_NoAlgs
20. gfca.07.03.4.29m_NoAlgs
21. gfca.07.04.4.11_NoAlgs
22. gfca.07.02.4.16m_NoAlgs
23. gfca.07.03.4.12m_NoAlgs
24. gfca.07.01.4.11m_NoAlgs
25. gfca.07.04.4.13m_NoAlgs

1. Graph the hyperbola.

   $9(y + 2)^2 - 4(x - 1)^2 = 36$

   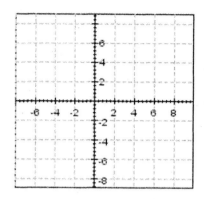

2. Solve the system of equations algebraically for real values of x and y.

   $\begin{cases} x^2 - 7x - y = -6 \\ x^2 - 7x + y = -6 \end{cases}$

   Select the correct answer.

   a. (6, 0), (1, 0)
   b. (0, 4), (0, 1)
   c. (4, 0), (1, 0)
   d. (0, 4), (6, 0)

3. Write the equation of the ellipse with foci at (4, 6), (8, 6), and b = 4.

   Select the correct answer.

   a. $\dfrac{(x - 6)^2}{20} + \dfrac{(y - 6)^2}{16} = 1$

   b. $\dfrac{(x - 6)^2}{20} + \dfrac{(y - 6)^2}{16} = -1$

   c. $\dfrac{(x - 6)^2}{16} + \dfrac{(y - 6)^2}{20} = 1$

   d. $\dfrac{(x - 6)^2}{20} - \dfrac{(y - 6)^2}{16} = 1$

4. Give the coordinates of the circle's center and its radius.

   $x^2 + y^2 - 4 = 0$

   Select the correct answer.

   a. (1, 1); $r = 2$
   b. (1, 1); $r = 4$
   c. (0, 0); $r = 2$
   d. (0, 0); $r = 4$

5. Find the equation, in the form $(x - h)^2 + (y - k)^2 = r^2$, of the circle passing through the given points.

   (0, 12), (5, 7), and (4, 10)

   Select the correct answer.

   a. $x^2 + (y - 7)^2 = 25$

   b. $x^2 - (y - 7)^2 = 25$

   c. $x^2 + (y + 7)^2 = 5$

   d. $x^2 + (y - 7)^2 = 5$

6. Find the equation of the parabola with vertex at (3, 1) and focus at (- 3, 1).

   Select the correct answer.

   a. $(y - 1)^2 = 24(x - 3)$

   b. $(y + 1)^2 = 24(x + 3)$

   c. $(y - 1)^2 = 6(x - 3)$

   d. $(y - 1)^2 = -24(x - 3)$

7. Write the equation of the hyperbola with center (2, 10), vertex (2, 12) and passing through $(26, 10 + 2\sqrt{10})$.

8. Write the equation of the ellipse that has its center at the origin with focus at (0, 5) and major axis equal to 20. Express the equation in standard form.

9. Graph the ellipse.
$$x^2 + 4y^2 - 4x + 8y + 4 = 0$$

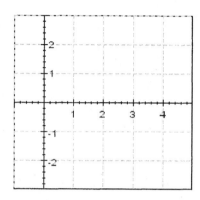

10. Find the equation of the circle graphed below.

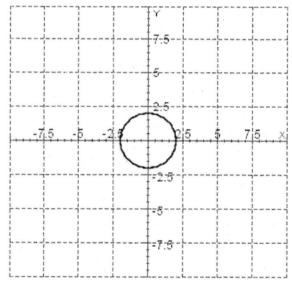

Select the correct answer.
a. $x^2 + y^2 = 4$

b. $y^2 = x^2 + 4$

c. $x^2 + y^2 = 2$

d. $x^2 + y^2 = 1$

11. Find the equation of the hyperbola with vertices (2, 0), (-2, 0) and focus (7, 0).

    Select the correct answer.

    a. $\dfrac{y^2}{4} - \dfrac{x^2}{45} = 1$

    b. $\dfrac{x^2}{4} + \dfrac{y^2}{45} = 1$

    c. $\dfrac{x^2}{4} - \dfrac{y^2}{45} = 1$

12. Solve the system of equations algebraically for real values of x and y.

    $$\begin{cases} x^2 - y^2 = -32 \\ 4x^2 + 3x^2 = 124 \end{cases}$$

13. Write the equation of the ellipse that has its center at the origin, focus at (3, 0), and major axis equal to 18.

    Select the correct answer.

    a. $\dfrac{x^2}{81} - \dfrac{y^2}{72} = 1$

    b. $\dfrac{x^2}{81} + \dfrac{y^2}{72} = 1$

    c. $\dfrac{x^2}{81} + \dfrac{y^2}{72} = -1$

    d. $\dfrac{x^2}{72} + \dfrac{y^2}{81} = 1$

14. Solve the system of equations by graphing.

$$\begin{cases} x^2 + y^2 = 2 \\ x + y = 2 \end{cases}$$

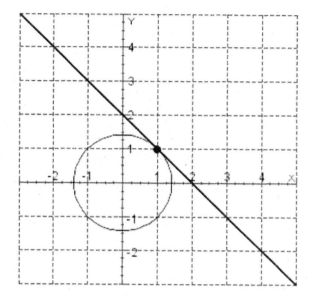

Select the correct answer.

a. (2, 2)
b. (2, 1)
c. (1, 2)
d. (1, 1)

15. Change the equation to standard form.

$$x^2 - 7y + 11 = -4x$$

16. Solve the system of equations algebraically for real values of x and y.

$$\begin{cases} \dfrac{1}{x} + \dfrac{6}{y} = 1 \\ \dfrac{2}{x} - \dfrac{1}{y} = \dfrac{1}{7} \end{cases}$$

17. Graph the hyperbola

$$4x^2 - 25y^2 = 100$$

Select the correct answer.

a.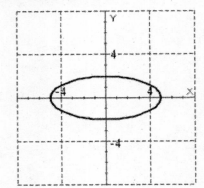

b.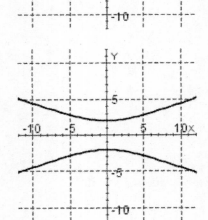

c.

18. Graph the following equation.

$x^2 - 2x + y^2 = 15$

Select the correct answer.

a.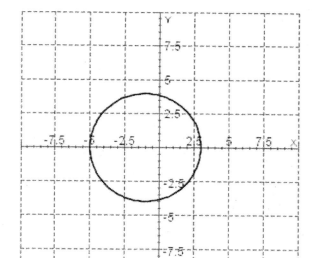

b.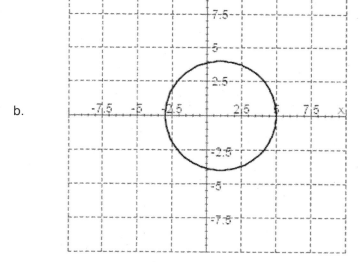

19. Find the equation of the hyperbola with foci (0, 13), (0, -13) and $\dfrac{c}{a} = \dfrac{13}{2}$.

   Select the correct answer.

   a. $\dfrac{y^2}{4} - \dfrac{x^2}{165} = 1$

   b. $\dfrac{(y-2)^2}{4} - \dfrac{x^2}{165} = 1$

   c. $\dfrac{y^3}{4} - \dfrac{x}{165} = 1$

20. Solve the system of equations algebraically for real values of x and y.

   $$\begin{cases} x^2 + 3y^2 = 200 \\ y = x^2 \end{cases}$$

   Select the correct answer.

   a. $(\sqrt{8}, 8), (\sqrt{11}, -11)$
   b. $(\sqrt{8}, 8), (-\sqrt{8}, 8)$
   c. $(\sqrt{8}, 11), (-\sqrt{11}, 8)$
   d. $(-\sqrt{8}, 11), (\sqrt{8}, 11)$

21. Write the equation of the circle with a radius of 7 and its center at the intersection of $3x + y = 3$ and $-2x - 4y = 8$. Express the equation in standard form.

22. Find the area of the fundamental rectangle of the hyperbola

   $64(x-5)^2 - 4(y-3)^2 = 256$

   Select the correct answer.

   a. 64 square units
   b. 16 square units
   c. 128 square units

23. Find the graph of the following ellipse.

$$\frac{x^2}{36} + \frac{(y+1)^2}{9} = 1$$

Select the correct answer.

a.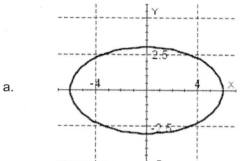

b.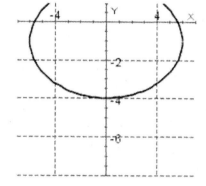

c.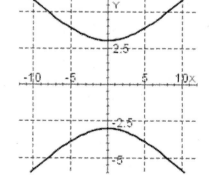

24. Solve the system of equations by graphing.

$$\begin{cases} y^2 + x^2 = 34 \\ y + x = 8 \end{cases}$$

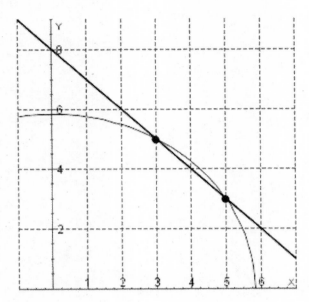

Select the correct answer.

a. (5, 5), (3, 3)
b. (2, 3), (3, 2)
c. (2, 5), (3, 2)
d. (5, 3), (3, 5)

25. To draw an ellipse that is 40 centimeters wide and 10 centimeters tall, how far apart should the two thumbtacks be?

Select the correct answer.

a. 10 cm
b. 38.7 cm
c. 38.2 cm
d. 40 cm

ANSWER KEY

Gustafson/Frisk - College Algebra 8E Chapter 7 Form G

1.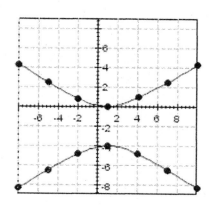

2. a

3. a

4. c

5. a

6. d

7. $\dfrac{(y-10)^2}{4} - \dfrac{(x-2)^2}{64} = 1$

8. $\dfrac{x^2}{75} + \dfrac{y^2}{100} = 1$

9.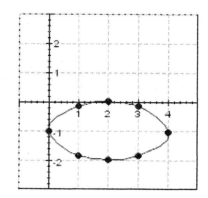

10. a

11. c

12. $(2, 6), (-2, 6), (2, -6), (-2, -6)$

ANSWER KEY

Gustafson/Frisk - College Algebra 8E Chapter 7 Form G

13. b

14. d

15. $(x+2)^2 = 7(y-1)$

16. $(7, 7)$

17. b

18. b

19. a

20. b

21. $(x-2)^2 + (y+3)^2 = 49$

22. a

23. b

24. d

25. b

List of Problem Codes for BCA Testing

Gustafson/Frisk - College Algebra 8E Chapter 7 Form G

1. gfca.07.03.4.32_NoAlgs
2. gfca.07.04.4.32m_NoAlgs
3. gfca.07.02.4.19m_NoAlgs
4. gfca.07.01.4.02m_NoAlgs
5. gfca.07.01.4.65m_NoAlgs
6. gfca.07.01.4.26m_NoAlgs
7. gfca.07.03.4.11_NoAlgs
8. gfca.07.02.4.13_NoAlgs
9. gfca.07.02.4.29_NoAlgs
10. gfca.07.01.4.11m_NoAlgs
11. gfca.07.03.4.07m_NoAlgs
12. gfca.07.04.4.35_NoAlgs
13. gfca.07.02.4.14m_NoAlgs
14. gfca.07.04.4.04m_NoAlgs
15. gfca.07.01.4.43_NoAlgs
16. gfca.07.04.4.37_NoAlgs
17. gfca.07.03.4.29m_NoAlgs
18. gfca.07.01.4.18m_NoAlgs
19. gfca.07.03.4.12m_NoAlgs
20. gfca.07.04.4.25m_NoAlgs
21. gfca.07.01.4.15_NoAlgs
22. gfca.07.03.4.19m_NoAlgs
23. gfca.07.02.4.27m_NoAlgs
24. gfca.07.04.4.06m_NoAlgs
25. gfca.07.02.4.08m_NoAlgs

1. Write the equation of the hyperbola with center (2, 10), vertex (2, 12) and passing through $(26, 10 + 2\sqrt{10})$.

2. Graph the following equation.

   $x^2 - 2x + y^2 = 15$

   Select the correct answer.

   a.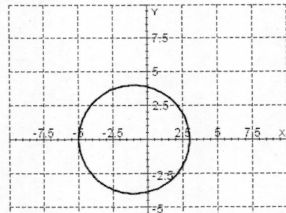

   b.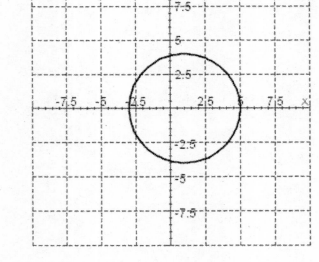

3. Graph the ellipse.

$$x^2 + 4y^2 - 4x + 8y + 4 = 0$$

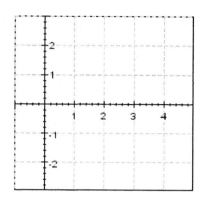

4. Find the equation, in the form $(x - h)^2 + (y - k)^2 = r^2$, of the circle passing through the given points.

(0, 12), (5, 7), and (4, 10)

Select the correct answer.

a. $x^2 + (y - 7)^2 = 25$

b. $x^2 - (y - 7)^2 = 25$

c. $x^2 + (y + 7)^2 = 5$

d. $x^2 + (y - 7)^2 = 5$

5. Solve the system of equations algebraically for real values of x and y.

$$\begin{cases} \dfrac{1}{x} + \dfrac{6}{y} = 1 \\ \dfrac{2}{x} - \dfrac{1}{y} = \dfrac{1}{7} \end{cases}$$

6. Solve the system of equations by graphing.

$$\begin{cases} x^2 + y^2 = 2 \\ x + y = 2 \end{cases}$$

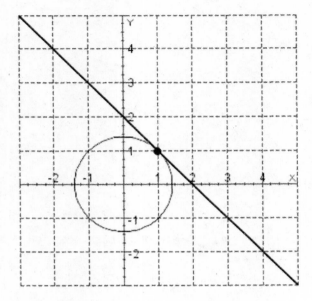

Select the correct answer.

a. (1, 2)
b. (1, 1)
c. (2, 2)
d. (2, 1)

7. Find the area of the fundamental rectangle of the hyperbola

$64(x - 5)^2 - 4(y - 3)^2 = 256$

Select the correct answer.

a. 64 square units
b. 16 square units
c. 128 square units

8. Find the equation of the circle graphed below.

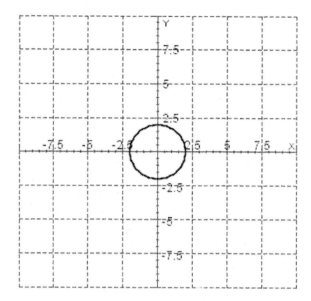

Select the correct answer.

a. $x^2 + y^2 = 4$

b. $y^2 = x^2 + 4$

c. $x^2 + y^2 = 2$

d. $x^2 + y^2 = 1$

9. Find the equation of the hyperbola with vertices (2, 0), (-2, 0) and focus (7, 0).

Select the correct answer.

a. $\dfrac{y^2}{4} - \dfrac{x^2}{45} = 1$

b. $\dfrac{x^2}{4} + \dfrac{y^2}{45} = 1$

c. $\dfrac{x^2}{4} - \dfrac{y^2}{45} = 1$

10. Solve the system of equations algebraically for real values of x and y.

$$\begin{cases} x^2 + 3y^2 = 200 \\ y = x^2 \end{cases}$$

Select the correct answer.

a. $(\sqrt{8}, 8), (\sqrt{11}, -11)$
b. $(\sqrt{8}, 8), (-\sqrt{8}, 8)$
c. $(\sqrt{8}, 11), (-\sqrt{11}, 8)$
d. $(-\sqrt{8}, 11), (-\sqrt{8}, 11)$

11. To draw an ellipse that is 40 centimeters wide and 10 centimeters tall, how far apart should the two thumbtacks be?

Select the correct answer.

a. 10 cm
b. 38.7 cm
c. 38.2 cm
d. 40 cm

12. Write the equation of the circle with a radius of 7 and its center at the intersection of $3x + y = 3$ and $-2x - 4y = 8$. Express the equation in standard form.

13. Solve the system of equations algebraically for real values of x and y.

$$\begin{cases} x^2 - 7x - y = -6 \\ x^2 - 7x + y = -6 \end{cases}$$

Select the correct answer.

a. (6, 0), (1, 0)
b. (0, 4), (0, 1)
c. (4, 0), (1, 0)
d. (0, 4), (6, 0)

14. Solve the system of equations algebraically for real values of x and y.

$$\begin{cases} x^2 - y^2 = -32 \\ 4x^2 + 3x^2 = 124 \end{cases}$$

15. Find the graph of the following ellipse.

$$\frac{x^2}{36} + \frac{(y+1)^2}{9} = 1$$

Select the correct answer.

a.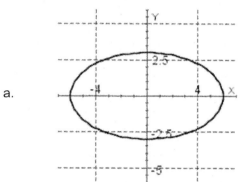

b.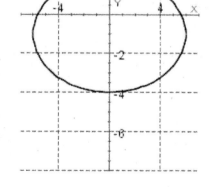

c.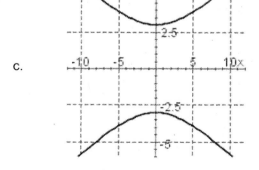

16. Write the equation of the ellipse with foci at (4, 6), (8, 6), and b = 4.

    Select the correct answer.

    a. $\dfrac{(x-6)^2}{20} + \dfrac{(y-6)^2}{16} = 1$

    b. $\dfrac{(x-6)^2}{20} + \dfrac{(y-6)^2}{16} = -1$

    c. $\dfrac{(x-6)^2}{16} + \dfrac{(y-6)^2}{20} = 1$

    d. $\dfrac{(x-6)^2}{20} - \dfrac{(y-6)^2}{16} = 1$

17. Solve the system of equations by graphing.

    $$\begin{cases} y^2 + x^2 = 34 \\ y + x = 8 \end{cases}$$

    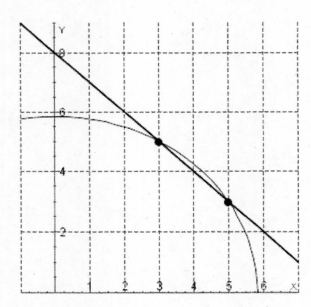

    Select the correct answer.

    a. (5, 5), (3, 3)
    b. (2, 3), (3, 2)
    c. (2, 5), (3, 2)
    d. (5, 3), (3, 5)

18. Find the equation of the hyperbola with foci (0, 13), (0, -13) and $\dfrac{c}{a} = \dfrac{13}{2}$.

Select the correct answer.

a. $\dfrac{y^2}{4} - \dfrac{x^2}{165} = 1$

b. $\dfrac{(y-2)^2}{4} - \dfrac{x^2}{165} = 1$

c. $\dfrac{y^3}{4} - \dfrac{x}{165} = 1$

19. Change the equation to standard form.

$x^2 - 7y + 11 = -4x$

20. Graph the hyperbola

$9(y+2)^2 - 4(x-1)^2 = 36$

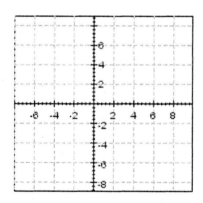

21. Graph the hyperbola

$$4x^2 - 25y^2 = 100$$

Select the correct answer.

a.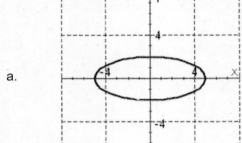

b.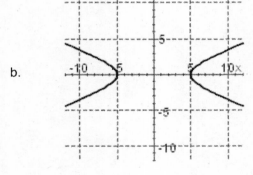

c.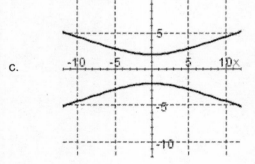

22. Write the equation of the ellipse that has its center at the origin, focus at (3, 0), and major axis equal to 18.

    Select the correct answer.

    a. $\dfrac{x^2}{72} + \dfrac{y^2}{81} = 1$

    b. $\dfrac{x^2}{81} - \dfrac{y^2}{72} = 1$

    c. $\dfrac{x^2}{81} + \dfrac{y^2}{72} = -1$

    d. $\dfrac{x^2}{81} + \dfrac{y^2}{72} = 1$

23. Give the coordinates of the circle's center and its radius.

    $x^2 + y^2 - 4 = 0$

    Select the correct answer.

    a. $(1, 1); r = 2$
    b. $(1, 1); r = 4$
    c. $(0, 0); r = 2$
    d. $(0, 0); r = 4$

24. Find the equation of the parabola with vertex at (3, 1) and focus at (- 3, 1).

    Select the correct answer.

    a. $(y - 1)^2 = 24(x - 3)$

    b. $(y + 1)^2 = 24(x + 3)$

    c. $(y - 1)^2 = 6(x - 3)$

    d. $(y - 1)^2 = -24(x - 3)$

25. Write the equation of the ellipse that has its center at the origin with focus at (0, 5) and major axis equal to 20. Express the equation in standard form.

ANSWER KEY

Gustafson/Frisk - College Algebra 8E Chapter 7 Form H

1. $\dfrac{(y-10)^2}{4} - \dfrac{(x-2)^2}{64} = 1$

2. b

3.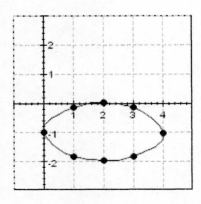

4. a

5. $(7, 7)$

6. b

7. a

8. a

9. c

10. b

11. b

12. $(x-2)^2 + (y+3)^2 = 49$

13. a

14. $(2, 6), (-2, 6), (2, -6), (-2, -6)$

15. b

16. a

17. d

18. a

ANSWER KEY

Gustafson/Frisk - College Algebra 8E Chapter 7 Form H

19. $(x + 2)^2 = 7(y - 1)$

20.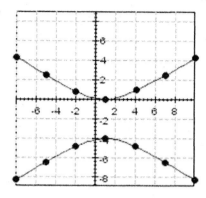

21. b

22. d

23. c

24. d

25. $\dfrac{x^2}{75} + \dfrac{y^2}{100} = 1$

List of Problem Codes for BCA Testing

Gustafson/Frisk - College Algebra 8E Chapter 7 Form H

1. gfca.07.03.4.11_NoAlgs
2. gfca.07.01.4.18m_NoAlgs
3. gfca.07.02.4.29_NoAlgs
4. gfca.07.01.4.65m_NoAlgs
5. gfca.07.04.4.37_NoAlgs
6. gfca.07.04.4.04m_NoAlgs
7. gfca.07.03.4.19m_NoAlgs
8. gfca.07.01.4.11m_NoAlgs
9. gfca.07.03.4.07m_NoAlgs
10. gfca.07.04.4.25m_NoAlgs
11. gfca.07.02.4.08m_NoAlgs
12. gfca.07.01.4.15_NoAlgs
13. gfca.07.04.4.32m_NoAlgs
14. gfca.07.04.4.35_NoAlgs
15. gfca.07.02.4.27m_NoAlgs
16. gfca.07.02.4.19m_NoAlgs
17. gfca.07.04.4.06m_NoAlgs
18. gfca.07.03.4.12m_NoAlgs
19. gfca.07.01.4.43_NoAlgs
20. gfca.07.03.4.32_NoAlgs
21. gfca.07.03.4.29m_NoAlgs
22. gfca.07.02.4.14m_NoAlgs
23. gfca.07.01.4.02m_NoAlgs
24. gfca.07.01.4.26m_NoAlgs
25. gfca.07.02.4.13_NoAlgs

Gustafson/Frisk - College Algebra 8E Chapter 8 Form A

1. Use the binomial theorem to expand the binomial.

    $(z + y)^3$

2. Use the binomial theorem to expand the binomial.

    $(a + 3b)^3$

3. Find the 5th term in the binomial expansion.

    $(x + d)^{17}$

4. Find the next term of the sequence.

    $c, cp, cp^2, cp^3, \ldots$

5. Find the sum of the first five terms of the sequence with the following general term.

    $5k$

6. Evaluate the sum.

    $$\sum_{k=1}^{9} 5k$$

7. The 5th term of an arithmetic sequence is 6, and the first term is -2. Find the common difference.

8. Insert three arithmetic means between 10 and 22.

9. Find the sum of the first *n* terms of the arithmetic sequence.

$$\sum_{n=1}^{26} \left(\frac{9}{4} n + 6 \right)$$

10. Write the first four terms of the geometric sequence with the given properties.

$$a = -3, \ r = \frac{1}{4}$$

11. Find the eighth term of the geometric sequence whose second and fourth terms are -12 and -108.

12. Find the sum of the indicated terms of the geometric sequence.

   -10, 40, -160 . . . (to 4 terms)

13. Change the decimal to a common fraction.

   $0.\overline{32}$

14. Evaluate the expression.

   $P(5, 0) \cdot C(3, 3)$

15. How many different seven-digit phone numbers can be used in one area code if no phone number begins with 0 or 1?

16. If there are 6 class periods in a school day, and a typical student takes 4 classes, how many different time patterns are possible for the student?

17. An ordinary die is rolled. Find the probability of rolling an odd number.

18. Find the probability of drawing a club on one draw from a card deck.

19. Find the probability of rolling a sum of 4 with one roll of three dice.

20. Assume that you draw one card from a standard card deck. Find the probability of the event.

    drawing a black card

21. Assume that you draw two cards from a card deck (standard playing deck of 52 cards), without replacement. Find the probability of the event.

    drawing two aces

22. Assume that you roll two dice once. Find the probability of the result.

    rolling a sum of 7 or 10

23. Assume a single roll of two dice. Find the probability of rolling a sum of 4.

24. If the odds in favor of victory are 13 to 6, find the probability of victory.

25. Find the odds against a couple having 5 baby boys in succession. (Assume $P(\text{boys}) = \frac{1}{2}$.)

    _____ to _____

ANSWER KEY

Gustafson/Frisk - College Algebra 8E Chapter 8 Form A

1. $z^3 + 3z^2 \cdot y + 3z \cdot y^2 + y^3$

2. $a^3 + 9a^2 \cdot b + 27a \cdot b^2 + (27b)^3$

3. $2380x^{13} \cdot d^4$

4. $c \cdot p^4$

5. 75

6. 225

7. 2

8. 13, 16, 19

9. 945.75

10. $-3, \dfrac{-3}{4}, \dfrac{-3}{16}, \dfrac{-3}{64}$

11. -8, 748

12. 510

13. $\dfrac{32}{99}$

14. 1

15. 8,000,000

16. 15

17. $\dfrac{1}{2}$

Page 1

ANSWER KEY

Gustafson/Frisk - College Algebra 8E Chapter 8 Form A

18. $\dfrac{1}{4}$

19. $\dfrac{1}{72}$

20. $\dfrac{1}{2}$

21. $\dfrac{1}{221}$

22. $\dfrac{1}{4}$

23. $\dfrac{1}{12}$

24. $\dfrac{13}{19}$

25. 31, 1

List of Problem Codes for BCA Testing

Gustafson/Frisk - College Algebra 8E Chapter 8 Form A

1. gfca.09.03.4.07_NoAlgs
2. gfca.09.02.4.30_NoAlgs
3. gfca.09.03.4.22m_NoAlgs
4. gfca.09.01.4.15_NoAlgs
5. gfca.09.01.4.39m_NoAlgs
6. gfca.09.03.4.14m_NoAlgs
7. gfca.09.03.4.11_NoAlgs
8. gfca.09.02.4.08_NoAlgs
9. gfca.09.03.4.10m_NoAlgs
10. gfca.09.01.4.38_NoAlgs
11. gfca.09.02.4.13m_NoAlgs
12. gfca.09.01.4.14m_NoAlgs
13. gfca.09.03.4.18m_NoAlgs
14. gfca.09.01.4.19m_NoAlgs
15. gfca.09.03.4.19_NoAlgs
16. gfca.09.01.4.35m_NoAlgs
17. gfca.09.01.4.27m_NoAlgs
18. gfca.09.01.4.41m_NoAlgs
19. gfca.09.03.4.16m_NoAlgs
20. gfca.09.02.4.20_NoAlgs
21. gfca.09.02.4.31m_NoAlgs
22. gfca.09.03.4.08m_NoAlgs
23. gfca.09.02.4.21m_NoAlgs
24. gfca.09.02.4.25m_NoAlgs
25. gfca.09.02.4.09m_NoAlgs

Gustafson/Frisk - College Algebra 8E Chapter 8 Form B

1. Use the binomial theorem to expand the binomial.

    $(a + 3b)^3$

2. An ordinary die is rolled. Find the probability of rolling an odd number.

3. If the odds in favor of victory are 13 to 6, find the probability of victory.

4. Evaluate the expression.

    $P(5, 0) \cdot C(3, 3)$

5. Find the odds against a couple having 5 baby boys in succession. (Assume $P(\text{boys}) = \frac{1}{2}$.)

    _____ to _____

6. If there are 6 class periods in a school day, and a typical student takes 4 classes, how many different time patterns are possible for the student?

7. Assume that you draw two cards from a card deck (standard playing deck of 52 cards), without replacement. Find the probability of the event.

    drawing two aces

8. Assume that you draw one card from a standard card deck. Find the probability of the event.

    drawing a black card

9. Find the probability of rolling a sum of 4 with one roll of three dice.

10. Find the sum of the first five terms of the sequence with the following general term.

    $5k$

11. Insert three arithmetic means between 10 and 22.

12. Change the decimal to a common fraction.

    $0.\overline{32}$

13. Find the eighth term of the geometric sequence whose second and fourth terms are -12 and -108.

14. Find the sum of the first *n* terms of the arithmetic sequence.

    $$\sum_{n=1}^{26} \left(\frac{9}{4}n + 6\right)$$

15. Write the first four terms of the geometric sequence with the given properties.

    $a = -3, \ r = \frac{1}{4}$

16. Find the probability of drawing a club on one draw from a card deck.

17. Find the next term of the sequence.

    $c, \ cp, \ cp^2, \ cp^3, \ \ldots$

18. Find the sum of the indicated terms of the geometric sequence.

    $-10, 40, -160 \ldots$ (to 4 terms)

19. The 5th term of an arithmetic sequence is 6, and the first term is -2. Find the common difference.

20. Assume that you roll two dice once. Find the probability of the result.

    rolling a sum of 7 or 10

21. How many different seven-digit phone numbers can be used in one area code if no phone number begins with 0 or 1?

22. Find the 5th term in the binomial expansion.

$(x+d)^{17}$

23. Assume a single roll of two dice. Find the probability of rolling a sum of 4.

24. Evaluate the sum.

$$\sum_{k=1}^{9} 5k$$

25. Use the binomial theorem to expand the binomial.

$(z+y)^3$

ANSWER KEY

Gustafson/Frisk - College Algebra 8E Chapter 8 Form B

1. $a^3 + 9a^2 \cdot b + 27a \cdot b^2 + (27b)^3$

2. $\dfrac{1}{2}$

3. $\dfrac{13}{19}$

4. 1

5. 31, 1

6. 15

7. $\dfrac{1}{221}$

8. $\dfrac{1}{2}$

9. $\dfrac{1}{72}$

10. 75

11. 13, 16, 19

12. $\dfrac{32}{99}$

13. −8, 748

14. 945.75

15. $-3, \dfrac{-3}{4}, \dfrac{-3}{16}, \dfrac{-3}{64}$

16. $\dfrac{1}{4}$

ANSWER KEY

Gustafson/Frisk - College Algebra 8E Chapter 8 Form B

17. $c \cdot p^4$

18. 510

19. 2

20. $\dfrac{1}{4}$

21. 8,000,000

22. $2380x^{13} \cdot d^4$

23. $\dfrac{1}{12}$

24. 225

25. $z^3 + 3z^2 \cdot y + 3z \cdot y^2 + y^3$

List of Problem Codes for BCA Testing

Gustafson/Frisk - College Algebra 8E Chapter 8 Form B

1. gfca.08.01.4.26_NoAlgs
2. gfca.08.07.4.12_NoAlgs
3. gfca.08.09.4.21_NoAlgs
4. gfca.08.06.4.23_NoAlgs
5. gfca.08.09.4.26_NoAlgs
6. gfca.08.06.4.56_NoAlgs
7. gfca.08.08.4.13_NoAlgs
8. gfca.08.08.4.09_NoAlgs
9. gfca.08.07.4.34_NoAlgs
10. gfca.08.02.4.18_NoAlgs
11. gfca.08.03.4.21_NoAlgs
12. gfca.08.04.4.36_NoAlgs
13. gfca.08.04.4.16_NoAlgs
14. gfca.08.03.4.27_NoAlgs
15. gfca.08.04.4.10_NoAlgs
16. gfca.08.07.4.22_NoAlgs
17. gfca.08.02.4.14_NoAlgs
18. gfca.08.04.4.25_NoAlgs
19. gfca.08.03.4.15_NoAlgs
20. gfca.08.08.4.17_NoAlgs
21. gfca.08.06.4.31_NoAlgs
22. gfca.08.01.4.43_NoAlgs
23. gfca.08.09.4.11_NoAlgs
24. gfca.08.02.4.37_NoAlgs
25. gfca.08.01.4.21_NoAlgs

1. Use the binomial theorem to expand the binomial.

   $(z + b)^3$

   Select the correct answer.

   a. $z^3 + 3z^2b + 3zb^2 + b^3$

   b. $z^3 + 2z^2b + 2zb^2 + b^3$

   c. $z^3 + b^3$

   d. $z^3 + z^2b + zb^2 + b^3$

2. Use the binomial theorem to expand the binomial.

   $(c + 2t)^3$

   Select the correct answer.

   a. $c^3 + 2t^3$

   b. $c^3 + 2c^2t + 4ct^2 + 8t^3$

   c. $c^3 + 6c^2t + 12ct^2 + 8t^3$

   d. $c^3 + 3c^2t + 3ct^2 + t^3$

3. Find the 5th term in the binomial expansion.

   $(x + d)^{17}$

   Select the correct answer.

   a. $1190x^{13}d^4$

   b. $12376x^{12}d^7$

   c. $28x^{12}d^7$

   d. $2380x^{13}d^4$

4. Find the next term of the sequence.

$$a, ap, ap^2, ap^3, \ldots$$

Select the correct answer.

a. ap^4

b. p^5

c. ap^2

d. pa^4

5. Find the sum of the first five terms of the sequence with the given general term.

    $4k$

    Select the correct answer.

    a. 24
    b. 20
    c. 69
    d. 60

6. Evaluate the sum.

    $$\sum_{k=1}^{5} 4k$$

    Select the correct answer.

    a. 20
    b. 57
    c. 120
    d. 60

Gustafson/Frisk - College Algebra 8E Chapter 8 Form C

7. The 7th term of an arithmetic sequence is 44, and the first term is -4. Find the common difference.

   Select the correct answer.

   a. -8
   b. 8
   c. 10
   d. -10

8. Insert three arithmetic means between 5 and 13.

   Select the correct answer.

   a. 7, 9, 11
   b. 10, 11, 12
   c. 6, 7, 8
   d. 11, 13, 15

9. Find the sum of the first n terms of the arithmetic sequence.

   $$\sum_{n=1}^{18}\left(\frac{9}{4}n + 9\right)$$

   Select the correct answer.

   a. $384\frac{3}{4}$

   b. $506\frac{1}{4}$

   c. $546\frac{3}{4}$

   d. $931\frac{1}{2}$

10. Write the first four terms of the geometric sequence with the given properties.

    $a = 5$, $r = \dfrac{1}{2}$

    Select the correct answer(s).

    a. $\dfrac{5}{15}$

    b. $\dfrac{5}{12}$

    c. $\dfrac{5}{10}$

    d. 5

    e. $\dfrac{5}{2}$

    f. $\dfrac{5}{8}$

    g. $\dfrac{5}{4}$

11. Find the eighth term of the geometric sequence whose second and fourth terms are 6 and 54.

    Select the correct answer.

    a. $a_8 = 1,458$
    b. $a_8 = 13,122$
    c. $a_8 = 486$
    d. $a_8 = 4,374$

12. Find the sum of the indicated terms of the geometric sequence.

    $-9, 27, -81 \ldots$ (to 4 terms)

    Select the correct answer.

    a. $s = 180$
    b. $s = 189$
    c. $s = 171$
    d. $s = 162$

13. Change the decimal to a common fraction.

    $0.\overline{25}$

    Select the correct answer.

    a. $\dfrac{25}{9}$

    b. $\dfrac{25}{99}$

    c. $\dfrac{25}{999}$

    d. $\dfrac{25}{90}$

14. Consider the equation.

    $1 + 3 + 5 + \ldots + 2n - 1 = 3n - 2$

    Is the equation true for all natural numbers n?

    Select the correct answer.

    a. no
    b. yes

15. Evaluate the expression.

    $P(6, 4) \cdot C(8, 5)$

    Select the correct answer.

    a. 100,800
    b. 10,080
    c. 2,419,200
    d. 20,160

16. How many different six-digit phone numbers can be used in one area code if no phone number begins with 0 or 1?

    Select the correct answer.

    a. 400,000
    b. 2,400,000
    c. 800,000

17. If there are 8 class periods in a school day, and a typical student takes 6 classes, how many different time patterns are possible for the student?

    Select the correct answer.

    a. 26
    b. 56
    c. 28

18. An ordinary die is rolled. Find the probability of rolling an odd number.

    Select the correct answer.

    a. $\frac{1}{4}$

    b. $\frac{1}{6}$

    c. $\frac{1}{3}$

    d. $\frac{1}{2}$

19. Find the probability of drawing a heart on one draw from a card deck.

    Select the correct answer.

    a. $\frac{1}{3}$

    b. $\frac{1}{2}$

    c. $\frac{1}{5}$

    d. $\frac{1}{4}$

20. Find the probability of rolling a sum of 4 with one roll of three dice.

    Select the correct answer.

    a. $\dfrac{5}{72}$

    b. $\dfrac{7}{72}$

    c. $\dfrac{1}{72}$

21. Assume that you draw one card from a standard card deck. Find the probability of the event.

    drawing a black card

    Select the correct answer.

    a. $\dfrac{3}{4}$

    b. $\dfrac{1}{2}$

    c. $\dfrac{1}{4}$

    d. $\dfrac{1}{3}$

22. Assume that you draw two cards from a card deck (standard playing deck of 52 cards), without replacement. Find the probability of the event.

    drawing two queens

    Select the correct answer.

    a. $\dfrac{1}{104}$

    b. $\dfrac{1}{221}$

    c. $\dfrac{1}{338}$

    d. $\dfrac{1}{289}$

23. Assume that you roll two dice once. Find the probability of the result.

    rolling a sum of 8 or 6

    Select the correct answer.

    a. $\dfrac{1}{14}$

    b. $\dfrac{5}{18}$

    c. $\dfrac{1}{15}$

    d. $\dfrac{15}{36}$

24. Assume a single roll of two dice. Find the probability of rolling a sum of 5.

    Select the correct answer.

    a. $\dfrac{1}{36}$

    b. $\dfrac{1}{8}$

    c. $\dfrac{1}{9}$

25. If the odds in favor of victory are 7 to 3, find the probability of victory.

    Select the correct answer.

    a. $\dfrac{7}{10}$

    b. $\dfrac{3}{7}$

    c. $\dfrac{3}{10}$

    d. $\dfrac{10}{3}$

ANSWER KEY

Gustafson/Frisk - College Algebra 8E Chapter 8 Form C

1. a
2. c
3. d
4. a
5. d
6. d
7. b
8. a
9. c
10. d,e,f,g
11. d
12. a
13. b
14. a
15. d
16. c
17. c
18. d
19. d
20. c
21. b
22. b
23. b
24. c
25. a

List of Problem Codes for BCA Testing

Gustafson/Frisk - College Algebra 8E Chapter 8 Form C

1. gfca.08.01.4.21m_NoAlgs
2. gfca.08.01.4.26m_NoAlgs
3. gfca.08.01.4.43m_NoAlgs
4. gfca.08.02.4.14m_NoAlgs
5. gfca.08.02.4.18m_NoAlgs
6. gfca.08.02.4.37m_NoAlgs
7. gfca.08.03.4.15m_NoAlgs
8. gfca.08.03.4.21m_NoAlgs
9. gfca.08.03.4.27m_NoAlgs
10. gfca.08.04.4.10m_NoAlgs
11. gfca.08.04.4.16m_NoAlgs
12. gfca.08.04.4.25m_NoAlgs
13. gfca.08.04.4.36m_NoAlgs
14. gfca.08.05.4.28m_NoAlgs
15. gfca.08.06.4.23m_NoAlgs
16. gfca.08.06.4.31m_NoAlgs
17. gfca.08.06.4.56m_NoAlgs
18. gfca.08.07.4.12m_NoAlgs
19. gfca.08.07.4.22m_NoAlgs
20. gfca.08.07.4.34m_NoAlgs
21. gfca.08.08.4.09m_NoAlgs
22. gfca.08.08.4.13m_NoAlgs
23. gfca.08.08.4.17m_NoAlgs
24. gfca.08.09.4.11m_NoAlgs
25. gfca.08.09.4.21m_NoAlgs

Gustafson/Frisk - College Algebra 8E Chapter 8 Form D

1. Consider the equation.

    $1 + 3 + 5 + \ldots + 2n - 1 = 3n - 2$

    Is the equation true for all natural numbers n?

    Select the correct answer.

    a. yes
    b. no

2. Evaluate the sum.

    $$\sum_{k=1}^{5} 4k$$

    Select the correct answer.

    a. 20
    b. 57
    c. 120
    d. 60

3. Evaluate the expression.

    $P(6, 4) \cdot C(8, 5)$

    Select the correct answer.

    a. 10,080
    b. 2,419,200
    c. 20,160
    d. 100,800

4. The 7th term of an arithmetic sequence is 44, and the first term is -4. Find the common difference.

    Select the correct answer.

    a. -8
    b. 8
    c. 10
    d. -10

5. Use the binomial theorem to expand the binomial.

$(c + 2t)^3$

Select the correct answer.

a. $c^3 + 2t^3$

b. $c^3 + 2c^2t + 4ct^2 + 8t^3$

c. $c^3 + 6c^2t + 12ct^2 + 8t^3$

d. $c^3 + 3c^2t + 3ct^2 + t^3$

6. Assume a single roll of two dice. Find the probability of rolling a sum of 5.

Select the correct answer.

a. $\dfrac{1}{36}$

b. $\dfrac{1}{8}$

c. $\dfrac{1}{9}$

7. Change the decimal to a common fraction.

$0.\overline{25}$

Select the correct answer.

a. $\dfrac{25}{9}$

b. $\dfrac{25}{99}$

c. $\dfrac{25}{999}$

d. $\dfrac{25}{90}$

8. Find the sum of the first five terms of the sequence with the given general term.

   $4k$

   Select the correct answer.

   a. 60
   b. 69
   c. 20
   d. 24

9. Find the sum of the first n terms of the arithmetic sequence.

   $$\sum_{n=1}^{18} \left(\frac{9}{4} n + 9 \right)$$

   Select the correct answer.

   a. $384 \frac{3}{4}$

   b. $506 \frac{1}{4}$

   c. $546 \frac{3}{4}$

   d. $931 \frac{1}{2}$

10. Assume that you roll two dice once. Find the probability of the result.

    rolling a sum of 8 or 6

    Select the correct answer.

    a. $\frac{5}{18}$

    b. $\frac{1}{15}$

    c. $\frac{1}{14}$

    d. $\frac{15}{36}$

11. Find the next term of the sequence.

$$a, ap, ap^2, ap^3, \ldots$$

Select the correct answer.

a. ap^4

b. p^5

c. ap^2

d. pa^4

12. If there are 8 class periods in a school day, and a typical student takes 6 classes, how many different time patterns are possible for the student?

Select the correct answer.

a. 26
b. 56
c. 28

13. Find the 5th term in the binomial expansion.

$$(x+d)^{17}$$

Select the correct answer.

a. $12376x^{12}d^7$

b. $28x^{12}d^7$

c. $2380x^{13}d^4$

d. $1190x^{13}d^4$

14. Find the probability of rolling a sum of 4 with one roll of three dice.

    Select the correct answer.

    a. $\dfrac{5}{72}$

    b. $\dfrac{7}{72}$

    c. $\dfrac{1}{72}$

15. If the odds in favor of victory are 7 to 3, find the probability of victory.

    Select the correct answer.

    a. $\dfrac{7}{10}$

    b. $\dfrac{3}{7}$

    c. $\dfrac{3}{10}$

    d. $\dfrac{10}{3}$

16. Find the probability of drawing a heart on one draw from a card deck.

    Select the correct answer.

    a. $\dfrac{1}{3}$

    b. $\dfrac{1}{2}$

    c. $\dfrac{1}{5}$

    d. $\dfrac{1}{4}$

17. Write the first four terms of the geometric sequence with the given properties.

$$a = 5, \; r = \frac{1}{2}$$

Select the correct answer(s).

a. $\dfrac{5}{15}$

b. $\dfrac{5}{12}$

c. $\dfrac{5}{10}$

d. 5

e. $\dfrac{5}{2}$

f. $\dfrac{5}{8}$

g. $\dfrac{5}{4}$

18. Find the eighth term of the geometric sequence whose second and fourth terms are 6 and 54.

Select the correct answer.

a. $a_8 = 13,122$
b. $a_8 = 1,458$
c. $a_8 = 486$
d. $a_8 = 4,374$

19. How many different six-digit phone numbers can be used in one area code if no phone number begins with 0 or 1?

Select the correct answer.

a. 400,000
b. 800,000
c. 2,400,000

20. Insert three arithmetic means between 5 and 13.

Select the correct answer.

a. 7, 9, 11
b. 10, 11, 12
c. 6, 7, 8
d. 11, 13, 15

21. Use the binomial theorem to expand the binomial.

$(z + b)^3$

Select the correct answer.

a. $z^3 + 3z^2b + 3zb^2 + b^3$

b. $z^3 + 2z^2b + 2zb^2 + b^3$

c. $z^3 + b^3$

d. $z^3 + z^2b + zb^2 + b^3$

22. An ordinary die is rolled. Find the probability of rolling an odd number.

Select the correct answer.

a. $\frac{1}{4}$

b. $\frac{1}{6}$

c. $\frac{1}{3}$

d. $\frac{1}{2}$

23. Assume that you draw two cards from a card deck (standard playing deck of 52 cards), without replacement. Find the probability of the event.

    drawing two queens

    Select the correct answer.

    a. $\dfrac{1}{104}$

    b. $\dfrac{1}{221}$

    c. $\dfrac{1}{338}$

    d. $\dfrac{1}{289}$

24. Find the sum of the indicated terms of the geometric sequence.

    -9, 27, -81 . . . (to 4 terms)

    Select the correct answer.

    a. $s = 162$
    b. $s = 180$
    c. $s = 171$
    d. $s = 189$

25. Assume that you draw one card from a standard card deck. Find the probability of the event.

    drawing a black card

    Select the correct answer.

    a. $\dfrac{3}{4}$

    b. $\dfrac{1}{2}$

    c. $\dfrac{1}{4}$

    d. $\dfrac{1}{3}$

ANSWER KEY

Gustafson/Frisk - College Algebra 8E Chapter 8 Form D

1. b
2. d
3. c
4. b
5. c
6. c
7. b
8. a
9. c
10. a
11. a
12. c
13. c
14. c
15. a
16. d
17. d,e,f,g
18. d
19. b
20. a
21. a
22. d
23. b
24. b
25. b

List of Problem Codes for BCA Testing

Gustafson/Frisk - College Algebra 8E Chapter 8 Form D

1. gfca.08.05.4.28m_NoAlgs
2. gfca.08.02.4.37m_NoAlgs
3. gfca.08.06.4.23m_NoAlgs
4. gfca.08.03.4.15m_NoAlgs
5. gfca.08.01.4.26m_NoAlgs
6. gfca.08.09.4.11m_NoAlgs
7. gfca.08.04.4.36m_NoAlgs
8. gfca.08.02.4.18m_NoAlgs
9. gfca.08.03.4.27m_NoAlgs
10. gfca.08.08.4.17m_NoAlgs
11. gfca.08.02.4.14m_NoAlgs
12. gfca.08.06.4.56m_NoAlgs
13. gfca.08.01.4.43m_NoAlgs
14. gfca.08.07.4.34m_NoAlgs
15. gfca.08.09.4.21m_NoAlgs
16. gfca.08.07.4.22m_NoAlgs
17. gfca.08.04.4.10m_NoAlgs
18. gfca.08.04.4.16m_NoAlgs
19. gfca.08.06.4.31m_NoAlgs
20. gfca.08.03.4.21m_NoAlgs
21. gfca.08.01.4.21m_NoAlgs
22. gfca.08.07.4.12m_NoAlgs
23. gfca.08.08.4.13m_NoAlgs
24. gfca.08.04.4.25m_NoAlgs
25. gfca.08.08.4.09m_NoAlgs

1. Find the 5th term in the binomial expansion.

   $(x + d)^{17}$

2. Change the decimal to a common fraction.

   $0.\overline{32}$

3. Assume that you draw two cards from a card deck (standard playing deck of 52 cards), without replacement. Find the probability of the event.

   drawing two aces

4. Find the odds against a couple having 5 baby boys in succession. (Assume $P(\text{boys}) = \frac{1}{2}$.)

   _____ to _____

5. Use the binomial theorem to expand the binomial.

   $(c + 2t)^3$

   Select the correct answer.

   a. $c^3 + 2c^2t + 4ct^2 + 8t^3$

   b. $c^3 + 6c^2t + 12ct^2 + 8t^3$

   c. $c^3 + 2t^3$

   d. $c^3 + 3c^2t + 3ct^2 + t^3$

6. The 5th term of an arithmetic sequence is 6, and the first term is -2. Find the common difference.

7. An ordinary die is rolled. Find the probability of rolling an odd number.

8. If the odds in favor of victory are 7 to 3, find the probability of victory.

   Select the correct answer.

   a. $\dfrac{3}{7}$

   b. $\dfrac{7}{10}$

   c. $\dfrac{3}{10}$

   d. $\dfrac{10}{3}$

9. Write the first four terms of the geometric sequence with the given properties.

   $a = 5,\ r = \dfrac{1}{2}$

   Select the correct answer(s).

   a. 5

   b. $\dfrac{5}{2}$

   c. $\dfrac{5}{12}$

   d. $\dfrac{5}{15}$

   e. $\dfrac{5}{8}$

   f. $\dfrac{5}{10}$

   g. $\dfrac{5}{4}$

10. If there are 8 class periods in a school day, and a typical student takes 6 classes, how many different time patterns are possible for the student?

    Select the correct answer.

    a. 26
    b. 56
    c. 28

11. Find the probability of drawing a heart on one draw from a card deck.

    Select the correct answer.

    a. $\dfrac{1}{4}$

    b. $\dfrac{1}{5}$

    c. $\dfrac{1}{3}$

    d. $\dfrac{1}{2}$

12. Find the sum of the indicated terms of the geometric sequence.

    - 9, 27, - 81 . . . (to 4 terms)

    Select the correct answer.

    a.   s = 189
    b.   s = 162
    c.   s = 171
    d.   s = 180

13. How many different seven-digit phone numbers can be used in one area code if no phone number begins with 0 or 1?

14. Find the sum of the first n terms of the arithmetic sequence.

    $$\sum_{n=1}^{26} \left(\frac{9}{4} n + 6 \right)$$

15. Evaluate the sum.

    $$\sum_{k=1}^{5} 4k$$

    Select the correct answer.

    a.   60
    b.   120
    c.   20
    d.   57

16. Find the probability of rolling a sum of 4 with one roll of three dice.

17. Find the eighth term of the geometric sequence whose second and fourth terms are -12 and -108.

18. Insert three arithmetic means between 5 and 13.

Select the correct answer.

a. 6, 7, 8
b. 7, 9, 11
c. 10, 11, 12
d. 11, 13, 15

19. Evaluate the expression.

$P(6, 4) \cdot C(8, 5)$

Select the correct answer.

a. 2,419,200
b. 20,160
c. 100,800
d. 10,080

20. Find the next term of the sequence.

$a, ap, ap^2, ap^3, \ldots$

Select the correct answer.

a. p^5

b. pa^4

c. ap^2

d. ap^4

21. Use the binomial theorem to expand the binomial.

$(z + y)^3$

22. Assume that you draw one card from a standard card deck. Find the probability of the event.

    drawing a black card

    Select the correct answer.

    a. $\dfrac{1}{4}$

    b. $\dfrac{1}{2}$

    c. $\dfrac{3}{4}$

    d. $\dfrac{1}{3}$

23. Assume that you roll two dice once. Find the probability of the result.

    rolling a sum of 8 or 6

    Select the correct answer.

    a. $\dfrac{5}{18}$

    b. $\dfrac{1}{15}$

    c. $\dfrac{1}{14}$

    d. $\dfrac{15}{36}$

24. Assume a single roll of two dice. Find the probability of rolling a sum of 4.

25. Find the sum of the first five terms of the sequence with the following general term.

    $5k$

ANSWER KEY

Gustafson/Frisk - College Algebra 8E Chapter 8 Form E

1. $2380x^{13} \cdot d^4$
2. $\dfrac{32}{99}$
3. $\dfrac{1}{221}$
4. 31.1
5. b
6. 2
7. $\dfrac{1}{2}$
8. b
9. a,b,e,g
10. c
11. a
12. d
13. 8,000,000
14. 945.75
15. a
16. $\dfrac{1}{72}$
17. -8,748
18. b
19. b
20. d
21. $z^3 + 3z^2 \cdot y + 3z \cdot y^2 + y^3$
22. b
23. a
24. $\dfrac{1}{12}$
25. 75

List of Problem Codes for BCA Testing

Gustafson/Frisk - College Algebra 8E Chapter 8 Form E

1. gfca.08.01.4.43_NoAlgs
2. gfca.08.04.4.36_NoAlgs
3. gfca.08.08.4.13_NoAlgs
4. gfca.08.09.4.26_NoAlgs
5. gfca.08.01.4.26m_NoAlgs
6. gfca.08.03.4.15_NoAlgs
7. gfca.08.07.4.12_NoAlgs
8. gfca.08.09.4.21m_NoAlgs
9. gfca.08.04.4.10m_NoAlgs
10. gfca.08.06.4.56m_NoAlgs
11. gfca.08.07.4.22m_NoAlgs
12. gfca.08.04.4.25m_NoAlgs
13. gfca.08.06.4.31_NoAlgs
14. gfca.08.03.4.27_NoAlgs
15. gfca.08.02.4.37m_NoAlgs
16. gfca.08.07.4.34_NoAlgs
17. gfca.08.04.4.16_NoAlgs
18. gfca.08.03.4.21m_NoAlgs
19. gfca.08.06.4.23m_NoAlgs
20. gfca.08.02.4.14m_NoAlgs
21. gfca.08.01.4.21_NoAlgs
22. gfca.08.08.4.09m_NoAlgs
23. gfca.08.08.4.17m_NoAlgs
24. gfca.08.09.4.11_NoAlgs
25. gfca.08.02.4.18_NoAlgs

Gustafson/Frisk - College Algebra 8E Chapter 8 Form F

1. The 5th term of an arithmetic sequence is 6, and the first term is -2. Find the common difference.

2. Find the probability of drawing a heart on one draw from a card deck.

   Select the correct answer.

   a. $\dfrac{1}{4}$

   b. $\dfrac{1}{5}$

   c. $\dfrac{1}{3}$

   d. $\dfrac{1}{2}$

3. Find the sum of the first n terms of the arithmetic sequence.

   $$\sum_{n=1}^{26} \left(\dfrac{9}{4} n + 6 \right)$$

4. Use the binomial theorem to expand the binomial.

   $(z + y)^3$

5. Assume a single roll of two dice. Find the probability of rolling a sum of 4.

6. Write the first four terms of the geometric sequence with the given properties.

   $a = 5$, $r = \dfrac{1}{2}$

   Select the correct answer(s).

   a.   5

   b. $\dfrac{5}{2}$

   c. $\dfrac{5}{12}$

   d. $\dfrac{5}{15}$

   e. $\dfrac{5}{8}$

   f. $\dfrac{5}{10}$

   g. $\dfrac{5}{4}$

7. Find the odds against a couple having 5 baby boys in succession. (Assume P (boys) = $\dfrac{1}{2}$.)

   _____ to _____

8. How many different seven-digit phone numbers can be used in one area code if no phone number begins with 0 or 1?

9. Change the decimal to a common fraction.

   $0.\overline{32}$

10. Insert three arithmetic means between 5 and 13.

    Select the correct answer.

    a. 6, 7, 8
    b. 7, 9, 11
    c. 10, 11, 12
    d. 11, 13, 15

11. Use the binomial theorem to expand the binomial.

    $(c + 2t)^3$

    Select the correct answer.

    a. $c^3 + 2c^2t + 4ct^2 + 8t^3$
    b. $c^3 + 6c^2t + 12ct^2 + 8t^3$
    c. $c^3 + 2t^3$
    d. $c^3 + 3c^2t + 3ct^2 + t^3$

12. Find the eighth term of the geometric sequence whose second and fourth terms are -12 and -108.

13. Assume that you draw two cards from a card deck (standard playing deck of 52 cards), without replacement. Find the probability of the event.

    drawing two aces

14. An ordinary die is rolled. Find the probability of rolling an odd number.

15. Find the 5th term in the binomial expansion.

    $(x + d)^{17}$

16. If there are 8 class periods in a school day, and a typical student takes 6 classes, how many different time patterns are possible for the student?

    Select the correct answer.

    a. 26
    b. 56
    c. 28

17. Find the probability of rolling a sum of 4 with one roll of three dice.

18. Evaluate the expression.

    $P(6, 4) \cdot C(8, 5)$

    Select the correct answer.

    a. 2,419,200
    b. 20,160
    c. 100,800
    d. 10,080

19. Find the sum of the first five terms of the sequence with the following general term.

    $5k$

20. Evaluate the sum.

    $$\sum_{k=1}^{5} 4k$$

    Select the correct answer.

    a. 57
    b. 20
    c. 120
    d. 60

21. Assume that you draw one card from a standard card deck. Find the probability of the event.

    drawing a black card

    Select the correct answer.

    a. $\dfrac{1}{4}$

    b. $\dfrac{3}{4}$

    c. $\dfrac{1}{3}$

    d. $\dfrac{1}{2}$

22. If the odds in favor of victory are 7 to 3, find the probability of victory.

    Select the correct answer.

    a. $\dfrac{3}{7}$

    b. $\dfrac{7}{10}$

    c. $\dfrac{3}{10}$

    d. $\dfrac{10}{3}$

23. Assume that you roll two dice once. Find the probability of the result.

    rolling a sum of 8 or 6

    Select the correct answer.

    a. $\dfrac{5}{18}$

    b. $\dfrac{1}{15}$

    c. $\dfrac{1}{14}$

    d. $\dfrac{15}{36}$

24. Find the sum of the indicated terms of the geometric sequence.

    - 9, 27, - 81 . . . (to 4 terms)

    Select the correct answer.

    a.   s = 189
    b.   s = 162
    c.   s = 171
    d.   s = 180

25. Find the next term of the sequence.

    $a, ap, ap^2, ap^3, \ldots$

    Select the correct answer.

    a. p^5

    b. pa^4

    c. ap^2

    d. ap^4

ANSWER KEY

Gustafson/Frisk - College Algebra 8E Chapter 8 Form F

1. 2

2. a

3. 945.75

4. $z^3 + 3z^2 \cdot y + 3z \cdot y^2 + y^3$

5. $\dfrac{1}{12}$

6. a,b,e,g

7. 31, 1

8. 8,000,000

9. $\dfrac{32}{99}$

10. b

11. b

12. −8,748

13. $\dfrac{1}{221}$

14. $\dfrac{1}{2}$

15. $2380x^{13} \cdot d^4$

16. c

17. $\dfrac{1}{72}$

18. b

19. 75

ANSWER KEY

Gustafson/Frisk - College Algebra 8E Chapter 8 Form F

20. d

21. d

22. b

23. a

24. d

25. d

List of problem Codes for BCA Testing

Gustafson/Frisk - College Algebra 8E Chapter 8 Form F

1. gfca.08.01.4.43_NoAlgs
2. gfca.08.04.4.36_NoAlgs
3. gfca.08.08.4.13_NoAlgs
4. gfca.08.09.4.26_NoAlgs
5. gfca.08.01.4.26m_NoAlgs
6. gfca.08.03.4.15_NoAlgs
7. gfca.08.07.4.12_NoAlgs
8. gfca.08.09.4.21m_NoAlgs
9. gfca.08.04.4.10m_NoAlgs
10. gfca.08.06.4.56m_NoAlgs
11. gfca.08.07.4.22m_NoAlgs
12. gfca.08.04.4.25m_NoAlgs
13. gfca.08.06.4.31_NoAlgs
14. gfca.08.03.4.27_NoAlgs
15. gfca.08.02.4.37m_NoAlgs
16. gfca.08.07.4.34_NoAlgs
17. gfca.08.04.4.16_NoAlgs
18. gfca.08.03.4.21m_NoAlgs
19. gfca.08.06.4.23m_NoAlgs
20. gfca.08.02.4.14m_NoAlgs
21. gfca.08.01.4.21_NoAlgs
22. gfca.08.08.4.09m_NoAlgs
23. gfca.08.08.4.17m_NoAlgs
24. gfca.08.09.4.11_NoAlgs
25. gfca.08.02.4.18_NoAlgs

Gustafson/Frisk - College Algebra 8E Chapter 8 Form G

1. Change the decimal to a common fraction.

   $0.\overline{32}$

2. Use the binomial theorem to expand the binomial.

   $(z + b)^3$

   Select the correct answer.

   a. $z^3 + z^2b + zb^2 + b^3$

   b. $z^3 + b^3$

   c. $z^3 + 3z^2b + 3zb^2 + b^3$

   d. $z^3 + 2z^2b + 2zb^2 + b^3$

3. Evaluate the expression.

   $P(6, 4) \cdot C(8, 5)$

   Select the correct answer.

   a. 2,419,200
   b. 20,160
   c. 100,800
   d. 10,080

4. Find the 5th term in the binomial expansion.

   $(x + d)^{17}$

   Select the correct answer.

   a. $28x^{12}d^7$

   b. $2380x^{13}d^4$

   c. $1190x^{13}d^4$

   d. $12376x^{12}d^7$

5. Use the binomial theorem to expand the binomial.

$(a + 3b)^3$

6. Consider the equation.

$1 + 3 + 5 + \ldots + 2n - 1 = 3n - 2$

Is the equation true for all natural numbers n?

Select the correct answer.

a. no
b. yes

7. Assume that you roll two dice once. Find the probability of the result.

rolling a sum of 7 or 10

8. Find the next term of the sequence.

$a, ap, ap^2, ap^3, \ldots$

Select the correct answer.

a. ap^4

b. p^5

c. pa^4

d. ap^2

9. Find the sum of the first n terms of the arithmetic sequence.

$$\sum_{n=1}^{26} \left(\frac{9}{4} n + 6 \right)$$

10. Find the probability of drawing a club on one draw from a card deck.

11. Insert three arithmetic means between 5 and 13.

    Select the correct answer.

    a. 7, 9, 11
    b. 11, 13, 15
    c. 10, 11, 12
    d. 6, 7, 8

12. If there are 8 class periods in a school day, and a typical student takes 6 classes, how many different time patterns are possible for the student?

    Select the correct answer.

    a. 28
    b. 56
    c. 26

13. Assume that you draw one card from a standard card deck. Find the probability of the event.

    drawing a black card

    Select the correct answer.

    a. $\dfrac{1}{2}$

    b. $\dfrac{3}{4}$

    c. $\dfrac{1}{4}$

    d. $\dfrac{1}{3}$

14. An ordinary die is rolled. Find the probability of rolling an odd number.

    Select the correct answer.

    a. $\dfrac{1}{6}$

    b. $\dfrac{1}{4}$

    c. $\dfrac{1}{3}$

    d. $\dfrac{1}{2}$

15. Evaluate the sum.

    $$\sum_{k=1}^{9} 5k$$

16. Find the probability of rolling a sum of 4 with one roll of three dice.

    Select the correct answer.

    a. $\dfrac{7}{72}$

    b. $\dfrac{5}{72}$

    c. $\dfrac{1}{72}$

17. Find the sum of the first five terms of the sequence with the given general term.

    $4k$

    Select the correct answer.

    a. 69
    b. 24
    c. 20
    d. 60

18. How many different seven-digit phone numbers can be used in one area code if no phone number begins with 0 or 1?

19. The 7th term of an arithmetic sequence is 44, and the first term is -4. Find the common difference.

    Select the correct answer.

    a. 10
    b. -10
    c. -8
    d. 8

20. Find the sum of the indicated terms of the geometric sequence.

    - 9, 27, - 81 . . . (to 4 terms)

    Select the correct answer.

    a. $s = 180$
    b. $s = 162$
    c. $s = 171$
    d. $s = 189$

21. Assume that you draw two cards from a card deck (standard playing deck of 52 cards), without replacement. Find the probability of the event.

    drawing two queens

    Select the correct answer.

    a. $\dfrac{1}{289}$

    b. $\dfrac{1}{221}$

    c. $\dfrac{1}{104}$

    d. $\dfrac{1}{338}$

22. If the odds in favor of victory are 13 to 6, find the probability of victory.

23. Write the first four terms of the geometric sequence with the given properties.

$a = 5$, $r = \dfrac{1}{2}$

Select the correct answer(s).

- a. $\dfrac{5}{2}$
- b. $\dfrac{5}{10}$
- c. 5
- d. $\dfrac{5}{12}$
- e. $\dfrac{5}{4}$
- f. $\dfrac{5}{8}$
- g. $\dfrac{5}{15}$

24. Find the odds in favor of a couple having 5 baby girls in succession. (Assume P (girls) = $\dfrac{1}{2}$.)

Select the correct answer.

- a.   31 to 32
- b.   1 to 31
- c.   1 to 5
- d.   5 to 1

25. Assume a single roll of two dice. Find the probability of rolling a sum of 5.

Select the correct answer.

- a. $\dfrac{1}{36}$
- b. $\dfrac{1}{9}$
- c. $\dfrac{1}{8}$

ANSWER KEY

Gustafson/Frisk - College Algebra 8E Chapter 8 Form G

1. $\dfrac{32}{99}$
2. c
3. b
4. b
5. $a^3 + 9a^2 \cdot b + 27a \cdot b^2 + (27b)^3$
6. a
7. $\dfrac{1}{4}$
8. a
9. 945.75
10. $\dfrac{1}{4}$
11. a
12. a
13. a
14. d
15. 225
16. c
17. d
18. 8,000,000
19. d
20. a
21. b
22. $\dfrac{13}{19}$
23. a,c,e,f
24. b
25. b

List of Problem Codes for BCA Testing

Gustafson/Frisk - College Algebra 8E Chapter 8 Form G

1. gfca.08.04.4.36_NoAlgs
2. gfca.08.01.4.21m_NoAlgs
3. gfca.08.06.4.23m_NoAlgs
4. gfca.08.01.4.43m_NoAlgs
5. gfca.08.01.4.26_NoAlgs
6. gfca.08.05.4.28m_NoAlgs
7. gfca.08.08.4.17_NoAlgs
8. gfca.08.02.4.14m_NoAlgs
9. gfca.08.03.4.27_NoAlgs
10. gfca.08.07.4.22_NoAlgs
11. gfca.08.03.4.21m_NoAlgs
12. gfca.08.06.4.56m_NoAlgs
13. gfca.08.08.4.09m_NoAlgs
14. gfca.08.07.4.12m_NoAlgs
15. gfca.08.02.4.37_NoAlgs
16. gfca.08.07.4.34m_NoAlgs
17. gfca.08.02.4.18m_NoAlgs
18. gfca.08.06.4.31_NoAlgs
19. gfca.08.03.4.15m_NoAlgs
20. gfca.08.04.4.25m_NoAlgs
21. gfca.08.08.4.13m_NoAlgs
22. gfca.08.09.4.21_NoAlgs
23. gfca.08.04.4.10m_NoAlgs
24. gfca.08.09.4.26m_NoAlgs
25. gfca.08.09.4.11m_NoAlgs

1. Find the sum of the first *n* terms of the arithmetic sequence.

$$\sum_{n=1}^{26}\left(\frac{9}{4}n+6\right)$$

2. Find the probability of rolling a sum of 4 with one roll of three dice.

   Select the correct answer.

   a. $\frac{7}{72}$

   b. $\frac{5}{72}$

   c. $\frac{1}{72}$

3. Find the odds in favor of a couple having 5 baby girls in succession. (Assume P (girls) = $\frac{1}{2}$.)

   Select the correct answer.

   a. 31 to 32
   b. 1 to 31
   c. 1 to 5
   d. 5 to 1

4. Change the decimal to a common fraction.

   $0.\overline{32}$

5. The 7th term of an arithmetic sequence is 44, and the first term is -4. Find the common difference.

   Select the correct answer.

   a. 10
   b. -10
   c. -8
   d. 8

6. Assume that you draw two cards from a card deck (standard playing deck of 52 cards), without replacement. Find the probability of the event.

   drawing two queens

   Select the correct answer.

   a. $\dfrac{1}{289}$

   b. $\dfrac{1}{221}$

   c. $\dfrac{1}{104}$

   d. $\dfrac{1}{338}$

7. Find the next term of the sequence.

   $a, ap, ap^2, ap^3, \ldots$

   Select the correct answer.

   a. ap^4

   b. p^5

   c. pa^4

   d. ap^2

8. Find the probability of drawing a club on one draw from a card deck.

9. Use the binomial theorem to expand the binomial.

   $(z + b)^3$

   Select the correct answer.

   a. $z^3 + z^2b + zb^2 + b^3$

   b. $z^3 + b^3$

   c. $z^3 + 3z^2b + 3zb^2 + b^3$

   d. $z^3 + 2z^2b + 2zb^2 + b^3$

10. Find the sum of the indicated terms of the geometric sequence.

    -9, 27, -81 . . . (to 4 terms)

    Select the correct answer.

    a. $s = 180$
    b. $s = 162$
    c. $s = 171$
    d. $s = 189$

11. Assume a single roll of two dice. Find the probability of rolling a sum of 5.

    Select the correct answer.

    a. $\dfrac{1}{36}$

    b. $\dfrac{1}{9}$

    c. $\dfrac{1}{8}$

12. Use the binomial theorem to expand the binomial.

    $(a + 3b)^3$

13. Evaluate the expression.

    $P(6, 4) \cdot C(8, 5)$

    Select the correct answer.

    a. 2,419,200
    b. 20,160
    c. 100,800
    d. 10,080

14. Assume that you roll two dice once. Find the probability of the result.

    rolling a sum of 7 or 10

15. Find the sum of the first five terms of the sequence with the given general term.

    $4k$

    Select the correct answer.

    a. 69
    b. 24
    c. 20
    d. 60

16. Write the first four terms of the geometric sequence with the given properties.

    $a = 5, r = \dfrac{1}{2}$

    Select the correct answer(s).

    a. $\dfrac{5}{2}$

    b. $\dfrac{5}{10}$

    c. 5

    d. $\dfrac{5}{12}$

    e. $\dfrac{5}{4}$

    f. $\dfrac{5}{8}$

    g. $\dfrac{5}{15}$

17. Assume that you draw one card from a standard card deck. Find the probability of the event.

drawing a black card

Select the correct answer.

a. $\dfrac{1}{2}$

b. $\dfrac{3}{4}$

c. $\dfrac{1}{4}$

d. $\dfrac{1}{3}$

18. Evaluate the sum.

$$\sum_{k=1}^{9} 5k$$

19. An ordinary die is rolled. Find the probability of rolling an odd number.

Select the correct answer.

a. $\dfrac{1}{6}$

b. $\dfrac{1}{4}$

c. $\dfrac{1}{3}$

d. $\dfrac{1}{2}$

20. If the odds in favor of victory are 13 to 6, find the probability of victory.

21. Find the 5th term in the binomial expansion.

    $(x + d)^{17}$

    Select the correct answer.

    a. $28x^{12}d^7$

    b. $2380x^{13}d^4$

    c. $1190x^{13}d^4$

    d. $12376x^{12}d^7$

22. How many different seven-digit phone numbers can be used in one area code if no phone number begins with 0 or 1?

23. Insert three arithmetic means between 5 and 13.

    Select the correct answer.

    a. 7, 9, 11
    b. 11, 13, 15
    c. 10, 11, 12
    d. 6, 7, 8

24. Consider the equation.

    $1 + 3 + 5 + ... + 2n - 1 = 3n - 2$

    Is the equation true for all natural numbers n?

    Select the correct answer.
    a. no
    b. yes

25. If there are 8 class periods in a school day, and a typical student takes 6 classes, how many different time patterns are possible for the student?

    Select the correct answer.

    a. 28
    b. 56
    c. 26

ANSWER KEY

Gustafson/Frisk - College Algebra 8E Chapter 8 Form H

1. 945.75

2. c

3. b

4. $\dfrac{32}{99}$

5. d

6. b

7. a

8. $\dfrac{1}{4}$

9. c

10. a

11. b

12. $a^3 + 9a^2 \cdot b + 27a \cdot b^2 + (27b)^3$

13. b

14. $\dfrac{1}{4}$

15. d

16. a,c,e,f

17. a

18. 225

19. d

20. $\dfrac{13}{19}$

21. b

ANSWER KEY

Gustafson/Frisk - College Algebra 8E Chapter 8 Form H

22. 8,000,000

23. a

24. a

25. a

List of Problem Codes for BCA Testing

Gustafson/Frisk - College Algebra 8E Chapter 8 Form H

1. gfca.08.03.4.27_NoAlgs
2. gfca.08.07.4.34m_NoAlgs
3. gfca.08.09.4.26m_NoAlgs
4. gfca.08.04.4.36_NoAlgs
5. gfca.08.03.4.15m_NoAlgs
6. gfca.08.08.4.13m_NoAlgs
7. gfca.08.02.4.14m_NoAlgs
8. gfca.08.07.4.22_NoAlgs
9. gfca.08.01.4.21m_NoAlgs
10. gfca.08.04.4.25m_NoAlgs
11. gfca.08.09.4.11m_NoAlgs
12. gfca.08.01.4.26_NoAlgs
13. gfca.08.06.4.23m_NoAlgs
14. gfca.08.08.4.17_NoAlgs
15. gfca.08.02.4.18m_NoAlgs
16. gfca.08.04.4.10m_NoAlgs
17. gfca.08.08.4.09m_NoAlgs
18. gfca.08.02.4.37_NoAlgs
19. gfca.08.07.4.12m_NoAlgs
20. gfca.08.09.4.21_NoAlgs
21. gfca.08.01.4.43m_NoAlgs
22. gfca.08.06.4.31_NoAlgs
23. gfca.08.03.4.21m_NoAlgs
24. gfca.08.05.4.28m_NoAlgs
25. gfca.08.06.4.56m_NoAlgs

Gustafson/Frisk - College Algebra 8E Chapter 9 Form A

1. Assume that $1,100 is deposited in an account in which interest is compounded 10 times a year, at an annual rate of 8%. Find the accumulated amount after 11 years, rounded to the nearest cent.

2. Find the present value of $100,000 due in 3 years, given an annual rate of 6% compounded quarterly. Round your answer to the nearest cent.

3. Assume that $1,500 is deposited in an account in which interest is compounded annually at a rate of 8%. Find the accumulated amount after 7 years, rounded to the nearest cent.

4. When the Fernandez family made reservations at the end of 1995 for the December 2001 New Year's celebration in Paris, they placed $4,700 into an account paying 6% interest, compounded monthly. What amount was available at the time of the celebration? Round your answer to the nearest cent.

5. The managers of a pension fund invested $2 million in government bonds paying 5.93% annual interest, compounded semiannually. After 3 years, what will the investment be worth?

6. Property in suburbs closer to the city is appreciating about 9% annually. If this trend continues, what will a $50,000 one-acre lot be worth in five years? Give the result to the nearest dollar.

7. Bank One offers a passbook account with a 4.75% annual rate, compounded quarterly. Bank Two offers a money market account at 5%, compounded monthly. What is the greater effective rate of these two banks? Give the greater effective rate, rounded to the nearest hundredth.

8. Craig borrows $1,110 for unexpected car repair costs. His bank writes a 70-day note at 13%, with interest compounded daily. What will Craig owe? Round your answer to the nearest cent.

9. Assume that $100 is deposited at the end of each year in an account in which interest is compounded annually at a rate of 5%. Find the accumulated amount after 4 years. Round your answer to the nearest cent.

10. Assume that $100 is deposited at the end of each year into an account in which interest is compounded annually at a rate of 4%. Find the accumulated amount after 15 years. Round to the nearest cent.

11. Assume that $500 is deposited at the end of each period in an account in which interest is compounded at the frequency $k = 6$, at an annual rate of 7%. Find the accumulated amount after 13 years. Round your answer to the nearest cent.

12. Find the amount of each regular payment to provide $10,000 in 8 years, at an annual rate $r = 7\%$, compounded semiannually. Round to the nearest cent.

13. Find the amount of each regular payment to provide $7,500 in 7 years, at an annual rate $r = 6\%$, compounded annually. Round to the nearest cent.

14. For next year's vacation, the Phelps family is saving $400 each month in an account paying 8.5% annual interest, compounded monthly. How much will be available a year from now? Round your answer to the nearest cent.

15. A company's new corporate headquarters will be completed in 5.5 years. At that time, $520,000 will be needed for office equipment. How much should be invested monthly to fund that expense? Assume 8.5% interest, compounded monthly. Round your answer to the nearest cent.

16. The last payment of a home mortgage is a balloon payment of $35,000, which the owner is scheduled to pay in 16 years. How much extra should he start including in each monthly payment to eliminate the balloon payment? His mortgage is at 8.6%, compounded monthly. Round to the nearest cent.

17. Find the present value of an annuity with annual payments of $3,670 at 5.65%, compounded annually for 10 years.

18. Find the periodic payment required to repay $10,180 repaid over 18 years, with monthly payments at a 12% annual rate. Round to the nearest cent.

19. Instead of making quarterly contributions of $730 to a retirement fund for the next 20 years, Jason would rather make only one contribution, now. How much should that be? Assume 6.5 % annual interest, compounded quarterly. Round to the nearest cent.

20. Instead of receiving an annuity of $10,160 each year for the next 15 years, Manuel would like a one-time payment, now. Assuming Manuel could invest the proceeds at 10.5%, what would be a fair amount? Round to the nearest cent.

21. The Jepsens are buying a $25,000 car and financing it over the next 2 years. They secure an 8.3% loan. What will their monthly payments be? Round to the nearest cent.

22. One lender offers two mortgages - a 10-year mortgage at 15%, and a 18-year mortgage at 9%. For each, find the monthly payment to repay $114,300.

23. One lender offers two mortgages - a 8-year mortgage at 13%, and a 23-year mortgage at 12%. For each, find the total of the monthly payments to repay $110,500.

24. As Jorge starts working now at the age of 23, he decides to make regular contributions to a savings account. He wants to accumulate enough by age 55 to fund an annuity of $4,700 per month until age 80. What should his monthly contributions be? Assume that both accounts pay 8.5%, compounded monthly. Round to the nearest cent.

25. Amy contributed $140 per month for 18 years to an account that paid 6% for the first 9 years, but 6.7% for the last 9 years. How much has she saved?

ANSWER KEY

Gustafson/Frisk - College Algebra 8E Chapter 9 Form A

1. 2,642.72

2. 83,638.74

3. 2,570.74

4. 6,730.61

5. $2,383

6. $76,931

7. 5.12

8. 1,138.41

9. 431.01

10. 2,002.36

11. 63,054.47

12. 476.85

13. 893.51

14. 4,991.49

15. 6,207.47

16. 85.33

17. $PV = \$27{,}465.13$

18. 115.23

19. 32,551.51

20. 75,121.26

ANSWER KEY

Gustafson/Frisk - College Algebra 8E Chapter 9 Form A

21. 1,134.11

22. $1,844.06, $1,070.36

23. $178,291.36, $325,890.92

24. 294.57

25. $57,135.47

List of Problem Codes for BCA Testing

Gustafson/Frisk - College Algebra 8E Chapter 9 Form A

1. gfca.09.01.4.23_NoAlgs
2. gfca.09.01.4.31_NoAlgs
3. gfca.09.01.4.15_NoAlgs
4. gfca.09.01.4.36_NoAlgs
5. gfca.09.01.4.38_NoAlgs
6. gfca.09.01.4.40_NoAlgs
7. gfca.09.01.4.42_NoAlgs
8. gfca.09.01.4.45_NoAlgs
9. gfca.09.02.4.08_NoAlgs
10. gfca.09.02.4.11_NoAlgs
11. gfca.09.02.4.15_NoAlgs
12. gfca.09.02.4.19_NoAlgs
13. gfca.09.02.4.20_NoAlgs
14. gfca.09.02.4.23_NoAlgs
15. gfca.09.02.4.27_NoAlgs
16. gfca.09.02.4.30_NoAlgs
17. gfca.09.03.4.07m_NoAlgs
18. gfca.09.03.4.09_NoAlgs
19. gfca.09.03.4.11_NoAlgs
20. gfca.09.03.4.13_NoAlgs
21. gfca.09.03.4.15_NoAlgs
22. gfca.09.03.4.17_NoAlgs
23. gfca.09.03.4.18_NoAlgs
24. gfca.09.03.4.19_NoAlgs
25. gfca.09.03.4.22_NoAlgs

1. A company's new corporate headquarters will be completed in 5.5 years. At that time, $520,000 will be needed for office equipment. How much should be invested monthly to fund that expense? Assume 8.5% interest, compounded monthly. Round your answer to the nearest cent.

2. For next year's vacation, the Phelps family is saving $400 each month in an account paying 8.5% annual interest, compounded monthly. How much will be available a year from now? Round your answer to the nearest cent.

3. As Jorge starts working now at the age of 23, he decides to make regular contributions to a savings account. He wants to accumulate enough by age 55 to fund an annuity of $4,700 per month until age 80. What should his monthly contributions be? Assume that both accounts pay 8.5%, compounded monthly. Round to the nearest cent.

4. Amy contributed $140 per month for 18 years to an account that paid 6% for the first 9 years, but 6.7% for the last 9 years. How much has she saved?

5. Instead of making quarterly contributions of $730 to a retirement fund for the next 20 years, Jason would rather make only one contribution, now. How much should that be? Assume 6.5 % annual interest, compounded quarterly. Round to the nearest cent.

6. One lender offers two mortgages - a 10-year mortgage at 15%, and a 18-year mortgage at 9%. For each, find the monthly payment to repay $114,300.

7. Assume that $100 is deposited at the end of each year in an account in which interest is compounded annually at a rate of 5%. Find the accumulated amount after 4 years. Round your answer to the nearest cent.

8. Assume that $100 is deposited at the end of each year into an account in which interest is compounded annually at a rate of 4%. Find the accumulated amount after 15 years. Round to the nearest cent.

9. Find the amount of each regular payment to provide $10,000 in 8 years, at an annual rate $r = 7\%$, compounded semiannually. Round to the nearest cent.

10. The Jepsens are buying a $25,000 car and financing it over the next 2 years. They secure an 8.3% loan. What will their monthly payments be? Round to the nearest cent.

11. Property in suburbs closer to the city is appreciating about 9% annually. If this trend continues, what will a $50,000 one-acre lot be worth in five years? Give the result to the nearest dollar.

12. The last payment of a home mortgage is a balloon payment of $35,000, which the owner is scheduled to pay in 16 years. How much extra should he start including in each monthly payment to eliminate the balloon payment? His mortgage is at 8.6%, compounded monthly. Round to the nearest cent.

13. Find the present value of $100,000 due in 3 years, given an annual rate of 6% compounded quarterly. Round your answer to the nearest cent.

14. Assume that $1,100 is deposited in an account in which interest is compounded 10 times a year, at an annual rate of 8%. Find the accumulated amount after 11 years, rounded to the nearest cent.

15. Assume that $1,500 is deposited in an account in which interest is compounded annually at a rate of 8%. Find the accumulated amount after 7 years, rounded to the nearest cent.

16. Assume that $500 is deposited at the end of each period in an account in which interest is compounded at the frequency $k = 6$, at an annual rate of 7%. Find the accumulated amount after 13 years. Round your answer to the nearest cent.

17. Bank One offers a passbook account with a 4.75% annual rate, compounded quarterly. Bank Two offers a money market account at 5%, compounded monthly. What is the greater effective rate of these two banks? Give the greater effective rate, rounded to the nearest hundredth.

18. When the Fernandez family made reservations at the end of 1995 for the December 2001 New Year's celebration in Paris, they placed $4,700 into an account paying 6% interest, compounded monthly. What amount was available at the time of the celebration? Round your answer to the nearest cent.

19. Find the amount of each regular payment to provide $7,500 in 7 years, at an annual rate $r = 6\%$, compounded annually. Round to the nearest cent.

20. One lender offers two mortgages - a 8-year mortgage at 13%, and a 23-year mortgage at 12%. For each, find the total of the monthly payments to repay $110,500.

21. Find the present value of an annuity with annual payments of $3,670 at 5.65%, compounded annually for 10 years.

22. Craig borrows $1,110 for unexpected car repair costs. His bank writes a 70-day note at 13%, with interest compounded daily. What will Craig owe? Round your answer to the nearest cent.

23. The managers of a pension fund invested $2 million in government bonds paying 5.93% annual interest, compounded semiannually. After 3 years, what will the investment be worth?

24. Find the periodic payment required to repay $10,180 repaid over 18 years, with monthly payments at a 12% annual rate. Round to the nearest cent.

25. Instead of receiving an annuity of $10,160 each year for the next 15 years, Manuel would like a one-time payment, now. Assuming Manuel could invest the proceeds at 10.5%, what would be a fair amount? Round to the nearest cent.

ANSWER KEY

Gustafson/Frisk - College Algebra 8E Chapter 9 Form B

1. 6,207.47

2. 4,991.49

3. 294.57

4. $57,135.47

5. 32,551.51

6. $1,844.06, $1,070.36

7. 431.01

8. 2,002.36

9. 476.85

10. 1,134.11

11. $76,931

12. 85.33

13. 83,638.74

14. 2,642.72

15. 2,570.74

16. 63,054.47

17. 5.12

18. 6,730.61

19. 893.51

ANSWER KEY

Gustafson/Frisk - College Algebra 8E Chapter 9 Form B

20. $178,291.36, $325,890.92

21. *PV* = $27,465.13

22. 1,138.41

23. $2.383

24. 115.23

25. 75,121.26

List of Problem Codes for BCA Testing

Gustafson/Frisk - College Algebra 8E Chapter 9 Form B

1. gfca.09.02.4.27_NoAlgs
2. gfca.09.02.4.23_NoAlgs
3. gfca.09.03.4.19_NoAlgs
4. gfca.09.03.4.22_NoAlgs
5. gfca.09.03.4.11_NoAlgs
6. gfca.09.03.4.17_NoAlgs
7. gfca.09.02.4.08_NoAlgs
8. gfca.09.02.4.11_NoAlgs
9. gfca.09.02.4.19_NoAlgs
10. gfca.09.03.4.15_NoAlgs
11. gfca.09.01.4.40_NoAlgs
12. gfca.09.02.4.30_NoAlgs
13. gfca.09.01.4.31_NoAlgs
14. gfca.09.01.4.23_NoAlgs
15. gfca.09.01.4.15_NoAlgs
16. gfca.09.02.4.15_NoAlgs
17. gfca.09.01.4.42_NoAlgs
18. gfca.09.01.4.36_NoAlgs
19. gfca.09.02.4.20_NoAlgs
20. gfca.09.03.4.18_NoAlgs
21. gfca.09.03.4.07m_NoAlgs
22. gfca.09.01.4.45_NoAlgs
23. gfca.09.01.4.38_NoAlgs
24. gfca.09.03.4.09_NoAlgs
25. gfca.09.03.4.13_NoAlgs

Gustafson/Frisk - College Algebra 8E Chapter 9 Form C

1. What is the effective rate, if after one year $200 grows to $224?

    Select the correct answer.

    a. 12 %
    b. 15 %
    c. 11 %
    d. 9 %

2. Assume that $1,400 is deposited in an account in which interest is compounded annually at a rate of 8%. Find the accumulated amount after 11 years.

    Select the correct answer.

    a. $3,164.29
    b. $3,664.29
    c. $3,264.29
    d. $3,364.29

3. Find the effective interest rate given an annual rate of 8% and a compounding frequency of 8.

    Select the correct answer.

    a. 8.29%
    b. 8.27%
    c. 8.34%
    d. 8.31%

4. At the birth of their child, the Fieldsons deposited $1,000 in an account paying 8% interest, compounded quarterly. How much will be available when the child turns 10?

    Select the correct answer.

    a. $2,508.44
    b. $2,208.04
    c. $2,107.50
    d. $2,308.58

5. When Jim retires in 13 years, he expects to live lavishly on the money in a retirement account that is earning 9% interest, compounded semiannually. If the account now contains $148,500, how much will be available at retirement?

   Select the correct answer.

   a. $466,390.83
   b. $466,490.83
   c. $466,491.17
   d. $467,390.95

6. Property values in the suburbs have been appreciating about 13% annually. If this trend continues, what will a $161,000 home be worth in four years? Give the result to the nearest dollar.

   Select the correct answer.

   a. $263,740
   b. $265,927
   c. $262,506
   d. $263,506

7. The gas utilities expect natural gas consumption to increase at 7.5% per year for the next decade. Monthly consumption for one county is currently 4.9 million cubic feet. What monthly demand for gas is expected in ten years?

   Select the correct answer.

   a. 11.20 million ft^3
   b. 10.22 million ft^3
   c. 10.10 million ft^3
   d. 11.10 million ft^3

8. Amy contributed $170 per month for 16 years to an account that paid 5% for the first 8 years, but 5.4% for the last 8 years. How much has she saved?

   Select the correct answer.

   a. $51,167.62
   b. $51,169.61
   c. $49,923.05
   d. $51,157.61

9. A credit union offers the two accounts shown in the table below. Find the effective rates.

	Annual rate	Compounding
Certificate of deposit	7.6	semiannually
Passbook	7	quarterly

Select the correct answer.

a. 8% for Certificate of deposit, 7.4% for Passbook
b. 7.2% for Certificate of deposit, 7.7% for Passbook
c. 8.1% for Certificate of deposit, 7.3% for Passbook
d. 7.8% for Certificate of deposit, 7.3% for Passbook
e. 7.7% for Certificate of deposit, 7.2% for Passbook

10. Assume that $150 is deposited at the end of each year in an account in which interest is compounded annually at a rate of 8%. Find the accumulated amount after 7 years.

Select the correct answer.

a. $1,338.42
b. $3,122.98
c. $1,488.42
d. $2,772.98

11. Assume that $400 is deposited at the end of each year in an account in which interest is compounded annually at a rate of 6%. Find the accumulated amount after 6 years.

Select the correct answer.

a. $2,769.38
b. $2,790.13
c. $2,810.03
d. $2,803.18

12. Assume that $200 is deposited at the end of each year into an account in which interest is compounded annually at a rate of 7%. Find the accumulated amount after 4 years.

Select the correct answer.

a. $859.84
b. $932.19
c. $921.04
d. $887.99

13. Find the amount of each regular payment to provide $18,500 in 13 years, at an annual rate $r = 6\%$, compounded semiannually.

    Select the correct answer.

    a. $492.01
    b. $492.31
    c. $494.56
    d. $479.86

14. Find the amount of each regular payment to provide $11,000 in 10 years, at an annual rate $r = 4\%$, compounded annually.

    Select the correct answer.

    a. $907.30
    b. $908.50
    c. $909.95
    d. $916.20

15. The managers of a company's pension fund invest the monthly employee contributions of $125,000 into a government fund paying 7.3%, compounded monthly. To what value (rounded to the nearest thousand) will the fund grow in 20 years?

    Select the correct answer.

    a. $67,535,000
    b. $67,546,000
    c. $67,562,000
    d. $67,540,000

16. A woman would like to receive a $150,000 lump-sum distribution from her retirement account when she retires in 17 years. She begins making monthly contributions now to an annuity paying 7%, compounded monthly. Find the amount of that contribution.

    Select the correct answer.

    a. $384.49
    b. $419.04
    c. $394.54
    d. $407.84

17. Jim will retire in 16 years. He will invest $200 each month for 8 years and then let the accumulated value continue to grow for the next 8 years. How much will be available at retirement? Assume 7%, compounded monthly.

    Select the correct answer.

    a. $44,803
    b. $44,773
    c. $44,834
    d. $44,814

18. What does the principal grow to after a specific time?

    Select the correct answer.

    a. amortization
    b. present value
    c. mortgage
    d. future value

19. Find the present value of an annuity with semiannual payments of $385 at 4.21%, compounded semiannually for 14 years.

    Select the correct answer.

    a. $PV = \$16,165.84$
    b. $PV = \$8,082.92$
    c. $PV = \$11,006.92$
    d. $PV = \$14,483.87$

20. Find the periodic payment required to repay $1,830 repaid in 18 monthly installments, at an annual rate of 18%.

    Select the correct answer.

    a. $P = \$163.67$
    b. $P = \$116.76$
    c. $P = \$374.96$
    d. $P = \$233.52$

21. To fund Jamie's lottery winnings of $17,300 per month for the next 16 years, the lottery commission needs to make a single deposit now. Assuming 9.1% compounded monthly, what should the deposit be?

    Select the correct answer.

    a. $1,746,460.39
    b. $1,746,472.49
    c. $1,746,437.29
    d. $1,746,482.79

22. What single amount deposited now into an account paying $9\frac{1}{3}$ % annual interest, compounded quarterly, would fund an annuity paying $3,520 quarterly for the next 23 years?

    Select the correct answer.

    a. $132,801.78
    b. $132,785.79
    c. $132,929.31
    d. $132,699.41

23. The Jepsens are buying a $21,600 car and financing it over the next 5 years. They secure a 8.6% loan. What will be the total of the monthly payments?

    Select the correct answer.

    a. $79,955.79
    b. $26,651.93
    c. $13,325.97
    d. $8,883.98

24. One lender offers two mortgages - a 8-year mortgage at 15%, and a 14-year mortgage at 12%. For each, find the total of the monthly payments to repay $145,000.

    Select the correct answer.

    a. 8-year: $249,788.04 , 14-year: $300,099.92
    b. 8-year: $249,800.04 , 14-year: $299,976.23
    c. 8-year: $249,756.70 , 14-year: $300,304.86
    d. 8-year: $251,000.92 , 14-year: $301,185.23

25. As Jorge starts working now at the age of 24, he decides to make regular contributions to a savings account. He wants to accumulate enough by age 55 to fund an annuity of $3,900 per month until age 80. What should his monthly contributions be? Assume that both accounts pay 8.25%, compounded monthly.

   Select the correct answer.

   a. $288.41
   b. $284.41
   c. $302.41
   d. $279.41

ANSWER KEY

Gustafson/Frisk - College Algebra 8E Chapter 9 Form C

1. a
2. c
3. a
4. b
5. a
6. c
7. c
8. d
9. e
10. a
11. b
12. d
13. d
14. d
15. d
16. a
17. d
18. d
19. b
20. b
21. a
22. b
23. b
24. b
25. a

List of Problem Codes for BCA Testing

Gustafson/Frisk - College Algebra 8E Chapter 9 Form C

1. gfca.09.01.4.14m_NoAlgs
2. gfca.09.01.4.19m_NoAlgs
3. gfca.09.01.4.27m_NoAlgs
4. gfca.09.01.4.35m_NoAlgs
5. gfca.09.01.4.37m_NoAlgs
6. gfca.09.01.4.39m_NoAlgs
7. gfca.09.01.4.41m_NoAlgs
8. gfca.09.03.4.22m_NoAlgs
9. gfca.09.01.4.44m_NoAlgs
10. gfca.09.02.4.07m_NoAlgs
11. gfca.09.02.4.09m_NoAlgs
12. gfca.09.02.4.13m_NoAlgs
13. gfca.09.02.4.21m_NoAlgs
14. gfca.09.02.4.22m_NoAlgs
15. gfca.09.02.4.25m_NoAlgs
16. gfca.09.02.4.28m_NoAlgs
17. gfca.09.02.4.31m_NoAlgs
18. gfca.09.03.4.01m_NoAlgs
19. gfca.09.03.4.08m_NoAlgs
20. gfca.09.03.4.10m_NoAlgs
21. gfca.09.03.4.12m_NoAlgs
22. gfca.09.03.4.14m_NoAlgs
23. gfca.09.03.4.16m_NoAlgs
24. gfca.09.03.4.18m_NoAlgs
25. gfca.09.03.4.19m_NoAlgs

Gustafson/Frisk - College Algebra 8E Chapter 9 Form D

1. Assume that $150 is deposited at the end of each year in an account in which interest is compounded annually at a rate of 8%. Find the accumulated amount after 7 years.

   Select the correct answer.

   a. $1,338.42
   b. $3,122.98
   c. $1,488.42
   d. $2,772.98

2. When Jim retires in 13 years, he expects to live lavishly on the money in a retirement account that is earning 9% interest, compounded semiannually. If the account now contains $148,500, how much will be available at retirement?

   Select the correct answer.

   a. $466,491.17
   b. $466,490.83
   c. $467,390.95
   d. $466,390.83

3. The gas utilities expect natural gas consumption to increase at 7.5% per year for the next decade. Monthly consumption for one county is currently 4.9 million cubic feet. What monthly demand for gas is expected in ten years?

   Select the correct answer.

   a. 11.20 million ft^3
   b. 10.22 million ft^3
   c. 10.10 million ft^3
   d. 11.10 million ft^3

4. Find the amount of each regular payment to provide $11,000 in 10 years, at an annual rate $r = 4\%$, compounded annually.

   Select the correct answer.

   a. $907.30
   b. $908.50
   c. $909.95
   d. $916.20

5. Amy contributed $170 per month for 16 years to an account that paid 5% for the first 8 years, but 5.4% for the last 8 years. How much has she saved?

   Select the correct answer.

   a. $51,167.62
   b. $51,169.61
   c. $49,923.05
   d. $51,157.61

6. Find the amount of each regular payment to provide $18,500 in 13 years, at an annual rate $r = 6\%$, compounded semiannually.

   Select the correct answer.

   a. $492.01
   b. $492.31
   c. $494.56
   d. $479.86

7. What does the principal grow to after a specific time?

   Select the correct answer.

   a. amortization
   b. present value
   c. mortgage
   d. future value

8. Assume that $1,400 is deposited in an account in which interest is compounded annually at a rate of 8%. Find the accumulated amount after 11 years.

   Select the correct answer.

   a. $3,164.29
   b. $3,664.29
   c. $3,264.29
   d. $3,364.29

9. At the birth of their child, the Fieldsons deposited $1,000 in an account paying 8% interest, compounded quarterly. How much will be available when the child turns 10?

   Select the correct answer.

   a. $2,508.44
   b. $2,208.04
   c. $2,107.50
   d. $2,308.58

10. Find the periodic payment required to repay $1,830 repaid in 18 monthly installments, at an annual rate of 18%.

    Select the correct answer.

    a. $P = \$163.67$
    b. $P = \$116.76$
    c. $P = \$374.96$
    d. $P = \$233.52$

11. Assume that $400 is deposited at the end of each year in an account in which interest is compounded annually at a rate of 6%. Find the accumulated amount after 6 years.

    Select the correct answer.

    a. $2,790.13
    b. $2,810.03
    c. $2,769.38
    d. $2,803.18

12. A woman would like to receive a $150,000 lump-sum distribution from her retirement account when she retires in 17 years. She begins making monthly contributions now to an annuity paying 7%, compounded monthly. Find the amount of that contribution.

    Select the correct answer.

    a. $384.49
    b. $419.04
    c. $394.54
    d. $407.84

13. The managers of a company's pension fund invest the monthly employee contributions of $125,000 into a government fund paying 7.3%, compounded monthly. To what value (rounded to the nearest thousand) will the fund grow in 20 years?

    Select the correct answer.

    a. $67,546,000
    b. $67,540,000
    c. $67,535,000
    d. $67,562,000

14. To fund Jamie's lottery winnings of $17,300 per month for the next 16 years, the lottery commission needs to make a single deposit now. Assuming 9.1% compounded monthly, what should the deposit be?

    Select the correct answer.

    a. $1,746,482.79
    b. $1,746,460.39
    c. $1,746,472.49
    d. $1,746,437.29

15. Find the effective interest rate given an annual rate of 8% and a compounding frequency of 8.

    Select the correct answer.

    a. 8.31%
    b. 8.27%
    c. 8.34%
    d. 8.29%

16. Find the present value of an annuity with semiannual payments of $385 at 4.21%, compounded semiannually for 14 years.

    Select the correct answer.

    a. $PV = \$14,483.87$
    b. $PV = \$8,082.92$
    c. $PV = \$16,165.84$
    d. $PV = \$11,006.92$

17. Property values in the suburbs have been appreciating about 13% annually. If this trend continues, what will a $161,000 home be worth in four years? Give the result to the nearest dollar.

    Select the correct answer.

    a. $263,506
    b. $263,740
    c. $265,927
    d. $262,506

18. Assume that $200 is deposited at the end of each year into an account in which interest is compounded annually at a rate of 7%. Find the accumulated amount after 4 years.

    Select the correct answer.

    a. $859.84
    b. $932.19
    c. $921.04
    d. $887.99

19. Jim will retire in 16 years. He will invest $200 each month for 8 years and then let the accumulated value continue to grow for the next 8 years. How much will be available at retirement? Assume 7%, compounded monthly.

    Select the correct answer.

    a. $44,834
    b. $44,773
    c. $44,803
    d. $44,814

20. The Jepsens are buying a $21,600 car and financing it over the next 5 years. They secure a 8.6% loan. What will be the total of the monthly payments?

    Select the correct answer.

    a. $13,325.97
    b. $8,883.98
    c. $79,955.79
    d. $26,651.93

Gustafson/Frisk - College Algebra 8E Chapter 9 Form D

21. One lender offers two mortgages - a 8-year mortgage at 15%, and a 14-year mortgage at 12%. For each, find the total of the monthly payments to repay $145,000.

    Select the correct answer.

    a. 8-year: $249,788.04 , 14-year: $300,099.92
    b. 8-year: $249,800.04 , 14-year: $299,976.23
    c. 8-year: $249,756.70 , 14-year: $300,304.86
    d. 8-year: $251,000.92 , 14-year: $301,185.23

22. What is the effective rate, if after one year $200 grows to $224?

    Select the correct answer.

    a. 12 %
    b. 15 %
    c. 11 %
    d. 9 %

23. A credit union offers the two accounts shown in the table below. Find the effective rates.

	Annual rate	Compounding
Certificate of deposit	7.6	semiannually
Passbook	7	quarterly

    Select the correct answer.

    a. 8% for Certificate of deposit, 7.4% for Passbook
    b. 7.2% for Certificate of deposit, 7.7% for Passbook
    c. 8.1% for Certificate of deposit, 7.3% for Passbook
    d. 7.8% for Certificate of deposit, 7.3% for Passbook
    e. 7.7% for Certificate of deposit, 7.2% for Passbook

24. As Jorge starts working now at the age of 24, he decides to make regular contributions to a savings account. He wants to accumulate enough by age 55 to fund an annuity of $3,900 per month until age 80. What should his monthly contributions be? Assume that both accounts pay 8.25%, compounded monthly.

Select the correct answer.

a. $284.41
b. $288.41
c. $279.41
d. $302.41

25. What single amount deposited now into an account paying $9\frac{1}{3}$ % annual interest, compounded quarterly, would fund an annuity paying $3,520 quarterly for the next 23 years?

Select the correct answer.

a. $132,801.78
b. $132,785.79
c. $132,929.31
d. $132,699.41

ANSWER KEY

Gustafson/Frisk - College Algebra 8E Chapter 9 Form D

1. a
2. d
3. c
4. d
5. d
6. d
7. d
8. c
9. b
10. b
11. a
12. a
13. b
14. b
15. d
16. b
17. d
18. d
19. d
20. d
21. b
22. a
23. e
24. b
25. b

List of Problem Codes for BCA Testing

Gustafson/Frisk - College Algebra 8E Chapter 9 Form D

1. gfca.09.02.4.07m_NoAlgs
2. gfca.09.01.4.37m_NoAlgs
3. gfca.09.01.4.41m_NoAlgs
4. gfca.09.02.4.22m_NoAlgs
5. gfca.09.03.4.22m_NoAlgs
6. gfca.09.02.4.21m_NoAlgs
7. gfca.09.03.4.01m_NoAlgs
8. gfca.09.01.4.19m_NoAlgs
9. gfca.09.01.4.35m_NoAlgs
10. gfca.09.03.4.10m_NoAlgs
11. gfca.09.02.4.09m_NoAlgs
12. gfca.09.02.4.28m_NoAlgs
13. gfca.09.02.4.25m_NoAlgs
14. gfca.09.03.4.12m_NoAlgs
15. gfca.09.01.4.27m_NoAlgs
16. gfca.09.03.4.08m_NoAlgs
17. gfca.09.01.4.39m_NoAlgs
18. gfca.09.02.4.13m_NoAlgs
19. gfca.09.02.4.31m_NoAlgs
20. gfca.09.03.4.16m_NoAlgs
21. gfca.09.03.4.18m_NoAlgs
22. gfca.09.01.4.14m_NoAlgs
23. gfca.09.01.4.44m_NoAlgs
24. gfca.09.03.4.19m_NoAlgs
25. gfca.09.03.4.14m_NoAlgs

Gustafson/Frisk - College Algebra 8E Chapter 9 Form E

1. Assume that $1,500 is deposited in an account in which interest is compounded annually at a rate of 8%. Find the accumulated amount after 7 years, rounded to the nearest cent.

2. Assume that $1,400 is deposited in an account in which interest is compounded annually at a rate of 8%. Find the accumulated amount after 11 years.

   Select the correct answer.

   a. $3,664.29
   b. $3,364.29
   c. $3,264.29
   d. $3,164.29

3. Assume that $1,100 is deposited in an account in which interest is compounded 10 times a year, at an annual rate of 8%. Find the accumulated amount after 11 years, rounded to the nearest cent.

4. At the birth of their child, the Fieldsons deposited $1,000 in an account paying 8% interest, compounded quarterly. How much will be available when the child turns 10?

   Select the correct answer.

   a. $2,208.04
   b. $2,107.50
   c. $2,308.58
   d. $2,508.44

5. The managers of a pension fund invested $2 million in governmentbonds paying 5.93% annual interest, compounded semiannually. After 3 years, what will the investment be worth?

6. Property values in the suburbs have been appreciating about 13% annually. If this trend continues, what will a $161,000 home be worth in four years? Give the result to the nearest dollar.

   Select the correct answer.

   a. $263,506
   b. $263,740
   c. $265,927
   d. $262,506

7. As Jorge starts working now at the age of 23, he decides to make regular contributions to a savings account. He wants to accumulate enough by age 55 to fund an annuity of $4,700 per month until age 80. What should his monthly contributions be? Assume that both accounts pay 8.5%, compounded monthly. Round to the nearest cent.

8. The gas utilities expect natural gas consumption to increase at 7.5% per year for the next decade. Monthly consumption for one county is currently 4.9 million cubic feet. What monthly demand for gas is expected in ten years?

    Select the correct answer.

    a. 11.20 million ft^3
    b. 10.22 million ft^3
    c. 11.10 million ft^3
    d. 10.10 million ft^3

9. Bank One offers a passbook account with a 4.75% annual rate, compounded quarterly. Bank Two offers a money market account at 5%, compounded monthly. What is the greater effective rate of these two banks? Give the greater effective rate, rounded to the nearest hundredth.

10. Assume that $100 is deposited at the end of each year in an account in which interest is compounded annually at a rate of 5%. Find the accumulated amount after 4 years. Round your answer to the nearest cent.

11. Assume that $400 is deposited at the end of each year in an account in which interest is compounded annually at a rate of 6%. Find the accumulated amount after 6 years.

    Select the correct answer.

    a. $2,769.38
    b. $2,803.18
    c. $2,810.03
    d. $2,790.13

12. Assume that $500 is deposited at the end of each period in an account in which interest is compounded at the frequency $k = 6$, at an annual rate of 7%. Find the accumulated amount after 13 years. Round your answer to the nearest cent.

13. Find the amount of each regular payment to provide $18,500 in 13 years, at an annual rate $r = 6\%$, compounded semiannually.

    Select the correct answer.

    a. $494.56
    b. $492.01
    c. $492.31
    d. $479.86

14. For next year's vacation, the Phelps family is saving $400 each month in an account paying 8.5% annual interest, compounded monthly. How much will be available a year from now? Round your answer to the nearest cent.

15. The managers of a company's pension fund invest the monthly employee contributions of $125,000 into a government fund paying 7.3%, compounded monthly. To what value (rounded to the nearest thousand) will the fund grow in 20 years?

    Select the correct answer.

    a. $67,546,000
    b. $67,540,000
    c. $67,535,000
    d. $67,562,000

16. The last payment of a home mortgage is a balloon payment of $35,000 which the owner is scheduled to pay in 16 years. How much extra should he start including in each monthly payment to eliminate the balloon payment? His mortgage is at 8.6%, compounded monthly. Round to the nearest cent.

17. Jim will retire in 16 years. He will invest $200 each month for 8 years and then let the accumulated value continue to grow for the next 8 years. How much will be available at retirement? Assume 7%, compounded monthly.

    Select the correct answer.

    a. $44,803
    b. $44,834
    c. $44,773
    d. $44,814

18. Find the present value of an annuity with annual payments of $3,380 at 5.15%, compounded annually for 16 years. Round to the nearest cent.

19. Find the present value of an annuity with semiannual payments of $385 at 4.21%, compounded semiannually for 14 years.

    Select the correct answer.

    a. $PV = \$14,483.87$
    b. $PV = \$8,082.92$
    c. $PV = \$11,006.92$
    d. $PV = \$16,165.84$

20. Instead of making quarterly contributions of $730 to a retirement fund for the next 20 years, Jason would rather make only one contribution, now. How much should that be? Assume 6.5 % annual interest, compounded quarterly. Round to the nearest cent.

21. To fund Jamie's lottery winnings of $17,300 per month for the next 16 years, the lottery commission needs to make a single deposit now. Assuming 9.1% compounded monthly, what should the deposit be?

    Select the correct answer.

    a. $1,746,472.49
    b. $1,746,437.29
    c. $1,746,460.39
    d. $1,746,482.79

22. Instead of receiving an annuity of $10,160 each year for the next 15 years, Manuel would like a one-time payment, now. Assuming Manuel could invest the proceeds at 10.5%, what would be a fair amount? Round to the nearest cent.

23. The Jepsens are buying a $21,600 car and financing it over the next 5 years. They secure a 8.6% loan. What will be the total of the monthly payments?

    Select the correct answer.

    a. $79,955.79
    b. $26,651.93
    c. $8,883.98
    d. $13,325.97

24. One lender offers two mortgages - a 8-year mortgage at 13%, and a 23-year mortgage at 12%. For each, find the total of the monthly payments to repay $110,500.

25. Amy contributed $170 per month for 16 years to an account that paid 5% for the first 8 years, but 5.4% for the last 8 years. How much has she saved?

Select the correct answer.

a. $51,169.61
b. $51,157.61
c. $51,167.62
d. $49,923.05

ANSWER KEY

Gustafson/Frisk - College Algebra 8E Chapter 9 Form E

1. 2,570.74
2. c
3. 2,642.72
4. a
5. $2.383
6. d
7. 294.57
8. d
9. 5.12
10. 431.01
11. d
12. 63,054.47
13. d
14. 4,991.49
15. b
16. 85.33
17. d
18. 36,243.68
19. b
20. 32,551.51
21. c
22. 75,121.26
23. b
24. $178,291.36, $325,890.92
25. b

List of Problem Codes for BCA Testing

Gustafson/Frisk - College Algebra 8E Chapter 9 Form E

1. gfca.09.01.4.15_NoAlgs
2. gfca.09.01.4.19m_NoAlgs
3. gfca.09.01.4.23_NoAlgs
4. gfca.09.01.4.35m_NoAlgs
5. gfca.09.01.4.38_NoAlgs
6. gfca.09.01.4.39m_NoAlgs
7. gfca.09.03.4.19_NoAlgs
8. gfca.09.01.4.41m_NoAlgs
9. gfca.09.01.4.42_NoAlgs
10. gfca.09.02.4.08_NoAlgs
11. gfca.09.02.4.09m_NoAlgs
12. gfca.09.02.4.15_NoAlgs
13. gfca.09.02.4.21m_NoAlgs
14. gfca.09.02.4.23_NoAlgs
15. gfca.09.02.4.25m_NoAlgs
16. gfca.09.02.4.30_NoAlgs
17. gfca.09.02.4.31m_NoAlgs
18. gfca.09.03.4.07_NoAlgs
19. gfca.09.03.4.08m_NoAlgs
20. gfca.09.03.4.11_NoAlgs
21. gfca.09.03.4.12m_NoAlgs
22. gfca.09.03.4.13_NoAlgs
23. gfca.09.03.4.16m_NoAlgs
24. gfca.09.03.4.18_NoAlgs
25. gfca.09.03.4.22m_NoAlgs

Gustafson/Frisk - College Algebra 8E Chapter 9 Form F

1. Jim will retire in 16 years. He will invest $200 each month for 8 years and then let the accumulated value continue to grow for the next 8 years. How much will be available at retirement? Assume 7%, compounded monthly.

   Select the correct answer.

   a. $44,803
   b. $44,834
   c. $44,773
   d. $44,814

2. Property values in the suburbs have been appreciating about 13% annually. If this trend continues, what will a $161,000 home be worth in four years? Give the result to the nearest dollar.

   Select the correct answer.

   a. $263,740
   b. $262,506
   c. $265,927
   d. $263,506

3. Assume that $500 is deposited at the end of each period in an account in which interest is compounded at the frequency $k = 6$, at an annual rate of 7%. Find the accumulated amount after 13 years. Round your answer to the nearest cent.

4. Bank One offers a passbook account with a 4.75% annual rate, compounded quarterly. Bank Two offers a money market account at 5%, compounded monthly. What is the greater effective rate of these two banks? Give the greater effective rate, rounded to the nearest hundredth.

5. Find the amount of each regular payment to provide $18,500 in 13 years, at an annual rate $r = 6\%$, compounded semiannually.

   Select the correct answer.

   a. $494.56
   b. $492.01
   c. $492.31
   d. $479.86

6. Assume that $1,500 is deposited in an account in which interest is compounded annually at a rate of 8%. Find the accumulated amount after 7 years, rounded to the nearest cent.

Page 1

7. The last payment of a home mortgage is a balloon payment of $35,000 which the owner is scheduled to pay in 16 years. How much extra should he start including in each monthly payment to eliminate the balloon payment? His mortgage is at 8.6%, compounded monthly. Round to the nearest cent.

8. To fund Jamie's lottery winnings of $17,300 per month for the next 16 years, the lottery commission needs to make a single deposit now. Assuming 9.1% compounded monthly, what should the deposit be?

    Select the correct answer.

    a. $1,746,472.49
    b. $1,746,437.29
    c. $1,746,460.39
    d. $1,746,482.79

9. Instead of receiving an annuity of $10,160 each year for the next 15 years, Manuel would like a one-time payment, now. Assuming Manuel could invest the proceeds at 10.5%, what would be a fair amount? Round to the nearest cent.

10. Find the present value of an annuity with semiannual payments of $385 at 4.21%, compounded semiannually for 14 years.

    Select the correct answer.

    a. $PV = \$14,483.87$
    b. $PV = \$8,082.92$
    c. $PV = \$11,006.92$
    d. $PV = \$16,165.84$

11. At the birth of their child, the Fieldsons deposited $1,000 in an account paying 8% interest, compounded quarterly. How much will be available when the child turns 10?

    Select the correct answer.

    a. $2,107.50
    b. $2,208.04
    c. $2,508.44
    d. $2,308.58

12. As Jorge starts working now at the age of 23, he decides to make regular contributions to a savings account. He wants to accumulate enough by age 55 to fund an annuity of $4,700 per month until age 80. What should his monthly contributions be? Assume that both accounts pay 8.5%, compounded monthly. Round to the nearest cent.

Gustafson/Frisk - College Algebra 8E Chapter 9 Form F

13. Instead of making quarterly contributions of $730 to a retirement fund for the next 20 years, Jason would rather make only one contribution, now. How much should that be? Assume 6.5 % annual interest, compounded quarterly. Round to the nearest cent.

14. The gas utilities expect natural gas consumption to increase at 7.5% per year for the next decade. Monthly consumption for one county is currently 4.9 million cubic feet. What monthly demand for gas is expected in ten years?

    Select the correct answer.

    a. 11.20 million ft^3
    b. 11.10 million ft^3
    c. 10.22 million ft^3
    d. 10.10 million ft^3

15. Assume that $100 is deposited at the end of each year in an account in which interest is compounded annually at a rate of 5%. Find the accumulated amount after 4 years. Round your answer to the nearest cent.

16. One lender offers two mortgages - a 8-year mortgage at 13%, and a 23-year mortgage at 12%. For each, find the total of the monthly payments to repay $110,500.

17. Assume that $1,100 is deposited in an account in which interest is compounded 10 times a year, at an annual rate of 8%. Find the accumulated amount after 11 years, rounded to the nearest cent.

18. The Jepsens are buying a $21,600 car and financing it over the next 5 years. They secure a 8.6% loan. What will be the total of the monthly payments?

    Select the correct answer.

    a. $79,955.79
    b. $26,651.93
    c. $8,883.98
    d. $13,325.97

19. Assume that $400 is deposited at the end of each year in an account in which interest is compounded annually at a rate of 6%. Find the accumulated amount after 6 years.

    Select the correct answer.

    a. $2,769.38
    b. $2,803.18
    c. $2,810.03
    d. $2,790.13

20. For next year's vacation, the Phelps family is saving $400 each month in an account paying 8.5% annual interest, compounded monthly. How much will be available a year from now? Round your answer to the nearest cent.

21. Assume that $1,400 is deposited in an account in which interest is compounded annually at a rate of 8%. Find the accumulated amount after 11 years.

    Select the correct answer.

    a. $3,364.29
    b. $3,664.29
    c. $3,164.29
    d. $3,264.29

22. Find the present value of an annuity with annual payments of $3,380 at 5.15%, compounded annually for 16 years. Round to the nearest cent.

23. Amy contributed $170 per month for 16 years to an account that paid 5% for the first 8 years, but 5.4% for the last 8 years. How much has she saved?

    Select the correct answer.

    a. $51,167.62
    b. $49,923.05
    c. $51,169.61
    d. $51,157.61

24. The managers of a pension fund invested $2 million in government bonds paying 5.93% annual interest, compounded semiannually. After 3 years, what will the investment be worth?

25. The managers of a company's pension fund invest the monthly employee contributions of $125,000 into a government fund paying 7.3%, compounded monthly. To what value (rounded to the nearest thousand) will the fund grow in 20 years?

Select the correct answer.

a. $67,540,000
b. $67,562,000
c. $67,546,000
d. $67,535,000

ANSWER KEY

Gustafson/Frisk - College Algebra 8E Chapter 9 Form F

1. d

2. b

3. 63,054.47

4. 5.12

5. d

6. 2,570.74

7. 85.33

8. c

9. 75,121.26

10. b

11. b

12. 294.57

13. 32,551.51

14. d

15. 431.01

16. $178,291.36, $325,890.92

17. 2,642.72

18. b

19. d

20. 4,991.49

21. d

22. 36,243.68

ANSWER KEY

Gustafson/Frisk - College Algebra 8E Chapter 9 Form F

23. d

24. $2.383

25. a

List of Problem Codes for BCA Testing

Gustafson/Frisk - College Algebra 8E Chapter 9 Form F

1. gfca.09.02.4.31m_NoAlgs
2. gfca.09.01.4.39m_NoAlgs
3. gfca.09.02.4.15_NoAlgs
4. gfca.09.01.4.42_NoAlgs
5. gfca.09.02.4.21m_NoAlgs
6. gfca.09.01.4.15_NoAlgs
7. gfca.09.02.4.30_NoAlgs
8. gfca.09.03.4.12m_NoAlgs
9. gfca.09.03.4.13_NoAlgs
10. gfca.09.03.4.08m_NoAlgs
11. gfca.09.01.4.35m_NoAlgs
12. gfca.09.03.4.19_NoAlgs
13. gfca.09.03.4.11_NoAlgs
14. gfca.09.01.4.41m_NoAlgs
15. gfca.09.02.4.08_NoAlgs
16. gfca.09.03.4.18_NoAlgs
17. gfca.09.01.4.23_NoAlgs
18. gfca.09.03.4.16m_NoAlgs
19. gfca.09.02.4.09m_NoAlgs
20. gfca.09.02.4.23_NoAlgs
21. gfca.09.01.4.19m_NoAlgs
22. gfca.09.03.4.07_NoAlgs
23. gfca.09.03.4.22m_NoAlgs
24. gfca.09.01.4.38_NoAlgs
25. gfca.09.02.4.25m_NoAlgs

Gustafson/Frisk - College Algebra 8E Chapter 9 Form G

1. What is the effective rate, if after one year $200 grows to $224?

    Select the correct answer.

    a. 15 %
    b. 9 %
    c. 12 %
    d. 11 %

2. Assume that $1,500 is deposited in an account in which interest is compounded annually at a rate of 8%. Find the accumulated amount after 7 years, rounded to the nearest cent.

3. Assume that $1,400 is deposited in an account in which interest is compounded annually at a rate of 8%. Find the accumulated amount after 11 years.

    Select the correct answer.

    a. $3,364.29
    b. $3,264.29
    c. $3,664.29
    d. $3,164.29

4. Find the effective interest rate given an annual rate of 8% and a compounding frequency of 8.

    Select the correct answer.

    a. 8.31%
    b. 8.27%
    c. 8.34%
    d. 8.29%

5. At the birth of their child, the Fieldsons deposited $1,000 in an account paying 8% interest, compounded quarterly. How much will be available when the child turns 10?

    Select the correct answer.

    a. $2,208.04
    b. $2,107.50
    c. $2,308.58
    d. $2,508.44

6. The managers of a pension fund invested $2 million in government bonds paying 5.93% annual interest, compounded semiannually. After 3 years, what will the investment be worth?

7. Property values in the suburbs have been appreciating about 13% annually. If this trend continues, what will a $161,000 home be worth in four years? Give the result to the nearest dollar.

   Select the correct answer.

   a. $263,506
   b. $262,506
   c. $265,927
   d. $263,740

8. The gas utilities expect natural gas consumption to increase at 7.5% per year for the next decade. Monthly consumption for one county is currently 4.9 million cubic feet. What monthly demand for gas is expected in ten years?

   Select the correct answer.

   a. 11.10 million ft^3
   b. 11.20 million ft^3
   c. 10.22 million ft^3
   d. 10.10 million ft^3

9. Assume that $100 is deposited at the end of each year in an account in which interest is compounded annually at a rate of 5%. Find the accumulated amount after 4 years. Round your answer to the nearest cent.

10. Assume that $400 is deposited at the end of each year in an account in which interest is compounded annually at a rate of 6%. Find the accumulated amount after 6 years.

    Select the correct answer.

    a. $2,769.38
    b. $2,803.18
    c. $2,810.03
    d. $2,790.13

11. Assume that $200 is deposited at the end of each year into an account in which interest is compounded annually at a rate of 7%. Find the accumulated amount after 4 years.

    Select the correct answer.

    a. $859.84
    b. $887.99
    c. $932.19
    d. $921.04

12. Find the amount of each regular payment to provide $7,500 in 7 years, at an annual rate $r = 6\%$, compounded annually. Round to the nearest cent.

13. Find the amount of each regular payment to provide $18,500 in 13 years, at an annual rate $r = 6\%$, compounded semiannually.

    Select the correct answer.

    a. $494.56
    b. $492.31
    c. $492.01
    d. $479.86

14. The managers of a company's pension fund invest the monthly employee contributions of $125,000 into a government fund paying 7.3%, compounded monthly. To what value (rounded to the nearest thousand) will the fund grow in 20 years?

    Select the correct answer.

    a. $67,540,000
    b. $67,562,000
    c. $67,546,000
    d. $67,535,000

15. The last payment of a home mortgage is a balloon payment of $35,000 which the owner is scheduled to pay in 16 years. How much extra should he start including in each monthly payment to eliminate the balloon payment? His mortgage is at 8.6%, compounded monthly. Round to the nearest cent.

16. Jim will retire in 16 years. He will invest $200 each month for 8 years and then let the accumulated value continue to grow for the next 8 years. How much will be available at retirement? Assume 7%, compounded monthly.

    Select the correct answer.

    a. $44,814
    b. $44,803
    c. $44,773
    d. $44,834

17. Find the present value of an annuity with annual payments of $3,380 at 5.15%, compounded annually for 16 years. Round to the nearest cent.

18. Find the present value of an annuity with semiannual payments of $385 at 4.21%, compounded semiannually for 14 years.

    Select the correct answer.

    a. $PV = \$11,006.92$
    b. $PV = \$8,082.92$
    c. $PV = \$16,165.84$
    d. $PV = \$14,483.87$

19. Find the periodic payment required to repay $1,830 repaid in 18 monthly installments, at an annual rate of 18%.

    Select the correct answer.

    a. $P = \$116.76$
    b. $P = \$163.67$
    c. $P = \$233.52$
    d. $P = \$374.96$

20. Instead of making quarterly contributions of $730 to a retirement fund for the next 20 years, Jason would rather make only one contribution, now. How much should that be? Assume 6.5 % annual interest, compounded quarterly. Round to the nearest cent.

21. What single amount deposited now into an account paying $9\frac{1}{3}$ % annual interest, compounded quarterly, would fund an annuity paying $3,520 quarterly for the next 23 years?

   Select the correct answer.

   a. $132,699.41
   b. $132,801.78
   c. $132,929.31
   d. $132,785.79

22. The Jepsens are buying a $21,600 car and financing it over the next 5 years. They secure a 8.6% loan. What will be the total of the monthly payments?

   Select the correct answer.

   a. $79,955.79
   b. $13,325.97
   c. $8,883.98
   d. $26,651.93

23. One lender offers two mortgages - a 8-year mortgage at 15%, and a 14-year mortgage at 12%. For each, find the total of the monthly payments to repay $145,000.

   Select the correct answer.

   a. 8-year: $249,800.04 , 14-year: $299,976.23
   b. 8-year: $249,788.04 , 14-year: $300,099.92
   c. 8-year: $251,000.92 , 14-year: $301,185.23
   d. 8-year: $249,756.70 , 14-year: $300,304.86

24. As Jorge starts working now at the age of 23, he decides to make regular contributions to a savings account. He wants to accumulate enough by age 55 to fund an annuity of $4,700 per month until age 80. What should his monthly contributions be? Assume that both accounts pay 8.5%, compounded monthly. Round to the nearest cent.

25. Amy contributed $170 per month for 16 years to an account that paid 5% for the first 8 years, but 5.4% for the last 8 years. How much has she saved?

   Select the correct answer.

   a. $51,157.61
   b. $51,167.62
   c. $49,923.05
   d. $51,169.61

ANSWER KEY

Gustafson/Frisk - College Algebra 8E Chapter 9 Form G

1. c

2. 2,570.74

3. b

4. d

5. a

6. $2,383

7. b

8. d

9. 431.01

10. d

11. b

12. 893.51

13. d

14. a

15. 85.33

16. a

17. 36,243.68

18. b

19. a

20. 32,551.51

21. d

ANSWER KEY

Gustafson/Frisk - College Algebra 8E Chapter 9 Form G

22. d

23. a

24. 294.57

25. a

List of Problem Codes for BCA Testing

Gustafson/Frisk - College Algebra 8E Chapter 9 Form G

1. gfca.09.01.4.14m_NoAlgs
2. gfca.09.01.4.15_NoAlgs
3. gfca.09.01.4.19m_NoAlgs
4. gfca.09.01.4.27m_NoAlgs
5. gfca.09.01.4.35m_NoAlgs
6. gfca.09.01.4.38_NoAlgs
7. gfca.09.01.4.39m_NoAlgs
8. gfca.09.01.4.41m_NoAlgs
9. gfca.09.02.4.08_NoAlgs
10. gfca.09.02.4.09m_NoAlgs
11. gfca.09.02.4.13m_NoAlgs
12. gfca.09.02.4.20_NoAlgs
13. gfca.09.02.4.21m_NoAlgs
14. gfca.09.02.4.25m_NoAlgs
15. gfca.09.02.4.30_NoAlgs
16. gfca.09.02.4.31m_NoAlgs
17. gfca.09.03.4.07_NoAlgs
18. gfca.09.03.4.08m_NoAlgs
19. gfca.09.03.4.10m_NoAlgs
20. gfca.09.03.4.11_NoAlgs
21. gfca.09.03.4.14m_NoAlgs
22. gfca.09.03.4.16m_NoAlgs
23. gfca.09.03.4.18m_NoAlgs
24. gfca.09.03.4.19_NoAlgs
25. gfca.09.03.4.22m_NoAlgs

Gustafson/Frisk - College Algebra 8E Chapter 9 Form H

1. Find the present value of an annuity with annual payments of $3,380 at 5.15%, compounded annually for 16 years. Round to the nearest cent.

2. The last payment of a home mortgage is a balloon payment of $35,000 which the owner is scheduled to pay in 16 years. How much extra should he start including in each monthly payment to eliminate the balloon payment? His mortgage is at 8.6%, compounded monthly. Round to the nearest cent.

3. Amy contributed $170 per month for 16 years to an account that paid 5% for the first 8 years, but 5.4% for the last 8 years. How much has she saved?

   Select the correct answer.

   a. $51,157.61
   b. $51,167.62
   c. $49,923.05
   d. $51,169.61

4. Assume that $1,500 is deposited in an account in which interest is compounded annually at a rate of 8%. Find the accumulated amount after 7 years, rounded to the nearest cent.

5. Property values in the suburbs have been appreciating about 13% annually. If this trend continues, what will a $161,000 home be worth in four years? Give the result to the nearest dollar.

   Select the correct answer.

   a. $263,506
   b. $262,506
   c. $265,927
   d. $263,740

6. What single amount deposited now into an account paying $9\frac{1}{3}$ % annual interest, compounded quarterly, would fund an annuity paying $3,520 quarterly for the next 23 years?

   Select the correct answer.

   a. $132,801.78
   b. $132,699.41
   c. $132,929.31
   d. $132,785.79

7. Instead of making quarterly contributions of $730 to a retirement fund for the next 20 years, Jason would rather make only one contribution, now. How much should that be? Assume 6.5 % annual interest, compounded quarterly. Round to the nearest cent.

8. Assume that $100 is deposited at the end of each year in an account in which interest is compounded annually at a rate of 5%. Find the accumulated amount after 4 years. Round your answer to the nearest cent.

9. Find the periodic payment required to repay $1,830 repaid in 18 monthly installments, at an annual rate of 18%.

    Select the correct answer.

    a. $P = \$116.76$
    b. $P = \$163.67$
    c. $P = \$233.52$
    d. $P = \$374.96$

10. The managers of a pension fund invested $2 million in government bonds paying 5.93% annual interest, compounded semiannually. After 3 years, what will the investment be worth?

11. Assume that $200 is deposited at the end of each year into an account in which interest is compounded annually at a rate of 7%. Find the accumulated amount after 4 years.

    Select the correct answer.

    a. $859.84
    b. $887.99
    c. $932.19
    d. $921.04

12. What is the effective rate, if after one year $200 grows to $224 ?

    Select the correct answer.

    a. 15 %
    b. 9 %
    c. 12 %
    d. 11 %

13. One lender offers two mortgages - a 8-year mortgage at 15%, and a 14-year mortgage at 12%. For each, find the total of the monthly payments to repay $145,000.

    Select the correct answer.

    a. 8-year: $249,756.70, 14-year: $300,304.86
    b. 8-year: $249,788.04, 14-year: $300,099.92
    c. 8-year: $249,800.04, 14-year: $299,976.23
    d. 8-year: $251,000.92, 14-year: $301,185.23

14. Assume that $1,400 is deposited in an account in which interest is compounded annually at a rate of 8%. Find the accumulated amount after 11 years.

    Select the correct answer.

    a. $3,364.29
    b. $3,264.29
    c. $3,664.29
    d. $3,164.29

15. As Jorge starts working now at the age of 23, he decides to make regular contributions to a savings account. He wants to accumulate enough by age 55 to fund an annuity of $4,700 per month until age 80. What should his monthly contributions be? Assume that both accounts pay 8.5%, compounded monthly. Round to the nearest cent.

16. At the birth of their child, the Fieldsons deposited $1,000 in an account paying 8% interest, compounded quarterly. How much will be available when the child turns 10?

    Select the correct answer.

    a. $2,308.58
    b. $2,208.04
    c. $2,107.50
    d. $2,508.44

17. Find the effective interest rate given an annual rate of 8% and a compounding frequency of 8.

    Select the correct answer.

    a. 8.31%
    b. 8.29%
    c. 8.34%
    d. 8.27%

18. The gas utilities expect natural gas consumption to increase at 7.5% per year for the next decade. Monthly consumption for one county is currently 4.9 million cubic feet. What monthly demand for gas is expected in ten years?

    Select the correct answer.

    a. 11.10 million ft^3
    b. 11.20 million ft^3
    c. 10.22 million ft^3
    d. 10.10 million ft^3

19. The Jepsens are buying a $21,600 car and financing it over the next 5 years. They secure a 8.6% loan. What will be the total of the monthly payments?

    Select the correct answer.

    a. $79,955.79
    b. $13,325.97
    c. $8,883.98
    d. $26,651.93

20. Find the amount of each regular payment to provide $7,500 in 7 years, at an annual rate r = 6%, compounded annually. Round to the nearest cent.

21. Jim will retire in 16 years. He will invest $200 each month for 8 years and then let the accumulated value continue to grow for the next 8 years. How much will be available at retirement? Assume 7%, compounded monthly.

    Select the correct answer.

    a. $44,814
    b. $44,803
    c. $44,773
    d. $44,834

22. Find the present value of an annuity with semiannual payments of $385 at 4.21%, compounded semiannually for 14 years.

    Select the correct answer.

    a. $PV = \$14,483.87$
    b. $PV = \$8,082.92$
    c. $PV = \$11,006.92$
    d. $PV = \$16,165.84$

23. Find the amount of each regular payment to provide $18,500 in 13 years, at an annual rate $r = 6\%$, compounded semiannually.

    Select the correct answer.

    a. $479.86
    b. $492.01
    c. $492.31
    d. $494.56

24. The managers of a company's pension fund invest the monthly employee contributions of $125,000 into a government fund paying 7.3%, compounded monthly. To what value (rounded to the nearest thousand) will the fund grow in 20 years?

    Select the correct answer.

    a. $67,540,000
    b. $67,562,000
    c. $67,546,000
    d. $67,535,000

25. Assume that $400 is deposited at the end of each year in an account in which interest is compounded annually at a rate of 6%. Find the accumulated amount after 6 years.

    Select the correct answer.

    a. $2,769.38
    b. $2,803.18
    c. $2,790.13
    d. $2,810.03

ANSWER KEY

Gustafson/Frisk - College Algebra 8E Chapter 9 Form H

1. 36,243.68
2. 85.33
3. a
4. 2,570.74
5. b
6. d
7. 32,551.51
8. 431.01
9. a
10. $2.383
11. b
12. c
13. c
14. b
15. 294.57
16. b
17. b
18. d
19. d
20. 893.51
21. a
23. a
24. a
25. c

List of Problem Codes for BCA Testing

Gustafson/Frisk - College Algebra 8E Chapter 9 Form H

1. gfca.09.03.4.07_NoAlgs
2. gfca.09.02.4.30_NoAlgs
3. gfca.09.03.4.22m_NoAlgs
4. gfca.09.01.4.15_NoAlgs
5. gfca.09.01.4.39m_NoAlgs
6. gfca.09.03.4.14m_NoAlgs
7. gfca.09.03.4.11_NoAlgs
8. gfca.09.02.4.08_NoAlgs
9. gfca.09.03.4.10m_NoAlgs
10. gfca.09.01.4.38_NoAlgs
11. gfca.09.02.4.13m_NoAlgs
12. gfca.09.01.4.14m_NoAlgs
13. gfca.09.03.4.18m_NoAlgs
14. gfca.09.01.4.19m_NoAlgs
15. gfca.09.03.4.19_NoAlgs
16. gfca.09.01.4.35m_NoAlgs
17. gfca.09.01.4.27m_NoAlgs
18. gfca.09.01.4.41m_NoAlgs
19. gfca.09.03.4.16m_NoAlgs
20. gfca.09.02.4.20_NoAlgs
21. gfca.09.02.4.31m_NoAlgs
22. gfca.09.03.4.08m_NoAlgs
23. gfca.09.02.4.21m_NoAlgs
24. gfca.09.02.4.25m_NoAlgs
25. gfca.09.02.4.09m_NoAlgs

1. Write the expression without using absolute value symbols.

   $|2|$

2. Find the distance between the following two points on the number line.

   -7 and 3

3. Let $x = -2$, $y = 0$, $z = 2$ and evaluate the expression.

   $$\frac{-(x^2 z^3)}{z^2 - y^2}$$

4. Graph the polynomial function.

   $f(x) = x^3 + x^2$

   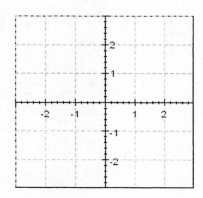

5. Assume that $1,500 is deposited in an account in which interest is compounded annually at a rate of 8%. Find the accumulated amount after 7 years, rounded to the nearest cent.

6. Rationalize the denominator and simplify.

   $$\frac{4}{\sqrt[3]{4}}$$

7. Rationalize the denominator.

$$\frac{7}{\sqrt{7}-2}$$

8. Perform the division and write the answer without using negative exponents.

$$\frac{77a^2b^3}{11ab^6}$$

9. Perform the operations and simplify. Assume that no denominators are 0.

$$\frac{-4}{2x-9y} + \frac{4}{2x-3z} - \frac{12z-36y}{(2x-9y)(2x-3z)}$$

10. Solve the equation.

    $3x + 9 = x + 13$

11. John drove to a distant city in 4 hours. When he returned, there was less traffic, and the trip took only 2 hours. If John averaged 22 mph faster on the return trip, how fast did he drive each way?

12. Solve the equation $x^2 - 6x - 55 = 0$ by completing the square.

13. A cyclist rides from DeKalb to Rockford, a distance of 120 miles. His return trip takes 1 hours longer, because his speed decreases by 10 miles per hour. How fast does he ride each way?

14. Do the operation and express the answer in *a + bi* form.

$$(3 + \sqrt{-16})(5 - \sqrt{-25})$$

15. Find all real solutions of the equation.

$$\sqrt{\sqrt{x+22} - \sqrt{x-29}} = \sqrt{3}$$

16. Solve the inequality and write the answer in interval notation.

$-4x - 13 > -9$

17. Solve the inequality. Write the answer in interval notation.

$$\frac{4}{x-7} \leq 2$$

18. Solve the equation for *x*.

$$\left| \frac{3x-20}{2} \right| = 13$$

19. Graph the equation.

$2(x - y) = 3x + 2$

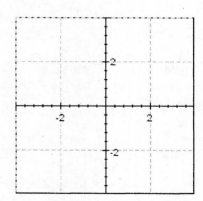

20. Use the slope-intercept form to write the equation of the line passing through the point P(12, 1) and having the slope $m = -\frac{1}{2}$. Express the answer in general form.

21. Find the equation in general form of the circle with center at (1, 1) and $r = 3$.

22. Let the function f be defined by $y = f(x)$, where x and $f(x)$ are real numbers. Find $f(2)$.

$f(x) = 93 - 2x^2$

23. Find the vertex of the parabola.

$y = 4x^2 + 12x + 19$

24. Let $f(x) = 3x$, $g(x) = x + 1$. Find the composite function.

$(g \circ f)(x)$

25. Find the inverse of the one-to-one function.

$y = 5x + 7$

26. An initial deposit of $200 earns 2% interest, compounded quarterly. How much will be in the account in 4 years?

27. Find the value of x.

$\log_6 x = -2$

28. Assume that x, y, z and b are positive numbers. Use the properties of logarithms to write the expression in terms of the logarithms of x, y, and z.

$$\log_b \sqrt[16]{\frac{x^7 y^8}{z^{16}}}$$

29. Solve the equation. If an answer is not exact, give the answer to four decimal places.

    $2 \log_2 x = 1 + \log_2 (x + 24)$

30. A partial solution set (3) is given for the equation. Find the complete solution set.

    $x^3 + 2x^2 - 9x - 18 = 0$

31. Find all rational roots of the equation.

    $x^5 - 5x^4 - 5x^3 + 25x^2 + 4x - 20 = 0$

32. Use a graphing calculator to find the real solution of the equation.

    $x^2 - 10x + 25 = 0$

33. Solve the system, if possible.

    $$\begin{cases} 4x + 7y + 2z = -36 \\ 7x - 3y + 10z = 135 \\ 5x - 5y - 8z = 19 \end{cases}$$

34. Solve the system by Gauss-Jordan elimination.

$$\begin{cases} x - 2y = -2 \\ y = -1 \end{cases}$$

35. Find the product.

$$\begin{bmatrix} 6 \\ -8 \\ -8 \end{bmatrix} \begin{bmatrix} 6 & -4 & -8 \end{bmatrix}$$

36. Evaluate the determinant.

$$\begin{vmatrix} 1 & -2 \\ -8 & 3 \end{vmatrix}$$

37. How many different seven-digit phone numbers can be used in one area code if no phone number begins with 0 or 1?

38. Find the probability of rolling a sum of 4 with one roll of three dice.

39. Assume that you draw two cards from a card deck (standard playing deck of 52 cards), without replacement. Find the probability of the event.

    drawing two aces

40. Decompose the fraction into partial fractions.

$$\frac{19x - 1}{x(x - 1)}$$

41. If the odds in favor of victory are 13 to 6, find the probability of victory.

42. Amy contributed $140 per month for 18 years to an account that paid 6% for the first 9 years, but 6.7% for the last 9 years. How much has she saved?

43. Graph the function.

$$f(x) = \frac{x^2}{x}$$

Note that the numerator and denominator of the fraction share a common factor.

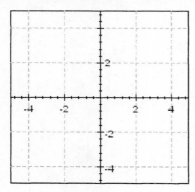

44. Two woodworkers, Tom and Carlos, earn a profit of $92 for making a table and $66 for making a chair. On average, Tom must work 3 hours and Carlos 2 hours to make a chair. Tom must work 2 hours and Carlos 6 hours to make a table. If neither wishes to work more than 42 hours per week, how many tables and how many chairs should they make each week to maximize their income? The information is summarized in the table below.

	Table	Chair	Time available
Profit (dollars)	92	66	
Tom's time (hours)	2	3	42
Carlos' time (hours)	6	2	42

45. Change the equation to standard form.

$$x^2 - 7y + 11 = -4x$$

46. Use the binomial theorem to expand the binomial.

$$(a + 3b)^3$$

47. Evaluate the sum.

$$\sum_{k=1}^{9} 5k$$

48. The 5th term of an arithmetic sequence is 6, and the first term is -2. Find the common difference.

49. Change the decimal to a common fraction.

$0.\overline{32}$

50. Graph the ellipse.

$x^2 + 4y^2 - 4x + 8y + 4 = 0$

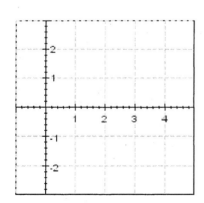

ANSWER KEY

Gustafson/Frisk - College Algebra 8E Final Exam Form A

1. 2

2. 10

3. (-8)

4.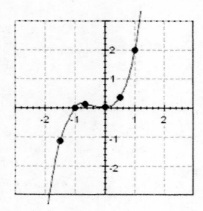

5. 2,570.74

6. $\sqrt[3]{16}$

7. $\dfrac{\left(7\sqrt{7}+14\right)}{3}$

8. $\dfrac{7a}{b^3}$

9. 0

10. 2

11. 22, 44

12. −5, 11

13. 30, 40

14. 35 + 5i

ANSWER KEY

Gustafson/Frisk - College Algebra 8E Final Exam Form A

15. 78

16. $(-\infty, -1)$

17. $(-\infty, 7) \cup [9, \infty)$

18. $\frac{46}{3}, -2$

19.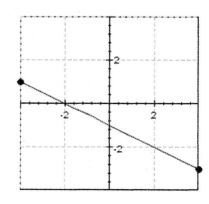

20. $x + 2y = 14$

21. $x^2 + y^2 - 2x - 2y - 7 = 0$

22. 85

23. $(-1.5, 10)$

24. $3x + 1$

25. $\frac{(x-7)}{5}$

26. $216.61

27. $\frac{1}{36}$

28. $\frac{7}{16} \cdot \log_b(x) + \frac{1}{2} \cdot \log_b(y) - \log_b(z)$

ANSWER KEY

Gustafson/Frisk - College Algebra 8E Final Exam Form A

29. $x = 8$

30. $3, -2, -3$

31. $1, 2, -1, -2, 5$

32. 5

33. $x = 5, y = -10, z = 7$

34. $(-4, -1)$

35. $\begin{pmatrix} 36 & -24 & -48 \\ -48 & 32 & 64 \\ -48 & 32 & 64 \end{pmatrix}$

36. -13

37. $8,000,000$

38. $\dfrac{1}{72}$

39. $\dfrac{1}{221}$

40. $\dfrac{1}{x} + \dfrac{18}{(x-1)}$

41. $\dfrac{13}{19}$

42. $\$57,135.47$

ANSWER KEY

Gustafson/Frisk - College Algebra 8E Final Exam Form A

43.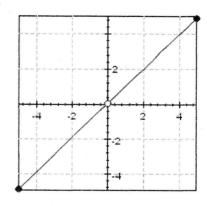

44. 3, 12, $1,068

45. $(x+2)^2 = 7(y-1)$

46. $a^3 + 9a^2 \cdot b + 27a \cdot b^2 + (27b)^3$

47. 225

48. 2

49. $\dfrac{32}{99}$

50.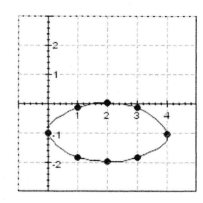

List of Problem Codes for BCA Testing

Gustafson/Frisk - College Algebra 8E Final Exam Form A

1. gfca.00.01.4.69_NoAlgs
2. gfca.00.01.4.85_NoAlgs
3. gfca.00.02.4.89_NoAlgs
4. gfca.03.03.4.12_NoAlgs
5. gfca.09.01.4.15_NoAlgs
6. gfca.00.03.4.97_NoAlgs
7. gfca.00.04.4.81_NoAlgs
8. gfca.00.04.4.99_NoAlgs
9. gfca.00.06.4.70_NoAlgs
10. gfca.01.01.4.18_NoAlgs
11. gfca.01.02.4.43_NoAlgs
12. gfca.01.03.4.39_NoAlgs
13. gfca.01.04.4.11_NoAlgs
14. gfca.01.05.4.25_NoAlgs
15. gfca.01.06.4.60_NoAlgs
16. gfca.01.07.4.17_NoAlgs
17. gfca.01.07.4.78_NoAlgs
18. gfca.01.08.4.25_NoAlgs
19. gfca.02.01.4.47_NoAlgs
20. gfca.02.03.4.30_NoAlgs
21. gfca.02.04.4.65_NoAlgs
22. gfca.03.01.4.40_NoAlgs
23. gfca.03.02.4.20_NoAlgs
24. gfca.03.06.4.43_NoAlgs
25. gfca.03.07.4.27_NoAlgs
26. gfca.04.01.4.67_NoAlgs
27. gfca.04.03.4.50_NoAlgs
28. gfca.04.05.4.36_NoAlgs
29. gfca.04.06.4.51_NoAlgs
30. gfca.05.01.4.23_NoAlgs
31. gfca.05.03.4.16_NoAlgs
32. gfca.05.04.4.24_NoAlgs
33. gfca.06.01.4.57_NoAlgs
34. gfca.06.02.4.41_NoAlgs

List of Problem Codes for BCA Testing

Gustafson/Frisk - College Algebra 8E Final Exam Form A

35. gfca.06.03.4.30_NoAlgs
36. gfca.06.05.4.09_NoAlgs
37. gfca.08.06.4.31_NoAlgs
38. gfca.08.07.4.34_NoAlgs
39. gfca.08.08.4.13_NoAlgs
40. gfca.06.06.4.05_NoAlgs
41. gfca.08.09.4.21_NoAlgs
42. gfca.09.03.4.22_NoAlgs
43. gfca.03.05.4.55_NoAlgs
44. gfca.06.08.4.21_NoAlgs
45. gfca.07.01.4.43_NoAlgs
46. gfca.08.01.4.26_NoAlgs
47. gfca.08.02.4.37_NoAlgs
48. gfca.08.03.4.15_NoAlgs
49. gfca.08.04.4.36_NoAlgs
50. gfca.07.02.4.29_NoAlgs

Gustafson/Frisk - College Algebra 8E Final Exam Form B

1. Solve the equation.

   $3x + 13 = x + 11$

   Select the correct answer.

   a. $x = 48$
   b. $x = -6$
   c. $x = -1$
   d. $x = 51$

2. John drove to a distant city in 4 hours. When he returned, there was less traffic, and the trip took only 2 hours. If John averaged 40 mph faster on the return trip, how fast did he drive each way?

   Select the correct answer.

   a. 44 mph, 84 mph
   b. 27 mph, 67 mph
   c. 51 mph, 91 mph
   d. 40 mph, 80 mph
   e. 48 mph, 88 mph

3. Solve the equation

   $x^2 - 6x - 40 = 0$ by completing the square.

   Select the correct answers.

   a. $x = 3, x = 8$
   b. $x = -4, x = 3$
   c. $x = 7, x = 10$
   d. $x = -4, x = 10$

4. A cyclist rides from DeKalb to Rockford, a distance of 120 miles. His return trip takes 1 hours longer, because his speed decreases by 10 miles per hour. How fast does he ride each way?

   Select the correct answer.

   a. 40 mph going and 30 mph returning
   b. 60 mph going and 30 mph returning
   c. 40 mph going and 40 mph returning
   d. 30 mph going and 120 mph returning

5. Do the operation and express the answer in *a + bi* form.

    $(5 + \sqrt{-25})(5 - \sqrt{-9})$

    Select the correct answer.

    a. $-5 + 10i$
    b. $40 - 5i$
    c. $-40 + 10i$
    d. $40 + 10i$

6. Find all real solutions of the equation.

    $\sqrt{\sqrt{x+22} - \sqrt{x-41}} = \sqrt{3}$

    Select the correct answer(s).

    a. $x = -22$
    b. $x = 22$
    c. $x = 122$
    d. $x = -122$

7. Write the expression without using absolute value symbols.

    $|-5|$

    Select the correct answer.

    a. -5
    b. 1
    c. 0
    d. 5

8. Let $x = -3$, $y = 0$, $z = 3$ and evaluate the expression.

    $$\frac{-(x^2 z^3)}{z^2 - y^2}$$

    Select the correct answer.

    a. 0
    b. -27
    c. -81

9. Rationalize the denominator and simplify.

$$\frac{3}{\sqrt[5]{3}}$$

Select the correct answer.

a. $\sqrt[5]{181}$

b. $\sqrt[6]{82}$

c. $\sqrt[5]{81}$

d. $\sqrt[5]{84}$

e. $\sqrt[10]{81}$

10. Rationalize the denominator.

$$\frac{7}{\sqrt{6} - 2}$$

Select the correct answer.

a. $\dfrac{7\sqrt{6}}{2}$

b. $\dfrac{7\sqrt{6} + \sqrt{14}}{2}$

c. $\dfrac{7\sqrt{6} + 14}{2}$

d. $\dfrac{7\sqrt{6} - 14}{2}$

11. Tell whether the function $y = x^5 + 8x^3$ is even or odd. If it is neither, so indicate.

Select the correct answer.

a. even
b. odd
c. neither

12. Perform the division and write the answer without using negative exponents.

$$\frac{50a^{-2}b^3}{10ab^6}$$

Select the correct answer.

a. $\dfrac{5}{ab^3}$

b. $5ab^3$

c. $\dfrac{5a}{b^3}$

d. $\dfrac{5a^2}{b^3}$

13. Perform the operations and simplify. Assume that no denominators are 0.

$$\frac{-8}{2x-5y} + \frac{8}{2x-9z} - \frac{72z-40y}{(2x-5y)(2x-9z)}$$

Select the correct answer.

a. $\dfrac{8}{2x-5y}$

b. 0

c. $\dfrac{8}{2x-9z}$

d. $\dfrac{72z-40y}{(2x-5y)(9z-2x)}$

14. Jim will retire in 16 years. He will invest $200 each month for 8 years and then let the accumulated value continue to grow for the next 8 years. How much will be available at retirement? Assume 7%, compounded monthly.

Select the correct answer.

a. $44,834
b. $44,803
c. $44,814
d. $44,773

15. A credit union offers the two accounts shown in the table below. Find the effective rates.

	Annual rate	Compounding
Certificate of deposit	7.6	semiannually
Passbook	7	quarterly

Select the correct answer.

a. 8% for Certificate of deposit, 7.4% for Passbook
b. 8.1% for Certificate of deposit, 7.3% for Passbook
c. 7.7% for Certificate of deposit, 7.2% for Passbook
d. 7.2% for Certificate of deposit, 7.7% for Passbook
e. 7.8% for Certificate of deposit, 7.3% for Passbook

16. Find the graph of the equation: $4(x - y) = 3x + 3$

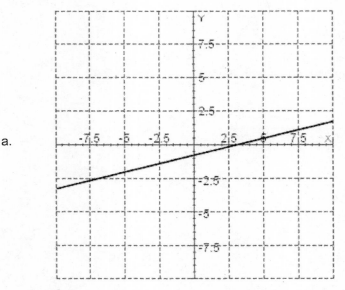

a.

b.

17. Use the slope-intercept form to write the equation of the line passing through the point $P(12, 0)$ and having the slope $m = -\dfrac{1}{4}$. Express the answer in general form.

Select the correct answer.

a. $x - 4y = -9$
b. $x + 4y = 12$
c. $x + 4y = -12$
d. $x - 4y = 12$

18. Find the equation in general form of the circle with center at $(3, 2)$ and $r = 6$

Select the correct answer.

a. $x^2 + y^2 + 6x + 4y - 23 = 0$
b. $x^2 + y^2 - 6x - 4y - 23 = 0$
c. $x^2 + y^2 - 6x - 4y - 49 = 0$
d. $x^2 + y^2 + 6x + 4y - 49 = 0$

19. Let the function f be defined by $y = f(x)$, where x and $f(x)$ are real numbers. Find $f(10)$.

$f(x) = 61 - 7x^2$

Select the correct answer.

a. $f(10) = -639$
b. $f(10) = -662$
c. $f(10) = 179$

20. Find the vertex of the parabola.

$y = 16x^2 + 40x + 32$

Select the correct answer.

a. $\left(\dfrac{5}{4}, 7\right)$

b. $\left(-\dfrac{5}{4}, 25\right)$

c. $\left(-\dfrac{5}{4}, 7\right)$

d. $\left(-\dfrac{5}{4}, 32\right)$

21. Assume that $1,400 is deposited in an account in which interest is compounded annually at a rate of 8%. Find the accumulated amount after 11 years.

    Select the correct answer.

    a. $3,364.29
    b. $3,264.29
    c. $3,664.29
    d. $3,164.29

22. Graph the rational function
    $$f(x) = \frac{4x^2}{5x}$$

    Select the correct answer.

    a.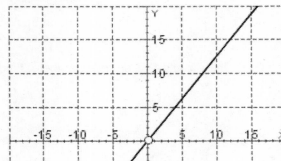

    b.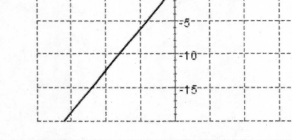

23. Let $f(x) = 3x$, $g(x) = x + 1$. Find the composite function.

    $(g \circ f)(x)$

    Select the correct answer.

    a. $(g \circ f)(x) = 3x + 1$
    b. $(g \circ f)(x) = 3x + 3$
    c. $(g \circ f)(x) = 9x$
    d. $(g \circ f)(x) = x + 2$

24. Find the inverse of the one-to-one function.

    $y = 6x + 3$

    Select the correct answer.

    a. $y = \dfrac{x - 3}{6}$

    b. $y = \dfrac{x - 6}{3}$

    c. $y = \dfrac{6}{x - 3}$

    d. $y = \dfrac{x + 3}{6}$

25. Solve the system, if possible.

    $$\begin{cases} 8x + 9y + 10z = -70 \\ 3x - 5y + 2z = 17 \\ 9x - 7y - 8z = -53 \end{cases}$$

    Select the correct answer.

    a. $x = 7$, $y = 6$, $z = -4$
    b. $x = -13$, $y = 0$, $z = -4$
    c. $x = -7$, $y = -6$, $z = 4$
    d. $x = -5$, $y = -7$, $z = -4$

26. An initial deposit of $800 earns 9% interest, compounded quarterly. How much will be in the account in 14 years?

    Select the correct answer.

    a. $8,933.71
    b. $2,781.22
    c. $99,764.00
    d. $1,092.39
    e. $2,768.56

27. Find the value of x.

    $\log_6 x = -2$

    Select the correct answer.

    a. $x = 6$
    b. $x = 36$
    c. $x = \dfrac{1}{36}$
    d. $x = 2$
    e. no solutions

28. Assume that x, y, z and b are positive numbers. Use the properties of logarithms to write the expression

    $$\log_b \sqrt[8]{\dfrac{x^7 y^4}{z^8}}$$

    in terms of the logarithms of x, y, and z.

    Select the correct answer.

    a. $\dfrac{7}{8} \log_b x + \log_b y - \log_b z$

    b. $56 \log_b x + 32 \log_b y - 64 \log_b z$

    c. $\dfrac{7}{8} \log_b x + \dfrac{1}{2} \log_b y - \log_b z$

    d. $\log_b x + \dfrac{1}{2} \log_b y - \log_b z$

29. Solve the equation.

    $2\log_2 x = 1 + \log_2(x + 40)$

    Select the correct answer.

    a. $x = 12$
    b. $x = 10$
    c. $x = 9$
    d. $x = -8$

30. A partial solution set (3) is given for the equation

    $x^3 + 4x^2 - 9x - 36 = 0$

    Find the complete solution set.

    Select the correct answer.

    a. $1, -4, 2$
    b. $3, 4, -3$
    c. $3, -4, -3$
    d. $3, -1, -3$

31. Find all rational roots of the equation.

    $x^5 - 4x^4 - 10x^3 + 40x^2 + 9x - 36 = 0$

    Select the correct answer(s).

    a. $x = 1$
    b. $x = -2$
    c. $x = -3$
    d. $x = -1$
    e. $x = 3$
    f. $x = 4$
    g. $x = 2$
    h. $x = -4$

32. Use a graphing calculator to find the solution of the equation.

    $x^2 - 8x + 16 = 0$

    Select the correct answer.

    a. $x = 4$
    b. $x = -16$
    c. $x = 8$
    d. $x = -4$
    e. $x = -8$

33. Solve the system by Gauss-Jordan elimination.

$$\begin{cases} x - 2y = 21 \\ y = -6 \end{cases}$$

Select the correct answer.

    a. $(10, -6)$
    b. $(9, -6)$
    c. $(-10, -6)$
    d. $(-9, -6)$

34. Change the equation to standard form.

$$x^2 - 9y + 61 = -8x$$

Select the correct answer.

    a. $(x + 4)^2 = 9(y - 5)$
    b. $(x - 4)^2 = 9(y + 5)$
    c. $(x + 4)^2 = y - 5$
    d. $(x + 4)^2 = 9(y + 5)$

35. The 7th term of an arithmetic sequence is 44, and the first term is -4. Find the common difference.

Select the correct answer.

    a.   8
    b.   -8
    c.   -10
    d.   10

36. Simplify the expression.

$$(x^3)^4 (x^2)^3$$

Select the correct answer.

    a. x^{12}
    b. x^{18}
    c. x^{10}

37. Find the product.

$$\begin{bmatrix} 1 \\ -9 \\ -9 \end{bmatrix} \begin{bmatrix} 4 & -1 & -9 \end{bmatrix}$$

Select the correct answer.

a. $\begin{bmatrix} 1 & 1 & -9 \\ 9 & 9 & 81 \\ 9 & 1 & 81 \end{bmatrix}$

b. 4

c. $\begin{bmatrix} 4 & -1 & -9 \\ -36 & 9 & 81 \\ -36 & 9 & 81 \end{bmatrix}$

d. $\begin{bmatrix} 4 & -1 & -9 \\ -36 & 9 & -36 \\ -9 & -1 & 4 \end{bmatrix}$

38. Evaluate the determinant.

$$\begin{vmatrix} 2 & -10 \\ -1 & 2 \end{vmatrix}$$

Select the correct answer.

a. $D = -6$
b. $D = 6$
c. $D = -14$
d. $D = 14$

39. Select the correct distance between the following two points on the number line.

-20 and 16

Select the correct answer.

a. 20
b. 37
c. 35
d. 16
e. 36

Page 12

40. Decompose the fraction into partial fractions.

$$\frac{19x + 17}{(x+1)(x-1)}$$

Select the correct answer.

a. $\dfrac{1}{x-1} + \dfrac{18}{x+1}$

b. $\dfrac{1}{x+1} + \dfrac{19}{x-1}$

c. $\dfrac{1}{x+1} + \dfrac{18}{x-1}$

d. $\dfrac{1}{x-18} + \dfrac{19}{x+1}$

41. Two woodworkers, Tom and Carlos, earn a profit of $81 for making a table and $77 for making a chair. On average, Tom must work 3 hours and Carlos 2 hours to make a chair. Tom must work 2 hours and Carlos 6 hours to make a table. If neither wishes to work more than 42 hours per week, how many tables and how many chairs should they make each week to maximize their income? The information is summarized in Illustration.

	Table	Chair	Time available
Profit (dollars)	81	77	
Tom's time (hours)	2	3	42
Carlos' time (hours)	6	2	42

Select the correct answer.

a. 3 tables and 12 chairs: $1,167
b. no tables and 21 chairs: $1,701
c. 12 tables and 3 chairs: $1,203

42. Amy contributed $170 per month for 16 years to an account that paid 5% for the first 8 years, but 5.4% for the last 8 years. How much has she saved?

Select the correct answer.

a. $49,923.05
b. $51,169.61
c. $51,157.61
d. $51,167.62

43. Find the graph of the following ellipse.

$$4x^2 + 9y^2 - 8x - 54y + 49 = 0$$

Select the correct answer.

a.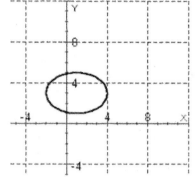

b.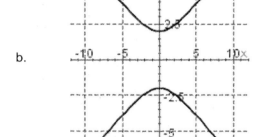

c.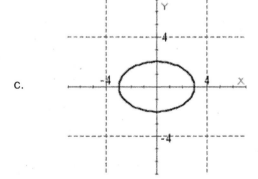

44. Use the binomial theorem to expand the binomial.

$(c + 2t)^3$

Select the correct answer.

a. $c^3 + 2c^2t + 4ct^2 + 8t^3$

b. $c^3 + 3c^2t + 3ct^2 + t^3$

c. $c^3 + 2t^3$

d. $c^3 + 6c^2t + 12ct^2 + 8t^3$

45. Evaluate the sum.

$$\sum_{k=1}^{5} 4k$$

Select the correct answer.

a. 120
b. 20
c. 57
d. 60

46. Change the decimal to a common fraction.

$0.\overline{25}$

Select the correct answer.

a. $\dfrac{25}{9}$

b. $\dfrac{25}{99}$

c. $\dfrac{25}{999}$

d. $\dfrac{25}{90}$

47. How many different six-digit phone numbers can be used in one area code if no phone number begins with 0 or 1?

Select the correct answer.

a. 400,000
b. 2,400,000
c. 800,000

48. Find the probability of rolling a sum of 4 with one roll of three dice.

Select the correct answer.

a. $\dfrac{5}{72}$

b. $\dfrac{7}{72}$

c. $\dfrac{1}{72}$

49. Assume that you draw two cards from a card deck (standard playing deck of 52 cards), without replacement. Find the probability of the event.

drawing two queens

Select the correct answer.

a. $\dfrac{1}{289}$

b. $\dfrac{1}{104}$

c. $\dfrac{1}{221}$

d. $\dfrac{1}{338}$

50. If the odds in favor of victory are 7 to 3, find the probability of victory.

Select the correct answer.

a. $\dfrac{3}{7}$

b. $\dfrac{3}{10}$

c. $\dfrac{10}{3}$

d. $\dfrac{7}{10}$

Gustafson/Frisk - College Algebra 8E Final Exam Form B

1. c
2. d
3. d
4. a
5. d
6. c
7. d
8. b
9. c
10. c
11. b
12. c
13. b
14. c
15. c
16. a
17. b
18. b
19. a
20. c
21. b
22. b
23. a
24. a
25. c
26. b
27. c

ANSWER KEY

Gustafson/Frisk - College Algebra 8E Final Exam Form B

28. c

29. b

30. c

31. a,c,d,e,f

32. a

33. b

34. a

35. a

36. b

37. c

38. a

39. e

40. c

41. a

42. c

43. a

44. d

45. d

46. b

47. c

48. c

49. c

50. d

List of Problem Codes for BCA Testing

Gustafson/Frisk - College Algebra 8E Final Exam Form B

1. gfca.01.01.4.18m_NoAlgs
2. gfca.01.02.4.43m_NoAlgs
3. gfca.01.03.4.39m_NoAlgs
4. gfca.01.04.4.11m_NoAlgs
5. gfca.01.05.4.25m_NoAlgs
6. gfca.01.06.4.60m_NoAlgs
7. gfca.00.01.4.69m_NoAlgs
8. gfca.00.02.4.89m_NoAlgs
9. gfca.00.03.4.97m_NoAlgs
10. gfca.00.04.4.81m_NoAlgs
11. gfca.03.03.4.21m_NoAlgs
12. gfca.00.04.4.99m_NoAlgs
13. gfca.00.06.4.70m_NoAlgs
14. gfca.09.02.4.31m_NoAlgs
15. gfca.02.01.4.47m_NoAlgs
16. gfca.02.03.4.30m_NoAlgs
17. gfca.02.04.4.65m_NoAlgs
18. gfca.03.01.4.40m_NoAlgs
19. gfca.03.02.4.20m_NoAlgs
20. gfca.03.05.4.55m_NoAlgs
21. gfca.03.06.4.43m_NoAlgs
22. gfca.03.07.4.27m_NoAlgs
23. gfca.06.01.4.57m_NoAlgs
24. gfca.04.01.4.67m_NoAlgs
25. gfca.04.03.4.50m_NoAlgs
26. gfca.04.05.4.36m_NoAlgs
27. gfca.04.06.4.51m_NoAlgs
28. gfca.05.01.4.23m_NoAlgs
29. gfca.05.03.4.16m_NoAlgs
30. gfca.05.04.4.24m_NoAlgs
31. gfca.06.02.4.41m_NoAlgs
32. gfca.07.01.4.43m_NoAlgs
33. gfca.08.03.4.15m_NoAlgs
34. gfca.06.03.4.30m_NoAlgs
35. gfca.06.05.4.09m_NoAlgs
36. gfca.06.06.4.07m_NoAlgs

List of Problem Codes for BCA Testing

Gustafson/Frisk - College Algebra 8E Final Exam Form B

37. gfca.06.08.4.21m_NoAlgs
38. gfca.07.02.4.29m_NoAlgs
39. gfca.08.01.4.26m_NoAlgs
40. gfca.08.02.4.37m_NoAlgs
41. gfca.08.04.4.36m_NoAlgs
42. gfca.08.06.4.31m_NoAlgs
43. gfca.08.07.4.34m_NoAlgs
44. gfca.08.08.4.13m_NoAlgs
45. gfca.08.09.4.21m_NoAlgs
46. gfca.00.02.4.41m_NoAlgs
47. gfca.00.01.4.85m_NoAlgs
48. gfca.09.03.4.22m_NoAlgs
49. gfca.09.01.4.19m_NoAlgs
50. gfca.09.01.4.44m_NoAlgs

1. Write the expression without using absolute value symbols.

   $|2|$

2. Find the distance between the following two points on the number line.

   -7 and 3

3. Simplify the expression.

   $(x^3)^4 (x^2)^3$

   Select the correct answer.

   a. x^{10}
   b. x^{12}
   c. x^{18}

4. Let $x = -2$, $y = 0$, $z = 2$ and evaluate the expression.

   $$\frac{-(x^2 z^3)}{z^2 - y^2}$$

5. Rationalize the denominator and simplify.

   $$\frac{4}{\sqrt[3]{4}}$$

6. Rationalize the denominator.

   $$\frac{7}{\sqrt{7} - 2}$$

7. Perform the division and write the answer without using negative exponents.

   $$\frac{77 a^{-2} b^3}{11 ab^6}$$

8. Assume that you draw two cards from a card deck (standard playing deck of 52 cards), without replacement. Find the probability of the event.

   drawing two aces

9. Perform the operations and simplify.

   $$\frac{-4}{2x - 9y} + \frac{4}{2x - 3z} - \frac{12z - 36y}{(2x - 9y)(2x - 3z)}$$

   Assume that no denominators are 0.

10. Solve the equation.

    $3x + 9 = x + 13$

11. John drove to a distant city in 4 hours. When he returned, there was less traffic, and the trip took only 2 hours. If John averaged 22 mph faster on the return trip, how fast did he drive each way?

12. Solve the equation $x^2 - 6x - 55 = 0$ by completing the square.

13. A cyclist rides from DeKalb to Rockford, a distance of 120 miles. His return trip takes 1 hour longer, because his speed decreases by 10 miles per hour. How fast does he ride each way?

14. Do the operation and express the answer in *a + bi* form.

    $(3 + \sqrt{-16})(5 - \sqrt{-25})$

15. Find all real solutions of the equation.

    $\sqrt{\sqrt{x + 22} - \sqrt{x - 29}} = \sqrt{3}$

16. Solve the inequality and write the answer in interval notation.

    $-4x - 13 > -9$

17. Solve the inequality. Write the answer in interval notation.

    $\dfrac{4}{x-7} \leq 2$

18. Solve the equation for x.

    $\left| \dfrac{3x - 20}{2} \right| = 13$

19. Amy contributed $140 per month for 18 years to an account that paid 6% for the first 9 years, but 6.7% for the last 9 years. How much has she saved?

20. Graph the equation.

    $2(x - y) = 3x + 2$

    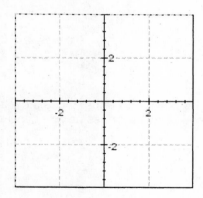

21. Use the slope-intercept form to write the equation of the line passing through the point $P(12, 1)$ and having the slope $m = -\dfrac{1}{2}$. Express the answer in general form.

22. Find the equation in general form of the circle with center at (1, 1) and $r = 3$.

23. Let the function f be defined by $y = f(x)$, where x and $f(x)$ are real numbers. Find $f(2)$.

$f(x) = 93 - 2x^2$

24. Find the vertex of the parabola.

$y = 4x^2 + 12x + 19$

25. How many different seven-digit phone numbers can be used in one area code if no phone number begins with 0 or 1?

26. Graph the function.

$f(x) = \dfrac{x^2}{x}$

Note that the numerator and denominator of the fraction share a common factor.

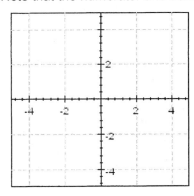

27. Let $f(x) = 3x$, $g(x) = x + 1$. Find the composite function.

$(g \circ f)(x)$

28. Find the inverse of the one-to-one function.

$y = 5x + 7$

29. An initial deposit of $ 200 earns 2% interest, compounded quarterly. How much will be in the account in 4 years?

30. Find the value of x.

$$\log_6 x = -2$$

31. Assume that x, y, z and b are positive numbers. Use the properties of logarithms to write the expression in terms of the logarithms of x, y, and z.

$$\log_b \sqrt[16]{\frac{x^7 y^8}{z^{16}}}$$

32. Solve the equation. If an answer is not exact, give the answer to four decimal places.

$$2 \log_2 x = 1 + \log_2 (x + 24)$$

33. Graph the polynomial function.

$$f(x) = x^3 + x^2$$

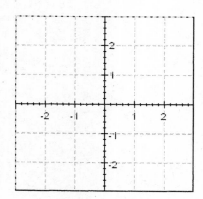

34. A partial solution set (3) is given for the equation. Find the complete solution set.

$$x^3 + 2x^2 - 9x - 18 = 0$$

35. Find all rational roots of the equation.

$$x^5 - 5x^4 - 5x^3 + 25x^2 + 4x - 20 = 0$$

36. Use a graphing calculator to find the real solution of the equation.

$$x^2 - 10x + 25 = 0$$

37. Solve the system, if possible.

$$\begin{cases} 4x + 7y + 2z = -36 \\ 7x - 3y + 10z = 135 \\ 5x - 5y - 8z = 19 \end{cases}$$

38. Solve the system by Gauss-Jordan elimination.

$$\begin{cases} x - 2y = -2 \\ y = -1 \end{cases}$$

39. Find the product.

$$\begin{bmatrix} 6 \\ -8 \\ -8 \end{bmatrix} \begin{bmatrix} 6 & -4 & -8 \end{bmatrix}$$

40. Evaluate the determinant.

$$\begin{vmatrix} 1 & -2 \\ -8 & 3 \end{vmatrix}$$

41. Decompose the fraction into partial fractions.

$$\frac{19x - 1}{x(x - 1)}$$

42. Find the probability of rolling a sum of 4 with one roll of three dice.

43. If the odds in favor of victory are 13 to 6, find the probability of victory.

44. Two woodworkers, Tom and Carlos, earn a profit of $92 for making a table and $66 for making a chair. On average, Tom must work 3 hours and Carlos 2 hours to make a chair. Tom must work 2 hours and Carlos 6 hours to make a table. If neither wishes to work more than 42 hours per week, how many tables and how many chairs should they make each week to maximize their income? The information is summarized in the table below.

	Table	Chair	Time available
Profit (dollars)	92	66	
Tom's time (hours)	2	3	42
Carlos' time (hours)	6	2	42

45. Change the equation to standard form.

$$x^2 - 7y + 11 = -4x$$

46. Graph the ellipse.

$$x^2 + 4y^2 - 4x + 8y + 4 = 0$$

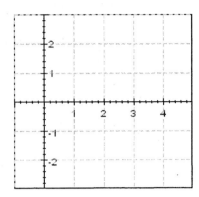

47. Use the binomial theorem to expand the binomial.

$$(a + 3b)^3$$

48. Evaluate the sum.

$$\sum_{k=1}^{9} 5k$$

49. The 5th term of an arithmetic sequence is 6, and the first term is -2. Find the common difference.

50. Change the decimal to a common fraction.

$0.\overline{32}$

ANSWER KEY

Gustafson/Frisk - College Algebra 8E Final Exam Form C

1. 2

2. 10

3. c

4. (-8)

5. $\sqrt[3]{16}$

6. $\dfrac{\left(7\sqrt{7}+14\right)}{3}$

7. $\dfrac{7a}{b^3}$

8. $\dfrac{1}{221}$

9. 0

10. 2

11. 22, 44

12. −5, 11

13. 30, 40

14. 35 + 5i

15. 78

16. $(-\infty, -1)$

17. $(-\infty, 7) \cup [9, \infty)$

18. $\dfrac{46}{3}$, −2

ANSWER KEY

Gustafson/Frisk - College Algebra 8E Final Exam Form C

19. $57,135.47

20.

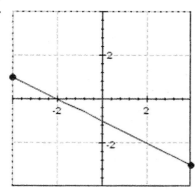

21. $x + 2y = 14$

22. $x^2 + y^2 - 2x - 2y - 7 = 0$

23. 85

24. $(-1.5, 10)$

25. 8,000,000

26.

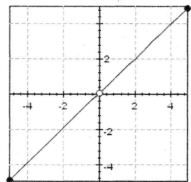

27. $3x + 1$

28. $\dfrac{(x-7)}{5}$

29. $216.61

30. $\dfrac{1}{36}$

31. $\dfrac{7}{16} \cdot \log_b(x) + \dfrac{1}{2} \cdot \log_b(y) - \log_b(z)$

32. $x = 8$

33.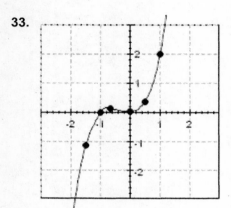

34. $3, -2, -3$

35. $1, 2, -1, -2, 5$

36. 5

37. $x = 5, y = -10, z = 7$

38. $(-4, -1)$

39. $\begin{pmatrix} 36 & -24 & -48 \\ -48 & 32 & 64 \\ -48 & 32 & 64 \end{pmatrix}$

40. -13

41. $\dfrac{1}{x} + \dfrac{18}{(x-1)}$

42. $\dfrac{1}{72}$

ANSWER KEY

Gustafson/Frisk - College Algebra 8E Final Exam Form C

43. $\dfrac{13}{19}$

44. 3, 12, $1,068

45. $(x+2)^2 = 7(y-1)$

46.
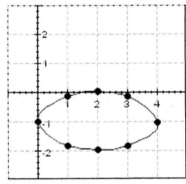

47. $a^3 + 9a^2 \cdot b + 27a \cdot b^2 + (27b)^3$

48. 225

49. 2

50. $\dfrac{32}{99}$

Page 4

List of Problem Codes for BCA Testing

Gustafson/Frisk - College Algebra 8E Final Exam Form C

1. gfca.00.01.4.69_NoAlgs
2. gfca.00.01.4.85_NoAlgs
3. gfca.00.02.4.41m_NoAlgs
4. gfca.00.02.4.89_NoAlgs
5. gfca.00.03.4.97_NoAlgs
6. gfca.00.04.4.81_NoAlgs
7. gfca.00.04.4.99_NoAlgs
8. gfca.08.08.4.13_NoAlgs
9. gfca.00.06.4.70_NoAlgs
10. gfca.01.01.4.18_NoAlgs
11. gfca.01.02.4.43_NoAlgs
12. gfca.01.03.4.39_NoAlgs
13. gfca.01.04.4.11_NoAlgs
14. gfca.01.05.4.25_NoAlgs
15. gfca.01.06.4.60_NoAlgs
16. gfca.01.07.4.17_NoAlgs
17. gfca.01.07.4.78_NoAlgs
18. gfca.01.08.4.25_NoAlgs
19. gfca.09.03.4.22_NoAlgs
20. gfca.02.01.4.47_NoAlgs
21. gfca.02.03.4.30_NoAlgs
22. gfca.02.04.4.65_NoAlgs
23. gfca.03.01.4.40_NoAlgs
24. gfca.03.02.4.20_NoAlgs
25. gfca.08.06.4.31_NoAlgs
26. gfca.03.05.4.55_NoAlgs
27. gfca.03.06.4.43_NoAlgs
28. gfca.03.07.4.27_NoAlgs
29. gfca.04.01.4.67_NoAlgs
30. gfca.04.03.4.50_NoAlgs
31. gfca.04.05.4.36_NoAlgs
32. gfca.04.06.4.51_NoAlgs
33. gfca.03.03.4.12_NoAlgs
34. gfca.05.01.4.23_NoAlgs
35. gfca.05.03.4.16_NoAlgs
36. gfca.05.04.4.24_NoAlgs

List of Problem Codes for BCA Testing

Gustafson/Frisk - College Algebra 8E Final Exam Form C

37. gfca.06.01.4.57_NoAlgs
38. gfca.06.02.4.41_NoAlgs
39. gfca.06.03.4.30_NoAlgs
40. gfca.06.05.4.09_NoAlgs
41. gfca.06.06.4.05_NoAlgs
42. gfca.08.07.4.34_NoAlgs
43. gfca.08.09.4.21_NoAlgs
44. gfca.06.08.4.21_NoAlgs
45. gfca.07.01.4.43_NoAlgs
46. gfca.07.02.4.29_NoAlgs
47. gfca.08.01.4.26_NoAlgs
48. gfca.08.02.4.37_NoAlgs
49. gfca.08.03.4.15_NoAlgs
50. gfca.08.04.4.36_NoAlgs